Accelerating Knowledge Sharing, Creativity, and Innovation Through Business Tourism

Muhammad Waseem Bari
Government College University, Faisalabad, Pakistan

Sadia Shaheen
Government College University, Faisalabad, Pakistan

Meng Fanchen
Beijing Institute of Technology, China

A volume in the Advances in Hospitality, Tourism, and the Services Industry (AHTSI) Book Series

IGI Global
DISSEMINATOR OF KNOWLEDGE

Published in the United States of America by
IGI Global
Business Science Reference (an imprint of IGI Global)
701 E. Chocolate Avenue
Hershey PA, USA 17033
Tel: 717-533-8845
Fax: 717-533-8661
E-mail: cust@igi-global.com
Web site: http://www.igi-global.com

Library of Congress Cataloging-in-Publication Data

Names: Bari, Muhammad Waseem, 1980- editor. | Shaheen, Sadia, 1985- editor. | Fanchen, Meng, 1967- editor.
Title: Accelerating knowledge sharing, creativity, and innovation through business tourism / Muhammad Waseem Bari, Sadia Shaheen, and Meng Fanchen, editors.
Description: Hershey : Business Science Reference, 2020. | Includes bibliographical references and index. | Summary: “”This book examines the importance of business tourism in the domain of social sciences and management sciences literature. It also highlights negatives and positives of business traveling specifically, the bright side of tourism and how to establish a logical link between business tourism and employee learning and knowledge transfer skills”--Provided by publisher”-- Provided by publisher.
Identifiers: LCCN 2019048521 (print) | LCCN 2019048522 (ebook) | ISBN 9781799831426 (hardcover) | ISBN 9781799831433 (paperback) | ISBN 9781799831440 (ebook)
Subjects: LCSH: Business travel--Social aspects. | Communication in organizations. | Organizational learning.
Classification: LCC G156.5.B86 A233 2020 (print) | LCC G156.5.B86 (ebook) | DDC 658.3/83--dc23
LC record available at https://lccn.loc.gov/2019048521
LC ebook record available at https://lccn.loc.gov/2019048522

This book is published in the IGI Global book series Advances in Hospitality, Tourism, and the Services Industry (AHTSI) (ISSN: 2475-6547; eISSN: 2475-6555)

British Cataloguing in Publication Data
A Cataloguing in Publication record for this book is available from the British Library.

Advances in Hospitality, Tourism, and the Services Industry (AHTSI) Book Series

Maximiliano Korstanje
University of Palermo, Argentina

ISSN:2475-6547
EISSN:2475-6555

Mission

Globally, the hospitality, travel, tourism, and services industries generate a significant percentage of revenue and represent a large portion of the business world. Even in tough economic times, these industries thrive as individuals continue to spend on leisure and recreation activities as well as services.

The Advances in Hospitality, Tourism, and the Services Industry (AHTSI) book series offers diverse publications relating to the management, promotion, and profitability of the leisure, recreation, and services industries. Highlighting current research pertaining to various topics within the realm of hospitality, travel, tourism, and services management, the titles found within the AHTSI book series are pertinent to the research and professional needs of managers, business practitioners, researchers, and upper-level students studying in the field.

Coverage

- Sustainable Tourism
- Hotel Management
- Service Management
- Destination Marketing and Management
- Leisure & Business Travel
- Service Design
- Food and Beverage Management
- Service Training
- Cruise Marketing and Sales
- Health and Wellness Tourism

IGI Global is currently accepting manuscripts for publication within this series. To submit a proposal for a volume in this series, please contact our Acquisition Editors at Acquisitions@igi-global.com or visit: http://www.igi-global.com/publish/.

The Advances in Hospitality, Tourism, and the Services Industry (AHTSI) Book Series (ISSN 2475-6547) is published by IGI Global, 701 E. Chocolate Avenue, Hershey, PA 17033-1240, USA, www.igi-global.com. This series is composed of titles available for purchase individually; each title is edited to be contextually exclusive from any other title within the series. For pricing and ordering information please visit http://www.igi-global.com/book-series/advances-hospitality-tourism-services-industry/121014. Postmaster: Send all address changes to above address.

Titles in this Series

For a list of additional titles in this series, please visit: http://www.igi-global.com/book-series/advances-hospitality-tourism-services-industry/121014

Global Entrepreneurial Trends in the Tourism and Hospitality Industry
Debasish Batabyal (Amity University, Kolkata, India)
Business Science Reference • © 2020 • 300pp • H/C (ISBN: 9781799826033) • US $225.00

Handbook of Research on the Impacts, Challenges, and Policy Responses to Overtourism
Cláudia Ribeiro de Almeida (University of Algarve, Portugal) Alfred Quintano (Independent Researcher, Malta) Moisés Simancas (University of La Laguna, Spain) Raquel Huete (University of Alicante, Spain) and Zélia Breda (University of Aveiro, Portugal)
Business Science Reference • © 2020 • 447pp • H/C (ISBN: 9781799822240) • US $285.00

Economic, Educational, and Touristic Development in Asia
Bryan Christiansen (Global Training Group, Ltd, UK) and Hakan Sezerel (Anadolu University, Turkey)
Business Science Reference • © 2020 • 241pp • H/C (ISBN: 9781799822394) • US $215.00

Industrial and Managerial Solutions for Tourism Enterprises
Atilla Akbaba (Izmir Katip Çelebi University, Turkey) and Volkan Altıntaş (Izmir Katip Celebi University, Turkey)
Business Science Reference • © 2020 • 336pp • H/C (ISBN: 9781799830306) • US $225.00

Handbook of Research on Smart Technology Applications in the Tourism Industry
Evrim Çeltek (Gaziosmanpasa University, Turkey)
Business Science Reference • © 2020 • 569pp • H/C (ISBN: 9781799819899) • US $295.00

Global Opportunities and Challenges for Rural and Mountain Tourism
Devkant Kala (University of Petroleum and Energy Studies, India) and Satish Chandra Bagri (Hemvati Nandan Bahuguna Garhwal University, India)
Business Science Reference • © 2020 • 354pp • H/C (ISBN: 9781799813026) • US $225.00

Organizational Behavior Challenges in the Tourism Industry
Şule Aydin (Nevsehir Haci Bektas Veli University, Turkey) Bekir Bora Dedeoglu (Nevsehir Haci Bektas Veli University, Turkey) and Ömer Çoban (Nevsehir Haci Bektas Veli University, Turkey)
Business Science Reference • © 2020 • 524pp • H/C (ISBN: 9781799814740) • US $225.00

IGI Global
DISSEMINATOR OF KNOWLEDGE

701 East Chocolate Avenue, Hershey, PA 17033, USA
Tel: 717-533-8845 x100 • Fax: 717-533-8661
E-Mail: cust@igi-global.com • www.igi-global.com

Table of Contents

Detailed Table of Contents

Gastronomy examines, researches, and applies the food and beverage culture of a society in the historical process. Gastronomy is also a field of science that includes elements of science and art. The aim of gastronomy education is to train experts with the knowledge and equipment needed by the sector. Creativity is an important parameter in the field of gastronomy. The concept of creativity can be defined in different ways. In a simple way, creativity is used in the meaning of creating, revealing, and discovering the unknown. The Torrance scale is one of the most widely used methods for measuring creativity. In this chapter, the effect of gender, the type of school graduated at high school, and pre-school education on creativity was examined for undergraduate students studying gastronomy at the universities. The importance of gastronomy in business tourism is increasing. The development of creative ideas and thoughts by gastronomy employees can positively affect gastronomy tourism.

Business tourism is type of tourism that involves work rather than enjoyment. It is becoming the most income-generating part of tourism. Business tourism has different types that are meetings, incentives, convention, and exhibition (MICE); congress tourism; event tourism; sport tourism; and business agencies. Meetings, incentives, convention, and exhibition is a detailed and comprehensive type of business tourism. The incentive traveling is that of business tourism that is offered by the company to their employees. Exhibitions are those types of traveling that are organized to introduced new products, services, and information to the business tourists. Convention consists of the congresses and conferences. Congresses are the larger-scale business meetings, and conferences are the smaller-scale business meetings while complexes are the combination of small exhibitions and conferences. There is a need to understand that leisure tourism is different from business tourism.

Preface

This book contributes to the advanced knowledge of creativity and innovation in business tourism. This book also attempts to give new insights and contributions in the existing literature of business tourism through the dissemination of information as well as empirical pieces of evidence from various parts of the world. This book is a fragment of ongoing empirical research in the relationship between tourism, creativity and innovation. The focus of this book is on advancing creativity and innovation through knowledge sharing on various platforms, geographical locations and diverse groups of intellectuals having diversified experience, knowledge and expertise. This book also focuses on knowledge sharing based on global competition, advance communication technologies, and changing customers' needs.

This book tells a story chapter by chapter for instance, what is tourism, how it is developing and what are the impacts of business tourism on individuals and groups' knowledge, creativity and innovation. Fundamentally, after reading this book the reader can understand how creativity and innovation would be enhanced through business tourism. After reading the present book, readers will understand the role of business tourism in knowledge sharing and creativity. The content of this book is very helpful for researchers, faculty, students, and business professionals. The present book "Accelerating Knowledge Sharing, Creativity, and Innovation through Business Tourism" is organized with fifteen chapters which provide insights into business tourism, creativity, and innovation as a whole.

The first chapter focuses on the key constraints of inbound business tourism in Pakistan, possible and suitable remedies to resolve these constraints. The importance of inbound business tourism from the perspective of the China-Pakistan economic corridor (CPEC).

The second chapter is about the analysis of Portuguese and Spanish congress tourists' profile. Professional and academic congresses play a relevant role in what came to be known as meetings tourism or business tourism. The congress tourists consume complementary activities, as well as accommodations services, within the destination, which makes them a very profitable segment. Therefore, many cities are including this market in their promotional strategies.

The third chapter investigates the relationship between learning strategies (cognitive learning strategies and behavioral learning strategies) and innovative work behavior.

The fourth chapter explores the role of authentic leaders and authentic relationships as a catalyst to create meaningful exchanges of knowledge and a genuine desire to maintain relationships after the business tourism activity ends.

The fifth chapter proposes a "Pluralistic Learning Theory for Project Knowledge Transfer" as a practical solution for all sorts of business tourism activities to transfer knowledge effectively.

The sixth chapter deals with the tourism industry in general from the Malaysian context. Tourism industry across the globe, gaining momentum and respected as an emerging business sector bringing the much-needed foreign exchange and employment opportunities for the countries. A Particular set of empirical results are reported to establish the effects of knowledge management on service innovation and performance of Malaysian tourism players. Moreover, emerging trends in the tourism industry highlighted as well.

The seventh chapter sheds the light on the link between corporate ethical values (CEV) and knowledge sharing behavior (KSB) by building and empirically examining a CEV-KSB model.

The eighth chapter explains the knowledge of traditional cultures, the history of religions, national religious traditions, rituals, ceremonies, and ceremonies, participating in them as spectators, acquaintance with unique architectural monuments, including the traditional methods of building religious buildings, with masterpieces of icon painting, sculptures, mosaics, and decoration.

The ninth chapter explores the comprehensive literature that helps in understanding the dynamics of social capital within an industrial cluster with mediation role of knowledge embeddedness.

The tenth chapter identifies the role of knowledge management processes and infrastructures as a determinant of financial and operational in the hotel industry, particularly how these knowledge management strategies have a positive impact on the hotels' performance.

The eleventh chapter is a case study carried for a focalized examination of knowledge management practices in the hospitality industry of Pakistan and highlighting the antecedents that impede the KM application to this industry.

The twelfth chapter highlights the various current state of tourism recommendation system models and discussed their preference criteria. As a part of that, the authors studied various important preference factors in the tourism recommendation system and categorized them based on their likeness.

The thirteenth chapter highlights the significance of business tourism in the global village where the advanced technologies are utilizing for knowledge sharing and implementing innovation to attract the potential customers effectively and enhance the profitability of tourism industry.

The fourteenth chapter explores the effect of gender, the type of school graduated at high school and pre-school education on creativity was examined for undergraduate students' study gastronomy at the universities. The development of creative ideas and thoughts by gastronomy employees can positively affect gastronomy tourism.

The fifteenth chapter explains how business tourism has become a famous and most income-generating part of tourism. In addition, also explains the different types of business tourism such as Meetings Incentives Convention and Exhibition (MICE), Congress tourism, Event tourism, Sport tourism, and Business Agencies.

Happy Reading!

Acknowledgment

I would like to express my special thanks of gratitude to my co-editors (Prof. Dr. Meng Fanchen and Dr. Sadia Shaheen) as well as the dean and director of my school (Prof. Dr. Hazoor Muhammad Sabir and Dr. Muhammad Abrar) who facilitated me to complete this book, I am really thankful to them. Secondly, I would also like to thank authors, my parents and friends who helped me a lot in finalizing this project within the limited time frame.

Thanking you,

Muhammad Waseem Bari
Lyallpur Business School, Government College University, Faisalabad, Pakistan

Chapter 1

Constraints of Business Tourism and Proposed Remedies:

China-Pakistan Economic Corridor-Based View

Muhammad Waseem Bari
Government College University, Faisalabad, Pakistan

Sadia Shaheen
Government College University, Faisalabad, Pakistan

ABSTRACT

Business tourism is the oldest form of tourism in human history. Business tourism is based on four types of activities i.e. meetings, incentives, conferences, and exhibitions (MICE). The inbound business tourism performs a significant role in the economic growth and development of any country. The objectives of this chapter are as follows. First, to highlight the key constraints of inbound business tourism in Pakistan. Second, possible and suitable remedies to resolve these constraints. Third, to explain the importance of inbound business tourism from the perspective of the China-Pakistan economic corridor (CPEC). The present chapter divides the constraints of business tourism in Pakistan into four levels, namely: state-level constraints, government level constraints, industrial level constraints, and individual-level constraints. This study proposes different remedies against different levels of constraints such as an establishment of an international platform of business tourism of Pakistan.

INTRODUCTION

After emerging the global village phenomenon, the world has been squeezed virtually (Pieterse, 2019). The information technology is providing global information at the doorsteps of the people. This global information brings the chances of business and career opportunities for this global village (Büyükbaykal, 2015). To seek these opportunities, the people travel to local and international destinations and stay over there as per their needs. In other words, digital information is a key driver of travel and tourism in the

DOI: 10.4018/978-1-7998-3142-6.ch001

world. Last few decades, the tourism industry has been growing rapidly and information technology is playing a significant role in this regard (Del Chiappa & Baggio, 2015). The tourism industry has a significant share of revenue in the total GDP in many countries. For instance, in 2018, Macau, Maldives, Grenada, Fiji, and Jamacia had a contribution of travel and tourism to GDP 72.2%, 66.4%, 56.6, 38.9, and 34% respectively (Knoema, 2018).

Leiper (1979) treated tourism as a system that is based on different elements such as tourists, transit routes, destination regions, generating regions, and the tourist industry. Leiper defined the tourism as a system including the optional travel and transitory stay of people away from their typical place of living for at least one night, with the exception of visits made for the basic purpose of gaining remuneration from focuses on the way (Leiper, 1979; Park, Kim, Kim, & Park, 2019). The literature explains the different types and forms of tourism. The types of tourism include domestic tourism and international tourism, international tourism is further catagorised into inbound and outbound tourism (Boniface, Cooper, & Cooper, 2016). As explained in Figure 1, tourism has several forms and sub-forms such as leisure and business tourism.

The focus of the present study is business tourism and constraints of business tourism in Pakistan, especially considering the case of China-Pakistan Economic Corridor (CPEC). Business tourism provides opportunities for interaction among entrepreneurs from different cultures and regions, and undoubtedly, business tourism, knowledge sharing, and innovation reciprocally support to each other (Ambler, Witzel, & Xi, 2016). Based on theories of interaction and networking (Pettenati & Cigognini, 2007), interaction among people of the same interests creates new knowledge, and new knowledge provides the base of innovation (Bari & Fanchen, 2017).

Prior to starting an in-depth discussion on business tourism, it is necessary to clarify the difference between business tourism and tourism business. The tourism business includes different sorts of business activities related to tourism such as transportation service, tour operator, accommodation, guided tours and tourist guide, hospitality, and destination management organization (Fess, 2019). On the other hand, business tourism has a different perspective. Swarbrooke & Horner (2012) referred to (Davidson, 1994) and defined business tourism as follows;

Business tourism is concerned with people traveling for purposes which are related to their work. As such it represents one of the oldest forms of tourism, man having traveled for this purpose of trade since very early times (Davidson, 1994).

The above definition clarifies that business tourism is the most popular and oldest form of tourism (Swarbrooke & Horner, 2012). The people have been traveling from one part to the other part of the world for the purposes of trade and employment since the civilization of the world. In the pre currency era, under the barter system, the people had been traveling from town to town and city to city for the exchange of goods and services. This sort of trade activity was also a source of information and wisdom sharing, knowledge and culture transfer and social values observation. Therefore, in business tourism, people travel, trade, and share knowledge, technology, and culture (Molz, 2013). However, the tourism business facilitates business tourism by providing facilities such as hoteling, food, traveling, etc.

Literature classifies the business tourism activities into four categories which include meetings, incentives, conferences, exhibitions (MICE) (Spencer & Bavuma, 2018). In other words, when a person travels for the purposes of business meetings, business incentives, business conferences, or business exhibitions are referred to as business tourism. Geographically, business tourism has two forms i.e.

national business tourism and international business tourism. When an individual trip and perform any activity from MICE within country refers to national business tourism, and when an individual travels and participates in any activity of MICE at international level/cross border, it refers to as international business tourism. The economic strength of any destination defines the inbound and outbound business tourism ratio (Ritchie & Crouch, 2003). Inbound business tourism means how many tourists visit a country, and how many tourists in that country travel to other countries referred to as outbound tourism. For example, Xinhua (2018) reported that in China inbound and outbound numbers of tourists reached 139 million and 131 million in 2017 respectively. According to the report of China Tourism Academy (CTA), outbound tourists of China have been increasing expenditures 5% yearly. Contrary, inbound tourists' expenditures are increasing by 3.6% yearly (Xinhua, 2018). The report further explains that inbound tourism in China is increasing due to increasing the tourist from the countries which are linked with One Belt-One Road project.

In a contemporary economic model, the tourism industry and especially inbound business tourism have significant importance. Every country is struggling not only to develop but also to sustain business tourism potential. The strong economies and developed states are successfully managing business tourism growth and activities (Elliott, 2002). However, mostly, underdeveloped and developing economies are still at the struggling phase to attract inbound business tourists specifically and tourism activities in general. There are multiple factors that impact business tourism such as social, cultural, economic, environmental, law and order, terrorism, and technological (Swanson & Edgell, 2013).

The above-mentioned factors impact on business tourism activities in different ways. For instance, a favorable climate of the destination boasts the chances of business activities such as international conferences and meetings. However, undesired changes in the destination environment such as hot weather, high winds, less greenery negatively impact on the happening of the business activities negatively. The turbulent economy such as the high unemployment rate, fewer business opportunities, negative balance of payment, etc., is another important factor to undermine the business tourism activities (Koens, 2012). In the presence of these types of negative economic indicators, inbound and outbound business tourism activities are affected simultaneously. Contemporary technology also plays an important role in inbound business tourism development. Internet and the latest digital devices facilitate business activities and business tourism. Lack of information technology can decrease the number of inbound business tourists. The historical and cultural richness of a destination also an important reason behind the growth of business tourism (Wang & Bramwell, 2012). Therefore, in most business conferences and exhibitions, organizers attach some historical places trips and cultural festivals with the conference/exhibition program.

Several business tourists travel in the research of innovative technologies, contemporary knowledge, and best business practices (Buhalis & Law, 2008; Connolly, Sigala, & Buhalis, 2001). If the destination has the potential of advanced technologies and business models, the number of inbound business tourists (researchers) may increase. Religious tourism also impacts business tourism activities (Ateljevic & Li, 2017). The travelers while going on pilgrimage to a religious destination (especially international destinations), they also arrange some business meetings or perform some trade activities within the same traveling budget. After World War II, traditional war has been transformed into an economic war. Therefore, most of the time terrorists hit economic activities across the world. Last few decades, it is observed that terrorist activities across the world have significantly affected business tourism in multiple ways. For instance, the visa policies of several countries especially, developed countries have been more tighten. Some developed countries have also disconnected or controlled the diplomatic relations with the severely affected countries by terrorism (Siegel, Brickman, Goldberg, & Pat-Horenczyk, 2019). Bomb blasting, kidnapping, and gun shooting have also created threats in the minds of business tourists.

Figure 1. Forms of tourism

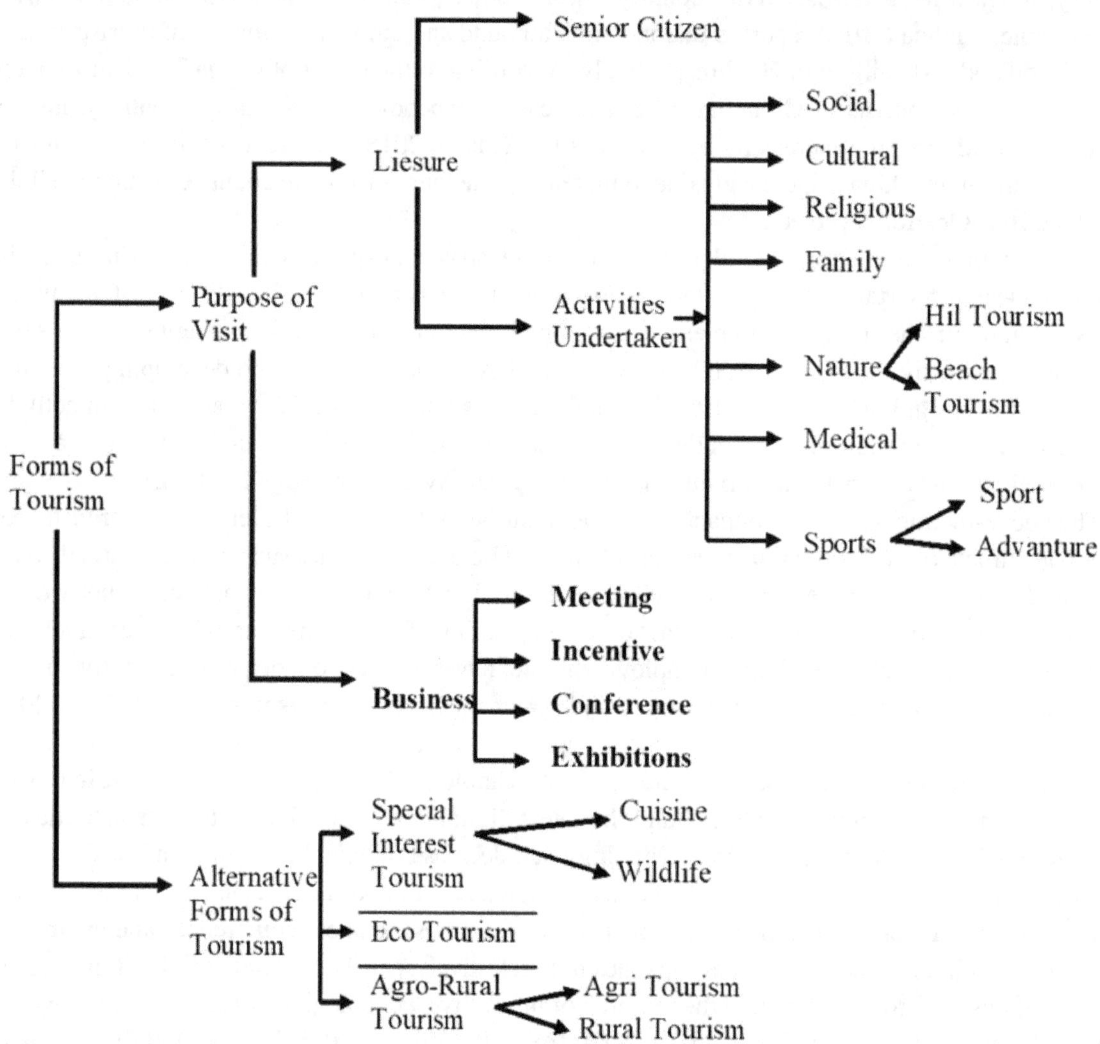

Belt and Road Initiative (BRI) is one of the biggest and global projects of the 21st century which is introduced by the Chinese government in 2013 (Lau, 2020). BRI is a striving infrastructural venture which connects 152 countries of different continents. The key objective behind this global infrastructure development under the umbrella of BRI is the promotion of international trade, increasing the opportunities of mutual business among member countries, tourism, transfer of innovative technology and skills, and poverty reduction (Voinescu & Moisoiu, 2019). Pakistan is a South Asian country and its major portion of the economy based on the agriculture sector. However, Pakistan is very rich in the perspective of mineral and natural resources (Ahmad, Öztürk, Ahmad, & Khan, 2015). P.R. China is situated at the west-east border of Pakistan. Pakistan shares its 535 kilometers border with China (Khan, 2013).

The BRI project has six economic corridors and CPEC is one of them (Malik, 2018). The CPEC is a great economic activity for Pakistan. In 2014, an agreement was signed between "Ministry of Planning, Reforms, and Development of Pakistan" and "National Reforms and Development Commission of China" for the investment of 46 billion US dollars by the Chinese government in 47 sub-projects of CPEC (Bari, Abrar, Bashir, Baig, & Fanchen, 2019). By nature, CPEC is a socioeconomic venture that consists of energy projects, telecommunication, dry ports, education, technical skills exchange, cotton industry, special economic zones, agriculture sector, law and security, and Chinese firms investment in Pakistan (Bari et al., 2019). Gwadar seaport has a central importance in the CPEC. Gwadar seaport connects several European and Asian countries through the sea and minimize their transportation cost and save time (Lee, Hu, Lee, Choi, & Shin, 2018). Since 2014, after the formal inauguration of CPEC, several countries are taking interest to join the CPEC which is the open indication of an increase in business tourism in Pakistan.

Although, the bright side of CPEC is presenting a positive image of Pakistan in front of the world. However, several countries such as India and America are not happy with CPEC. Therefore, these countries are creating serious hurdles in front of Pakistan's prosperity, economic growth, and soft image. On the other side, the weak social and economic infrastructure of Pakistan also a key reason behind the less business tourism in Pakistan. In the next sections of this chapter, the following points will be discussed. First, constraints of business tourism in Pakistan by considering the perspective of CPEC. Second, what is suitable and doable remedies to enhance and sustain business tourism in Pakistan?

CONSTRAINTS OF BUSINESS TOURISM IN PAKISTAN

As discussed earlier, business tourism consists of four types of activities namely, meetings, incentives, conferences, and exhibitions. While persuing these business tourism activities, four different levels of constraints of business tourism are faced in Pakistan;

1. State Level Constraints
2. Government Level Constraints
3. Industrial Level Constraints
4. Individual Level Constraints

State Level Constraints

Law and Order

One of the biggest hurdles in front of the inbound business tourism of Pakistan is an unstable situation of law and order in Pakistan (Muhammad, 2014). Like many other developing countries, Pakistan since its birth (1947) has been facing the issue of law and order. Pakistani security agencies and police are professionally not much proficient to provide sufficient security services to foreign business tourists. There is no special training for security forces on how to deal with international business tourists. Karachi is the biggest and metropolitan city of Pakistan. Maximum chances of inbound and outbound business tourism activities of Pakistan are expected in Karachi. However, in the last three decades, the situation of law and order in Karachi was very poor (Abbas, Shouping, Sidra, & Sharif, 2018). Karachi was considered very dangerous, that why not only international business tourists bus also local ones were frightened while traveling to Karachi. Last few years, the situation of law and order is getting better and its hope that in near future it will be much better. In 2016, the state minister of parliamentary affairs explained that poor situation of law and order, and terrorism is a key reason behind the decrease in the number of inbound international tourists (The News, 2016). However, the minister clarifies that international tourist is increasing due to recent betterment in the situation of law and order (Alasttal & Burdey, 2017).

The two states, China and Pakistan are treating each other friendly and brotherly since their birth. The CPEC is the bright evidence of this friendship. However, China and Pakistan have strict in their visa policies as compare to the past. The reason behind this strictness is some illegal activities performed by Chinese citizens in Karachi and other parts of Pakistan (Yasir, 2017). The strict measures regarding visa policy impact negatively on inbound and outbound business tourism industry of Pakistan.

Terrorism

Last four to five decades, the world is seriously facing the issue of terrorism. Post 9/11, Pakistan is the country which is highly affected after Afghanistan (Puri, 2017). More than 70 thousand people have died, and billions of rupees have been lost during the war of terrorism in Pakistan. From 2005 to 2017, Pakistan faced incidents of bomb blasting, target killing, kidnapping, etc. as a daily routine. The international media portrayed in the world that Pakistan is in war. The terrorists that were normally funded by the Indian intelligence agency (RAW) destroyed the peace of Pakistan (Lieven, 2017). These terrorism activities minimize the inbound tourism of Pakistan, especially, business tourism. The business tourism is associated with leisure tourism. Northern areas of Pakistan such as Sawat Valley, Naran, and Kagan are the most popular sites for leisure tourism in Pakistan. The people of northern areas of Pakistan faced terrorist activities and control over the area last 15 years for which another major reason for a decrease in the number of inbound tourists in Pakistan.

The CPEC, where brought the economic and infrastructure development opportunities of Pakistan, there, also increased the external security threats to Pakistan. India and America are not ready to accept the completion and success of CPEC (Chawla, 2017). Therefore, they both are continuously creating the security and economic threats in different parts of Pakistan such as terrorist activities in Balochistan (a province of Pakistan, where CPEC is developing). Although, security forces are arresting and breaking the terrorist networks in Pakistan and continuously improving the peace. However, still, it is a negative impact on the inbound business tourism of Pakistan.

Image of the State

By birth, Pakistan is a Muslim country and now it has become the first nuclear power of the Muslim world. Unfortunately, after 9/11, the western world associated the link of terrorism with Islam and focused on Afghanistan (Taliban) by considering it as a fundamentalist Islamic terrorist state (Rabasa, Waxman, Larson, & Marcum, 2004). Pakistan and Afghanistan have a common border and both are Muslim states. Several Taliban across the border of Pakistan sometimes for the purpose of safety form US troops. Therefore, the image of Pakistan as a state had been developed as a state of terrorists' home. Several western countries banned Pakistani product access in international markets and stop the business activities in Pakistan as well (Kronstadt, 2009). With the launching of CPEC, Pakistan has an opportunity to change the image of the state positively. A soft image of Pakistan is necessary for the attraction of inbound business tourism in the long run. However, India and many other countries are creating hurdles in the success of CPEC and the development of the soft image of Pakistan. Better economic conditions of Pakistan can improve the positive image of Pakistan.

Political Instability

Instability in the political system of Pakistan is another important issue that negatively impacts on business and business tourism activities in Pakistan (Sönmez, 1998). Since 1947, Pakistan has been experienced a very unstable and incoherent political environment. Except for two political governments, all others did not complete their constitutional period as a government. Several times, Army interfered and controlled the affairs of the state. Even, Pakistan faced the Martial law three times in the last 72 years (Siddiqa, 2019). All these types of activities are the cause of instability in the business environment in Pakistan. The policies developed for the betterment of business and business-related activities are stopped by every upcoming ruler.

Nowadays, the affairs of CPEC are also stuck up due to the change of political government. The government of PTI (Pakistan Tehreek-e- Insaf) has stopped the further development of CPEC projects and asked China for the time to review the agreements that were done by previous political governments (Boni, 2019). Obviously, this decision will develop a negative perception and uncertainty in the business community at the national level and the international investor will also reconsider their decisions of investment in Pakistan. Therefore, business tourism will also be impacted negatively due to the negative business environment. Moreover, a regular fight for power between the army and civil government is a serious issue for business tourism development.

Government Level Constraints

Government Priorities and Business Tourism

The development of the tourism industry is a top priority of the developing and developed countries of the world in the 21st century. Most of the governments are prioritizing the tourism industry while developing the economic development policy of the country. The business tourism is the mother of all tourism types and forms (Swarbrooke & Horner, 2012). Therefore, no one country is ready to ignore business tourism especially, inbound business tourism. The governments through their policies do special arrangements for business tourists such as easy visa availability, airport welcome services, security, access to information, etc. However, the case of Pakistan is totally different. No one government in the last 72 years did a serious effort for the development of business tourism in Pakistan. In fact, in the history of Pakistan, the tourism industry never is a focused point of any government. Even, in the last three decades, the tourism industry has been declined. This is the first time, the current government (PTI) took serious steps for the development tourism industry in Pakistan. However, business tourism is still required special attention not only for the development of the tourism industry but also for the betterment of Pakistan's economy.

Currently, in Pakistan, CPEC is creating inbound business tourism activities (Kanwal et al., 2019). Several entrepreneurs, especially from China, are visiting and interested to visit Pakistan in the future for business activities. The current government also providing a few facilities such as easy visa facilities and information about Pakistan from the perspective of business tourism. However, tourists are not completely satisfied with these tourism facilities in Pakistan, especially in local cities such as Faisalabad, Gujranwala, Sialkot, Peshawar, and Hyderabad.

Pakistan Tourism Development Corporation

Pakistan tourism development corporation (PTDC) is considered the backbone of tourism activities in Pakistan. The PTDC has full authority after the dissolution of the ministry of tourism due to the 18th amendment in the constitution of Pakistan. However, professionally, PTDC is a very weak organization. There is a shortage of professional staff in PTDC (Baloch, 2007). Mostly, bureaucrats are managing PTDC and several officers on the basis of favoritism and nepotism in PTDC. The ratio of experienced staff is not sufficient. Although, PTDC has the most attractive points for tourism activities with a handsome budget, however, corruption and non-professionalism are destroying everything. With a handsome financial budget, a monopolistic position in the market, and full government machinery, PTDC still believes in traditional strategies and local operational standards of tourism.

Along with all the above-mentioned issues with PTDC, it is still failed to focus on the business tourism perspective. Although, the current government has launched the new visa policy to encourage the tourists. However, its application at ground level is still ineffective. PTDC has no special arrangements for business tourists at all levels of a business tour. Due to CPEC, many business tourists are visiting since 2014 and most of them are from China. The Chinses business tourists are not satisfied with the PTDC services. Therefore, the Chinese are constructing their own hotels at Islamabad.

Poor Planning for Business Tourism

The planning for any project plays a significant role in its success (Gunduz & Yahya, 2018). The government of Pakistan has not any comprehensive strategic planning for the development of inbound business tourism in Pakistan. The corporate bodies such as chambers of commerce, different investment boards, and PTDC which have the joint responsibility of business tourism development lacks mutual coordination. At the provincial level, governments also have no common policy and mutual coordination regarding the development of business tourism. There is not a single event such as business exhibition and business fairs in Pakistan at the international level throughout the year for the attraction of international business tourists. The national and provincial governments of Pakistan are still failed in the development of knowledge and technological parks for the interests of international tourists. There is also a high deficiency of tourism-related educational institutes both in government and private sectors of Pakistan. The universities of Pakistan are also lacking in the development of the institutes of tourism development. Since 1947, the government machinery has been relying on a few naturally developed points for tourism. Now, these spots (northern areas) have been over sold out and tourists do not feel comfortable during their visits. The government of Pakistan never planned seriously for the development of tourists' spots in different areas of Pakistan. The government of Pakistan has no planning regarding sustainable tourism activities in Pakistan. The government is also not up to date regarding the future needs of tourism.

Ignorance About Available Tourism Infrastructure

Business tourism and leisure tourism reciprocally promote each other (Wu & Wu, 2019). However, the government of Pakistan only focuses on conventional tourism such as hills tracking, etc. The government of Pakistan has not focused on the forms of leisure tourism such as sports tourism, beach tourism, religious tourism, and water tourism. Pakistan has beautiful beach areas, beautiful lakes for water tourism, rich historical culture, and the center of Sikh religion (Fakhar, 2010). All these types of tourism naturally available to Pakistan and the government just need to develop the proper infrastructure and facilities for the promotion of these types of tourism. In some cases, if government plans regarding tourism development in some new areas, the views of local communities at destinations are not considered important. Moreover, the government just focuses on the development of the point of tourism but the facilities such as hospitals, pure drinking water, hotels, washing sites and cleaning system for the destination are not developed properly. In some places, these facilities are developed but there is no proper control of services quality and facilities for the tourists. While developing the infrastructure for new tourism destinations, the officials do not perform adequate research on the infrastructure of international tourism destinations.

Industrial Level Constraints

The inbound business tourism of Pakistan highly depends on the true efforts of the Pakistan industrial sector. In the history of Pakistan, the industrial sector never performed serious efforts to encourage business tourism in Pakistan. There is no well-known platform for the business community in Pakistan, where business tourism activities can be performed. Even, the industries are not awarded about the importance of inbound business tourism. Instead of creating inbound business tourism activities, the business community prefers outbound business tourism. The exporters of Pakistan travel throughout the world for participation in the world-renown business exhibitions to get the orders (Jost & Sandhu, 2000). However, the success rate in getting the export orders is not very good due to tough international competition in these exhibitions. Moreover, these businesspeople look like bagging for orders due to brandless products/ unpopular brands (Alam, Almotairi, & Gaadar, 2013).

Several organizations (exporters) invite foreign businesspeople individually for their own business products sale and marketing purposes. This way, these organizations spend more money on foreign business tourists and chances of success of business deals are also decrease. Moreover, the overall relevant industry and country have no benefits from such types of individual business tour invitations. Several industrial products are linked with each other's' and business exhibitions provide this opportunity to all inbound business tourists (Swarbrooke & Horner, 2012).

Individual Level Constraints

One of the key constraints of inbound business tourism development in Pakistan is the incapability of individual businesspersons. To enhance inbound business tourism, the social competence of Pakistani business individuals is quite necessary (Smallbone, Kitching, & Athayde, 2010). As discussed earlier in this chapter, the state image, government policies, and industrial efforts are not enough and supportive from the perspective of inbound business tourism. Therefore, it is important for the individual businesspeople/ entrepreneurs to enhance their social competence by considering social activities such as social perception, impression management, social adaptability, persuasiveness, expressiveness (Meutia, 2013).

This way not only individual inbound tourism will be increased but also the image of the state will be positively changed. A high number of businesspersons of Pakistan are not well educated and several are illiterate. Therefore, the English language also a big barrier to the development of social and business ties with the international business community (Hyder & Lussier, 2016). Similarly, foreign business tourists also have concerns regarding language barriers. Cultural differences also impact on inbound business tourism activities (Yang & Wong, 2012). Interactive behaviors and living styles of Pakistani business citizens. Religious rigidness is also a hurdle in the development of business tourism in Pakistan. Pakistani people are very rigid towards culture and religion. They did not provide sufficient liberty to foreign tourists to enjoy their tour as per their lifestyles (Avraham & Ketter, 2016).

PROPOSED REMEDIES TO DEVELOP BUSINESS TOURISM IN PAKISTAN

The present government of Pakistan has been taking several initiatives to uplift the tourism industry of Pakistan. For instance, relaxation and easiness in visa processing, multiple activities at national and international levels to develop the soft and positive image of Pakistan, development of current and new destinations of tourism in Pakistan, providing the awareness of tourism to the local community, etc. However, Pakistan still needs to take several measures at the state, government, industry, and individual levels. The present study proposes the following remedies to improve the inbound business tourism in Pakistan.

State Level Remedies

The major hurdle in the development of inbound business tourism in Pakistan is the unfavorable situation of law and order, negative perception about Pakistan as a state, and terrorist activities. Currently, the security forces of Pakistan have significantly improved the law and order situation in Pakistan and controlled the terrorist activities in Pakistan. Pakistan has struggled a lot in the last 20 years for peace in Pakistan. However, still, it is a long journey to cover. Pakistan needs to develop a positive image in front of international tourism. The stable political system in Pakistan can play a significant role in this regard because a political government can present a soft image of the state more effectively. Currently, several mega projects are running under the umbrella of CPEC, such as Gawader Seaport, specialized industrial zones, and eastern and western routes of CPEC in Pakistan. Several Chinese business organizations are working on different projects in Pakistan. A positive and peaceful environment for Chinese workers and employers can develop a positive perception about Pakistan and these foreign workers can perform the role of ambassadors of Pakistan. This way, inbound business tourism can improve in Pakistan and in the future, the business community of the countries that are interested in CPEC such as Russia, Afghanistan, and Iran can visit Pakistan.

Government Level Remedies

A peaceful environment is useless if the civil government did not develop and apply effective business tourism policies. The government should focus on the following recommendations.

- The government should give the task to all its ambassadors for the development of international inbound business tourism.
- The ambassadors should arrange the seminars/meetings with the foreign business community and explain to them the business opportunities in Pakistan.
- The government should not only relax the visa policy for the business community but also ensure its implementation.
- The special arrangement should be made at airports to facilitate the business tourists
- The government should activate PTDC and deploy the professional in PTDC. The PTDC should develop a separate section to deal with business tourists only.
- The government should promote the hotel industry in Pakistan and through policy international standards, the quality of hotel services should also be maintained.
- The transport standard in Pakistan is also very poor, the foreign tourist did not feel comfortable while using an unorganized transport service. The government should do special arrangements for tourists in Pakistan.
- The professional bodies such as chambers of commerce should be activated in the perspective of business tourism. The chambers of commerce should promote and support business tourism activities in Pakistan.
- Pakistan's government should develop new destinations and forms of business tourism in Pakistan because leisure tourism in interconnected with business tourism.
- The government should also explain the importance of business tourism to the business community of Pakistan.
- For the promotion of inbound business tourism, the government should facilitate the Pakistani business community for outbound business tourism.

Industrial Level Remedies

Business organizations and industries are the main key holders and beneficiaries of business tourism. However, in Pakistan industries are not well aware of the importance and impact of inbound business tourism. The industrial sector of Pakistan and big organizations should consider the following steps for the development of business tourism in Pakistan.

- All Pakistan industries should develop a platform/ legal body at the international level for the promotion of business tourism activities in Pakistan.
- That business tourism development platform should develop the infrastructure to facilitate the inbound/outbound business tourism activities.
- That business tourism development platform should also discuss the problems related to business tourism development with government institutions and should try to solve them.
- Different industries in Pakistan such as textiles, leather products, agricultural products, sports, etc. should arrange business fairs and business trips through the platform of business tourism development.
- The business tourism development platform should also arrange business tourism awareness seminars for the business community.
- Through these seminars/workshops, the business community can be educated that how a businessman/woman can get benefits from business tourism activities.

- The business community of Pakistan should also learn that the money incurred on business tourism is not an expense, but it is an investment.

Individual Level Remedies

For the development of business tourism in Pakistan, individual-level motivation is very important. The following points are very important in this regard.

- The individual businesspersons should learn other nations' culture.
- The CPEC is a big business and economic concern of Pakistan and China is the backbone of CPEC. Therefore, Chinese social and organizational culture is very important to learn for Pakistani businesspeople.
- Pakistan's individual businesspeople should develop their international social networking through different social media platforms.
- Businesspersons/ entrepreneurs of Pakistan should understand the importance of information technology and e-commerce.
- Pakistani entrepreneurs should also promote the activities of outbound business tourism because face to face interaction will enhance the chances of inbound business tourism and business activities in Pakistan.
- The business community of Pakistan should participate in international business conferences. International business conferences and fairs are big sources of social and instrumental networking.
- The government should launch a business tourism portal based on information and opportunities for business tourism and business activities.

CONCLUSION

After reviewing the different levels of constraints (state-level constraints, government level constraints, industrial level constraints, and individual-level constraints) of business tourism in Pakistan from the perspective of CPEC, the present study proposes multiple remedies to control these constraints. Specifically, the present study proposes a business tourism development body at the national level. This body will be helpful not only to solve the constraints of business tourism in Pakistan but also to promote and motivate the Pakistani entrepreneurs for outbound business tourism. Secondly, the government of Pakistan should also control the religious impact on tourism activities in Pakistan. The citizens of Pakistan should be little liberal and social towards international tourists in Pakistan. These activities will be helpful for the promotion soft image of Pakistan.

REFERENCES

Abbas, S., Shouping, L., Sidra, F., & Sharif, A. (2018). Impact of Crime on Socio-Economic Development: A Study of Karachi. *Malaysian Journal of Social Sciences and Humanities*, *3*(3), 148–159.

Ahmad, H., Öztürk, M., Ahmad, W., & Khan, S. M. (2015). Status of natural resources in the uplands of the Swat Valley Pakistan. In *Climate change impacts on high-altitude ecosystems* (pp. 49–98). Springer. doi:10.1007/978-3-319-12859-7_2

Alam, A., Almotairi, M., & Gaadar, K. (2013). Nation branding: An effective tool to enhance fore going direct investment (FDI) in Pakistan. *Research Journal of International Studies*, *25*, 134–141.

Alasttal, M. N. M., & Burdey, M. B. (2017). An Exploratory Analysis Of Pakistan Tourism Market: Its Present Scenario, Issues, Challenges And Future Prospects. *Grassroots, 50*(3).

Ambler, T., Witzel, M., & Xi, C. (2016). *Doing business in China*. Routledge.

Ateljevic, J., & Li, L. (2017). Tourism entrepreneurship–concepts and issues. In *Tourism and entrepreneurship* (pp. 30–53). Routledge. doi:10.4324/9780080942728

Avraham, E., & Ketter, E. (2016). *Tourism marketing for developing countries: Battling stereotypes and crises in Asia, Africa and the Middle East.* Springer. doi:10.1057/9781137342157

Baloch, Q. B. (2007). Managing tourism in Pakistan: A case study of Chitral valley. *Journal of Managerial Sciences*, *2*(2), 169–190.

Bari, M. W., Abrar, M., Bashir, M., Baig, S. A., & Fanchen, M. (2019). Soft Issues During Cross-Border Mergers and Acquisitions and Industry Performance, China–Pakistan Economic Corridor Based View. *SAGE Open*, *9*(2). doi:10.1177/2158244019845180

Bari, M. W., & Fanchen, M. (2017). Personal Interaction Drives Innovation: Instrumental Guanxi-Based Knowledge Café Approach. In Handbook of research on tacit knowledge management for organizational success (Vol. 1, pp. 177–200). Hershey, PA: IGI Publishers.

Boni, F. (2019). *Sino-Pakistani Relations: Politics*. Military and Regional Dynamics. doi:10.4324/9780429431982

Boniface, B., Cooper, R., & Cooper, C. (2016). *Worldwide destinations: The geography of travel and tourism*. Routledge. doi:10.4324/9781315697666

Buhalis, D., & Law, R. (2008). Progress in information technology and tourism management: 20 years on and 10 years after the Internet—The state of eTourism research. *Tourism Management*, *29*(4), 609–623. doi:10.1016/j.tourman.2008.01.005

Büyükbaykal, C. I. (2015). Communication technologies and education in the information age. *Procedia: Social and Behavioral Sciences*, *174*, 636–640. doi:10.1016/j.sbspro.2015.01.594

Chawla, M. I. (2017). One belt one road summit 2017 and its implications for CPEC: An overview. *South Asian Studies*, *32*(2), 277–284.

Connolly, D. J., Sigala, M., & Buhalis, D. (2001). Major trends and IT issues facing the hospitality industry in the new economy: A review of the 5th annual pan-European hospitality technology exhibition and conference (EURHOTEC 2000). *International Journal of Tourism Research, 3*(4), 325–327. doi:10.1002/jtr.308

Davidson, R. (1994). *Business Travel*. Addison Wesley Longman.

Del Chiappa, G., & Baggio, R. (2015). Knowledge transfer in smart tourism destinations: Analyzing the effects of a network structure. *Journal of Destination Marketing & Management, 4*(3), 145–150. doi:10.1016/j.jdmm.2015.02.001

Elliott, J. (2002). *Tourism: Politics and public sector management*. Routledge. doi:10.4324/9780203416136

Fakhar, A. H. (2010). Factors affecting tourism, tourism potential and strategies For development as an Industry in Pakistan.

Fess, C. (2019). Beginners Guide to Starting a Tourism Business. Tourism Tattler. Retrieved from https://www.tourismtattler.com/articles/business-and-finance/beginners-guide-to-starting-a-tourism-business/6681

Gunduz, M., & Yahya, A. M. A. (2018). Analysis of project success factors in construction industry. *Technological and Economic Development of Economy, 24*(1), 67–80. doi:10.3846/20294913.2015.1074129

Hyder, S., & Lussier, R. N. (2016). Why businesses succeed or fail: A study on small businesses in Pakistan. *Journal of Entrepreneurship in Emerging Economies, 8*(1), 82–100. doi:10.1108/JEEE-03-2015-0020

Jost, P. M., & Sandhu, H. S. (2000). *The hawala alternative remittance system and its role in money laundering*. Interpol Lyon.

Kanwal, S., Pitafi, A. H., Pitafi, A., Nadeem, M. A., Younis, A., & Chong, R. (2019). China–Pakistan Economic Corridor (CPEC) development projects and entrepreneurial potential of locals. *Journal of Public Affairs*.

Khan, S. A. (2013). Geo-economic imperatives of Gwadar Sea Port and Kashgar economic zone for Pakistan and China. *IPRI Journal, 13*(2), 87–100.

Knoema. (2018). *Contribution of travel and tourism to GDP as a share of GDP*. Retrieved from https://knoema.com/atlas/topics/Tourism/Travel-and-Tourism-Total-Contribution-to-GDP/Contribution-of-travel-and-tourism-to-GDP-percent-of-GDP

Koens, K. (2012). Competition, cooperation and collaboration: business relations and power in township tourism. In *Slum Tourism* (pp. 101–118). Routledge.

Kronstadt, K. A. (2009). *Pakistan-US relations*. Library of Congress Washington Dc Congressional Research Service.

Lau, R. K.-S. (2020). Africa-China Relations in the Context of Belt and Road Initiative: Realizing African-Chinese Dreams for Common Development? In China's Globalization and the Belt and Road Initiative (pp. 157–178). Springer.

Lee, P. T.-W., Hu, Z.-H., Lee, S.-J., Choi, K.-S., & Shin, S.-H. (2018). Research trends and agenda on the Belt and Road (B&R) initiative with a focus on maritime transport. *Maritime Policy & Management, 45*(3), 282–300. doi:10.1080/03088839.2017.1400189

Leiper, N. (1979). The framework of tourism. *Annals of Tourism Research, 6*(4), 390–407. doi:10.1016/0160-7383(79)90003-3

Lieven, A. (2017). Counter-insurgency in Pakistan: The role of legitimacy. *Small Wars & Insurgencies, 28*(1), 166–190. doi:10.1080/09592318.2016.1266128

Malik, A. R. (2018). The China–Pakistan Economic Corridor (CPEC): A Game Changer for Pakistan's Economy. In China's Global Rebalancing and the New Silk Road (pp. 69–83). Springer.

Meutia, M. (2013). Entrepreneurial Social Competence and Entrepreneurial Orientation to Build SME's Business Network and Business Performance. *International Journal of Social Science and Humanity, 3*(4), 381–385. doi:10.7763/IJSSH.2013.V3.266

Molz, J. G. (2013). Social networking technologies and the moral economy of alternative tourism: The case of couchsurfing.org. *Annals of Tourism Research, 43*, 210–230. doi:10.1016/j.annals.2013.08.001

Muhammad, S. B. (2014). Dynamics of the nationalist insurgency in the Balochistan province of Pakistan. Institute of World Economy and International Relations, 69.

TourismNotes. (2018). Tourism. Retrieved from https://tourismnotes.com/travel-tourism/

Park, D., Kim, J., Kim, W. G., & Park, H. (2019). Does distance matter? Examining the distance effect on tourists' multi-attraction travel behaviors. *Journal of Travel & Tourism Marketing, 36*(6), 693–710. doi:10.1080/10548408.2019.1624243

Pettenati, M. C., & Cigognini, M. E. (2007). Social networking theories and tools to support connectivist learning activities. *International Journal of Web-Based Learning and Teaching Technologies, 2*(3), 42–60. doi:10.4018/jwltt.2007070103

Pieterse, J. N. (2019). *Globalization and culture: Global mélange*. Rowman & Littlefield Publishers.

Puri, S. (2017). Brinkmanship, not COIN, in Pakistan's post-9/11 Internal War. *Small Wars & Insurgencies, 28*(1), 218–232. doi:10.1080/09592318.2016.1263471

Rabasa, A., Waxman, M., Larson, E. V., & Marcum, C. Y. (2004). *The Muslim world after 9/11*. Rand Corporation. doi:10.7249/RB151

Ritchie, J. R. B., & Crouch, G. I. (2003). *The competitive destination: A sustainable tourism perspective*. Cabi. doi:10.1079/9780851996646.0000

Siddiqa, A. (2019). Pakistan—From Hybrid-Democracy to Hybrid-Martial Law. *Journal of South Asian and Middle Eastern Studies, 42*(2), 52–72.

Siegel, A., Brickman, S., Goldberg, Z., & Pat-Horenczyk, R. (2019). Preventing Future Terrorism: Intervening on Youth Radicalization. In An International Perspective on Disasters and Children's Mental Health (pp. 391–418). Springer.

Smallbone, D., Kitching, J., & Athayde, R. (2010). Ethnic diversity, entrepreneurship and competitiveness in a global city. *International Small Business Journal*, *28*(2), 174–190. doi:10.1177/0266242609355856

Sönmez, S. F. (1998). Tourism, terrorism, and political instability. *Annals of Tourism Research*, *25*(2), 416–456. doi:10.1016/S0160-7383(97)00093-5

Spencer, J. P., & Bavuma, Z. (2018). How important are mice to the tourism economy? *The Business & Management Review*, *9*(4), 125–134.

Swanson, J. R., & Edgell, D. L. Sr. (2013). *Tourism policy and planning: Yesterday, today, and tomorrow*. Routledge.

Swarbrooke, J., & Horner, S. (2012). *Business travel and tourism*. Routledge. doi:10.4324/9780080490601

The News. (2016, March 17). 5,634 foreign tourists visited Pakistan in 2015, NA told.

Voinescu, R., & Moisoiu, C. (2019). Belt and Road Initiative and Possible Implications for Central and Eastern Europe Countries. *Global Economic Observer*, *7*(1), 195–204.

Wang, Y., & Bramwell, B. (2012). Heritage protection and tourism development priorities in Hangzhou, China: A political economy and governance perspective. *Tourism Management*, *33*(4), 988–998. doi:10.1016/j.tourman.2011.10.010

Wu, T.-P., & Wu, H.-C. (2019). The link between tourism activities and economic growth: Evidence from China's provinces. *Tourism and Hospitality Research*, *19*(1), 3–14. doi:10.1177/1467358416683769

Xinhua. (2018). China's inbound, outbound tourism markets expand in 2017: reports. Retrieved from http://usa.chinadaily.com.cn/a/201806/27/WS5b3371e aa3 1033 49141df4cf.html

Yang, Y., & Wong, K. K. F. (2012). The influence of cultural distance on China inbound tourism flows A panel data gravity model approach. *Asian Geographer*, *29*(1), 21–37. doi:10.1080/10225706.2012.662314

Yasir, M. (2017). Is Crime Rate Increasing in Pakistan as Chinese Immigrants Increase? Propakistani. Retrieved from https://propakistani.pk/problog/welcome-china-crimecorruption-pakistan/

KEY TERMS AND DEFINITIONS

Belt and Road Initiative (BRI): The Belt and Road Initiative (BRI) is a strategy to develop global infrastructure introduced and adopted by P.R. China with a big investment in more than 152 states and international organizations of Asia, Africa, the Middle East, Europe, and America.

Business Tourism: Business tourism is specialized area of overall tourism. During business tour, the tourist not only performs different business-related activities but also receives financial benefits. There are four types of activities (Meetings, Incentives, Conferences, Exhibitions) which are performed by a business tourist.

China Pakistan Economic Corridor (CPEC): CPEC is a mixture of infrastructure-based projects which are presently under structure across Pakistan. Under the umbrella of CPEC, a big network of railways, highways, and energy plants is under construction. CPEC was started in May 2013 with the initial investment of 46 billion US $ by Chinese and Pakistani governments.

Inbound and Outbound Tourism: When a country or destination receives tourists, it refers to inbound tourism and when tourists of country travel to other countries/ destinations, it refers to outbound tourism for that country.

Leisure Tourism: Leisure tourism refers to take a break from daily routine life. Leisure tourists usually enjoy beauty and nice hotels and resorts, enjoy atmosphere of beaches. The most popular activities of leisure tourism include social tours, cultural tours, religious tours, family tours, sports tours, and medical tours.

Religious Tourism: Religious tourism which is also denoted as faith tourism refers to a type of tourism where persons travel in groups or sometimes individually for missionary, pilgrimage, and internal (spiritual) leisure.

Tourism: Tourism refers to travel for the purpose of happiness or business. More generally, tourism denotes to outdoor activities during holidays at domestic and international levels.

Chapter 2
Spanish and Portuguese MICE Tourists' Profile

Noelia Araújo Vila
https://orcid.org/0000-0002-3395-8536
University of Vigo, Spain

Lucília Cardoso
CITUR – Centre for Tourism Research, Development, and Innovation, Portugal

Arthur Filipe de Araújo
Lusófona do Porto University, Portugal

Francisco Dias
CITUR – Centre for Tourism Research, Development, and Innovation, Portugal

ABSTRACT

Professional and academic congresses play a relevant role on what came to be known as meetings tourism or business tourism. This market has grown significantly in the last years and is expected to maintain, or even increase, such growth rates in the short range. Moreover, congress tourists consume complementary activities, as well as accommodations services, within the destination, which make them a very profitable segment. For these reasons, many cities are including this market in their promotional strategies. To illustrate this scenario, the chapter presents an analysis of Portuguese and Spanish congress tourists' profile. Results corroborate that congress attendees do some sightseeing, visit local cultural attractions, and taste local cuisine while in the destination. In other words, their consumption goes beyond the participation in the congress per se, and consequently, their average daily expenditure is significantly higher than that of the average tourist.

INTRODUCTION

The MICE (meetings, incentives, conventions and exhibitions) segment has been one of the fastest growing tourism trends in the last decade (Chiang *et al.*, 2012). The term includes all kinds of business

DOI: 10.4018/978-1-7998-3142-6.ch002

events, and in a broader sense, also cultural, sports, academic or corporate summits, symposia, conventions, congresses and workshops (Lau, 2009). Conferences and congresses are understood as events that gather people to discuss and interchange information about a certain subject, potentially aiming to generate knowledge and find a solution to a certain challenge. Other specific goals might include motivating and inspiring participants, releasing new products and communicating the latest research findings on a particular subject (Rogers, 2008).

Conferences have become one of the main boosters of tourism worldwide. Consequently, they have attracted significant amounts of investments, which ultimately create jobs and generate income to the host communities. Therefore, conferences are important business opportunities that greatly benefit the whole economy, not least because, as event participants also tend to engage in leisure activities while in the destination, their average daily expense is significantly higher than that of the average tourist. Moreover, they help reduce seasonality and regenerate destinations, spread knowledge and facilitate creativity and innovation (UNWTO, 2014). Big events also entail an effective use of resources and standardized services. Therefore, they boost many local businesses, which benefit from scale economies. Events also benefit the host place in the long run, as they tend to improve destination image (Lau, 2009). Moreover, conferences are usually a starting point for stablishing auspicious relationships among attendees, as many new ideas for joint projects are discussed. Consequently, some of positive impacts of conferences are projected in the future, and cannot be measured as its outcomes, although they really are.

Due to the mentioned positive outcomes, destinations worldwide have invested 1.14€ billion in the MICE segment in 2018, 28% of which coming from Europe (Global Business Travel Association - GBTA, 2019). In the last years, many places have invested in adequate facilities and services to cater for this segment. These destinations normally aim to integrate business tourism with elements of leisure travel, which is attained through MICE facilities' partnerships with hospitality, entertainment and transportation providers (Wan, 2011).

The latest trends within the MICE segment are consequences of the increasingly significant role played by technology in providing attractive experiences. In this vein, the demand for unconventional venues and activities, such as tree houses, rooftops and museums, must increase, especially in Europe. Accordingly, new collaborative event formats, which directly involve participants in the creation of their own experience, must also be developed (CWT Meetings & Events, 2019). Destinations and businesses willing to thrive in this segment must be in tune with these trends. Namely, they should increase security, provide personalised experiences that reinforce human relations, and use new technologies to provide immersive experiences (Hosteltur, 2019).

Regarding the event tourists, Swarbrooke and Horner (2001) define them as people who attend any type of events, including entrepreneurs, professionals, scholars, members of a certain group and people associated with governance. According to ICCA (International Congress and Convention Association, 2019), these visitors spend 53% more than leisure travellers in the visited places. In absolute numbers, according to the MICE Travel Report, carried out by IBTM Americas, this translates into a daily expenditure of U$239.90 in the context of the United States.

The addressed scenario evidences the relevance of events tourism within the global tourism market. For this reason, the present chapter is dedicated to this topic, with emphasis on the one of the sub-groups it encompasses: congress tourism. To achieve a better understanding of this specific market, its consumption patterns were characterised through a survey with 74 Spanish and Portuguese congress tourists. Results show that this public effectively spends more than the average tourist due to the complementary leisure activities and services – outside the event – that they consume whilst in the destination, and that their expenditure pattern is significantly related to several sociodemographic characteristics.

BUSINESS TOURISM

Within the context of the tourism system, events are a key element to both tourist generating regions – as they motivate travellers to visit host cities and regions – and destination regions – as they play a relevant role in destinations' development and marketing plans (Leiper, 1990). In this context, events are not only drivers of attraction, but mainly, key elements of destination marketing, which make them particularly important in the current scenario of global competitiveness amongst tourist destinations. Events also foment the development of the hospitality industry, transports, auxiliary services and even attractions, which improve places' tourist offer, allowing them to overcome the limited scope of leisure tourism.

Events planned for tourism development purposes aim to achieve specific objectives. What was once made mostly by individual and community initiatives has become the realm of professionals and entrepreneurs. In this context, there are four main types of events with relevance for local tourism industries: i) Business events, (or the MICE sector), which require convention and exhibition centres, as well as hotels, resorts and restaurants; ii) Sports events, which require purpose build facilities, such as sports parks, arenas and stadiums; iii) Festivals and cultural celebrations, which are less dependent on facilities, as they usually take place in parks, streets, theatres and concert rooms; and iv) Entertainment events, such as concerts, which are normally organised by private entrepreneurs and can take place in many different environments (Getz & Page, 2016).

Meetings or business tourism is part of a broader market, which encompasses conventions, congresses, incentive trips and corporate meetings. The segment has been defined by the United Nations' World Tourism Organization (UNWTO) as the activities of people who travel and stay outside their habitual area for business purposes. The definition excludes, however, regular trips made by workers or entrepreneurs to the same places, as those become part of their habitual area (Wootton, 1995). In a broader and more detailed definition, business tourism is understood as trips made by workers or entrepreneurs with the goal of participating in business or work meetings, attending congresses and conferences related to their professional field, visiting fairs, either as expositor or attendee, visiting partner companies or clients, doing benchmarking research, etc. (Gouvêa & dos Santos, 2005). Such definition, in turn, includes every displacement made for professional or business-related reasons.

Business tourism has been featured in travel and tourism textbooks since the 1990's (Lickorish et al., 1991). In these books, as well as within the industry, different terms, such as "professional tourism", "MICE (meetings, incentives, conventions and exhibitions)" and "events industry", are often employed interchangeably. To balance the meanings of all these terms, Eurostat (2000) establishes two key rules: they must exclude all event participants that are not tourists (local participants), and offer a coherent classification of what business tourism, or professional/business trips (see Figure 1), include.

Figure 1. Events and business tourism delimitation criterium
Source. European Commission. Eurostat (2000).

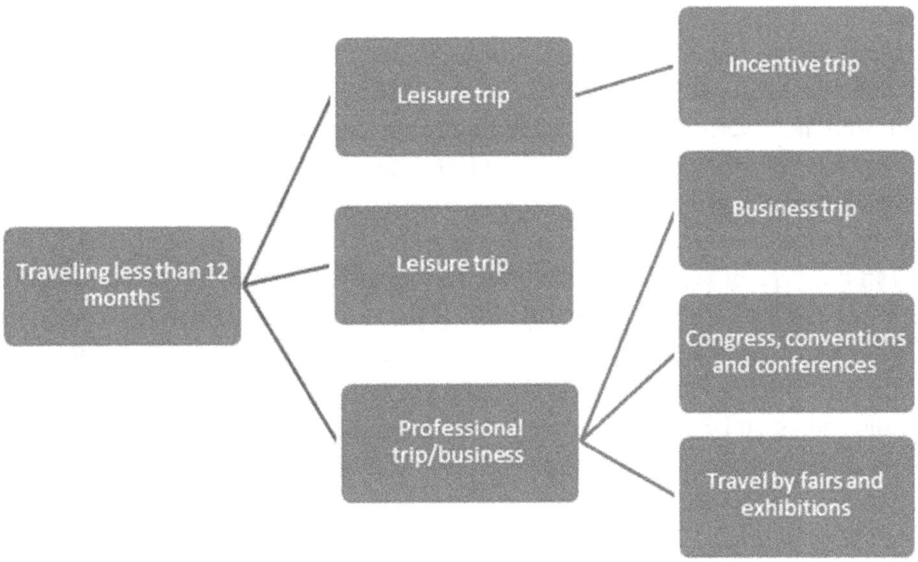

According to Smagina (2017), the business tourism marketing encompasses a set of multidirectional services. The market can be segmented in two groups according to the activities that are carried out: industry meetings (Meetings Industry/MICE) and corporate travellers' industry (travel industry). Business travel, in turn, can be divided into two industries, and its activities are the most important elements for the development and implementation of such category. Moreover, two distinct sectors may be described and estimated according to characteristics such as the objective of the global MICE industry, innovation, and entrepreneurial cooperation.

The main reason why business tourism has received so much attention from practitioners and researchers is the significance of the economic benefits it generates to host cities. For this reason, although the average stay of business tourists is shorter than that of leisure tourists, their average daily expenditure is far superior. Besides, business tourism has other characteristics that make it particularly advantageous from the economic stance. For instance, it is a less seasonal type of tourism, as its demand is not tied to weather conditions or to the school vacation period (Richards, 1992).

Other benefits of business tourism are listed by Clark (2008), who classifies them into two types: primary and secondary benefits. Primary benefits include the event's incorporation into the city's strategy, benefits to the economic sector or company involved, private and/or public investments, impacts to destination image (and the consequent attraction of visitors), dynamization of commerce and expansion of local economy (through the event's supply chain) and positive environmental impacts. The secondary benefits include infrastructure and equipment built or acquired for the event (which serve local population afterwards), commonly referred to as the event's legacy, impacts in the work market (which is not limited to temporary work) and on the real estate market, the destination's global positioning and knowledge transfer benefits. In sum, business tourism is one of the main drivers of local tourism development and an important generator of income, jobs and investments (especially foreign). Besides the economic benefits, the business tourism industry creates opportunities for knowledge interchange, creation of networks and development of capabilities, which makes it an important motor of intellectual development and regional cooperation (Smagina, 2017).

Due to these advantages, although business tourism is often referred to as an emergent segment, many cities already have a very mature offer to cater for it (Flamarich & Duro, 2011), aiming to capitalise on the potential benefits. Such cities also actively work in order to attract events. The strategies that they employ to this end, however, might vary according to the specific type of business tourism in question. According to Bigné et al. (2000) and Díaz (1995), the segment encompasses two main markets: the associative (congresses and conferences) and the corporate (company meetings). The latter can be attracted through promotional initiatives from the very own tourism companies (mainly hotels), as they are interested in this type of tourism, which brings about an additional demand for both their events/meetings facilities and their rooms. The promotion of the associative market, however, cannot rely on the private sector alone, since it requires initiatives from the academic or professional field. Moreover, the facilities they use, that is, congress and convention centres, are often public (Villena, 2014).

In face of the addressed benefits of business tourism, and the consequent initiatives of destinations to capitalise on it, the segment has been growing significantly since the 1980's (Peterson & Belchambers, 1990). Moreover, although business tourism has been affected by the changes in the national and global economies in the early 1990's, it continued to grow above average afterwards. Accordingly, the expenditure pattern of business tourists has also grown faster than that of leisure tourists. In sum, travel and tourism is an enormous industry, which has generated 10 billion dollars in the global economy in 2017; and business trips are an important part of such industry (Binimelis & Ordinas, 2003), as they contributed with 1.23 billion dollars in the same year (12.3%). Notwithstanding, this market is expected to grow even further within the next years, and its total expenditure is forecasted to increase 7% in 2019 (Statista, 2019). In this context, China will probably keep leading the global market, while other countries, such as India, must experience a significant growth.

The fast-paced globalisation of the world economy has generated new possibilities to the service market, both in developed countries and in emerging economies. Consequently, services have become a driver of regional development, and are already the fastest growing sector in international trade (Javalgi & White 2002). Naturally, as part of this process, the tourism industry has also been internationalised, which deeply changed the global dynamics of both leisure and business tourism. This ever-changing environment has attracted the attention of many researches, which have paid special attention to understanding the sectors' evolution, learning from the world leaders, identifying the main actors and analysing how they affect other stakeholders. Not enough attention has been paid, however, to the impact of business travel in particular countries or regions. The present study aims to contribute to filling this gap by analysing the profile of Portuguese and Spanish MICE tourists.

MICE TOURISM

In order to understand the concept of meetings tourism, first, the term "meeting" needs to be precisely defined. In the MICE tourism context, meetings refer to gatherings of at least 10 persons, for a minimum of 4 hours, in a pre-established place, with the purpose of carrying out a particular activity, which is often related to business endeavours (WTO, 2006). In this context, meetings tourism is understood as the activities of people travelling to outside their habitual surroundings, for at least 24 hours (with overnight), with the goal of participating in a meeting (according to the previously mentioned definition).

Although it was initially addressed in the 1990's, the concept of MICE, or MI (meetings industry) as known today, was introduced in 2006. The definitions were the result of a joint decision by the International Congress & Convention Association (ICCA), the Meeting Professionals International (MPI), the Reeds Travel Exhibitions and the UNWTO, which aimed to standardise concepts and create a stronger image for the sector (Davidson & Cope 2003; Rogers, 2008). MICE travellers, in turn, are "personas que realizan viajes personales o financiados por el empleador para asistir a reuniones, incentivos, convenciones y exposiciones, tanto en el hogar como en el extranjero" (Chiang, King & Nguyen, 2012, p. 104). Although MICE is most frequently described as segment of business tourism, it may also be related to the leisure component. However, since MICE is defined in very distinct ways according to the source, it is difficult to precisely define its volume in terms of tourist arrivals and spending (Centre for the Promotion of Imports from Developing Countries, 2016). Business travel, on the other hand, is conceptualised in a more consistent fashion, and therefore, can be more easily evaluated. In this context, according to the UNWTO (n.d.), business travellers accounted for 14% of global international tourist arrivals in 2012.

The UNWTO classifies meetings into 5 types, which are followingly addressed:

- Conventions and corporate/business meetings

Meetings normally organised or sponsored by a company, whereas attendees are the company's own employees and/or those of its clients, suppliers or partners. Expenses are typically covered by the company, since attendance is usually mandatory. Formal and informative company meetings, deliberation meetings and some exhibitions are examples of events included in this category.

- Association congresses

This is the type of meetings in which the present chapter focuses. Congresses are defined as non-business-oriented gatherings of big groups of people with the goal of discussing a certain topic, which can be of a professional, cultural, sports, academic or some other nature. Congresses typically last several days and usually include parallel sessions. Moreover, they usually take place with a regular frequency, typically every year or every two years.

- Fairs and commercial expositions

Commercial or cultural events that gather members of a certain business, professional or social sector, with the propose of promoting products or services and facilitating networking amongst members. These events typically consist of an exposition area where expositors show off their products or services to attendees.

- Incentive trips

Trips offered by companies to employees who have achieved certain objectives, such as sales or productivity targets, or whose work was considered exceptionally good. They aim to recognise and reward employees' good performances and to motivate the personnel to strive and achieve good results.

- Other meetings

This group includes all the meetings that fit in the previously mentioned criteria – gathering at least 10 persons in a pre-established place, for a minimum of four hours to discuss a certain topic – which do not fit in any of the four previous groups.

All these types of business events create a link between business and culture. International business programs affect the development of cultural tourism, which in turn, increase destinations' attractiveness. Moreover, the MICE industry encompasses commerce, transports, finance and travel. It was long being acknowledged as a profitable activity for destination regions, and consequently, an increasing number of destinations have actively worked towards capitalising on it. The industry is growing in a particularly fast pace in Asia, where new challenges and opportunities are constantly arising. In this context, new facilities, hotels, casinos resorts and theme parks are being created throughout the continent. This new product development, combined with the increasing global interest in Asia's economic development, translates into a significant predicted growth for the following years (Smagina, 2017).

According to the Centre for the Promotion of Imports from Developing Countries (2016), the MICE industry is a driver of economic, social and cultural development, as it:

- Adds value to destinations and improves local tourism economies, especially in the low season;
- Creates jobs and improves workforce qualification;
- Increases commerce through the promotion of international collaboration and the support to professional networks;
- Attracts investment by promoting host cities and regions.
- Creates opportunities for the regeneration of destinations.
- Spreads knowledge, improves creativity and fosters innovation.

Within the context of city tourism, meetings tourism is an important catalyser of socioeconomic growth, as well as a tertiary activity of great relevance for understanding economic and territorial logics (Ponce, 2006). In order to capitalise on this potential market, cities have to fulfil three basic criteria, according to the Trends Watch Report 2019: being safe and politically table (66%), offering good value for money (65%) and having good quality services (59%) (IBTM, 2019). Moreover, cities must also fulfil some more specific requirements, namely: availability of meeting rooms; good communication infrastructure and access network; an attractive city image; satisfactory institutional support; competitive prices; and good climate (Espejo & Ponce, 1999).

Naturally, big cities have an edge in fulfilling those requirements, and some cities are exceptionally successful in attracting events. This is the case of Barcelona, Paris, Vienna, Berlin and London, which made up ICCA's ranking's (of cities that host the most events) top 5 for three consecutive years, from 2015 to 2017. The positions among them, however, have changed. In 2017, Barcelona went back to the top for the first time since 2004; and in 2018, Madrid replaced London (which dropped to the 7th position) on the top 5.

Table 1. ICCA's cities ranking: 2017 and 2018

	2017		2018	
	City	Nº of meetings	City	Nº of meetings
1º	Barcelona	195	Paris	212
2º	Paris	190	Vienna	172
3º	Vienna	190	Madrid	165
4º	Berlin	185	Barcelona	163
5º	London	177	Berlin	162
6º	Singapore	160	Lisbon	152
7º	Madrid	153	London	150
8º	Prague	151	Singapore	145
9º	Lisbon	149	Prague	136
10º	Soul	142	Bangkok	135

Source. The authors, based on ICCA (2018)

Besides having two representatives (Barcelona and Madrid) in the cities' ranking, Spain also stands out in the countries' ranking, as it occupied the 4th position in 2017, and the 3rd in 2018. Although business tourism represents only 6% of Spanish tourism, it is a very profitable segment, not less because the average daily expenditure of business tourists in Spain is 208€ euros, while sun and beach tourists, for instance, spend an average of 146€ daily. Moreover, this segment has grown about 63% since 2013[1]. In this context, the Spanish MICE sector generated 6,000 million euros of revenue in 2017.

Table 2. ICCA's countries' ranking: 2017 and 2018

	2017		2018	
	Country	Nº of meetings	Country	Nº of meetings
1º	United States	941	United States	947
2º	Germany	682	Germany	642
3º	United Kingdom	592	Spain	595
4º	Spain	564	France	579
5º	Italy	515	United Kingdom	574
6º	France	506	Italy	522
7º	Japan	414	Japan	492
8º	China	376	China	449
9º	Canada	360	Netherlands	355
10º	Netherlands	307	Canada	315

Source. The authors, based on ICCA (2018)

The global expectations for the congress tourism segment are also positive, as the Trends Watch Report 2019 (IBTM, 2019) predicts a greater growth of professional associations (e.g., doctors, engineers, computer specialists, etc.). The report also identifies other trends that favour the segment, such as an increasing use of new technologies (especially, augmented reality) in events, and a greater demand for activities focused on wellness and cultural experiences. Moreover, the FMI (International Monetary Fund) predicts a global economic growth of 3.7%, which helps to minimize the sector's uncertainty.

OBJECTIVE AND METHODOLOGY

Given the importance of business travel & MICE, and the benefits this segment brings about to host destinations, the present study aims to characterise the profile of one of the sub-segments encompassed by this market: congress tourist. To this end, data was collected through a survey questionnaire consisting of 14 questions, which were divided in 5 parts: I) Transport and accommodation; II) Leisure activities carried out in the destination; III) Means through which the participant learned about the event (aiming to assess the promotional methods' effectiveness); IV) Daily average expenditure in the destination; and V) Sociodemographic profile. The survey was operationalised through an online platform and sent to scholars, specialists and professionals in Spain and Portugal, who have attended a congress within the last 24 months. The sample included 74 observations, whereas 34 were from Spain and 40 were from Portugal. The collected data was first analysed through descriptive statistics, in order to trace a general characterisation of the sample and their congress attendance behaviour. Next, bi-variate analysis, namely Pearson chi-square tests, was employed to explore possible relationships between variables.

RESULTS

Sample Characterization

As mentioned in the methodology section, a total of 74 valid questionnaires have been retrieved. The sample was relatively balanced in terms of age, with 54% of men and 46% of women. The predominant age range was 35-55 years (35.1%), followed by 25-34 and 45-54. The sample consists mostly of highly educated people, as the majority of respondents hold a Ph.D. tittle (63.5%), and most of the remaining hold a master's degree. Accordingly, fully dedicated Professor is the most frequent main occupation (48.6%), followed by Professors who are also researchers (16.2%) (a common situation in Spanish and Portuguese universities, formally referred to in the former as *PDI – Personal Docente e Investigador* (Teaching and Research Staff)). The sample's demographic characteristics are summarised in Table 3.

Table 3. Sample's demographic profile

Sociodemographic variable	n (sample=74)	%
Gender		
Male	40	54.1%
Female	34	45.9%
Age		
18-24 years	3	4.1%
25-34 years	20	27.0%
35-44 years	26	35.1%
45-54 years	19	25.7%
55-65 years	5	6.8%
65 years or more	1	1.4%
Education		
Graduate	1	1.4%
Master	26	35.1%
Ph.D.	47	63.5%
Main occupation		
Student	10	13.5%
Professor	34	48.6%
Researcher		9.5%
Professor + researcher	12	16.2%
Professor + student	2	2.7%
Researcher + student	4	5.4%
Professor + researcher + student	3	4.1%
Nationality		
Portuguese	40	54.1%
Spanish	34	45.9%

Participants attend an average of 2.5 congresses per year, whereas men attend slightly more (2.6) than women (2.3). The most common number for both genders is 2 (46.4% of men and 40.9% of women). Regarding the different age groups, people between 45 and 55 years old attend the most events per year, namely 3. Meanwhile, most of the remaining groups attend 2 congresses on average, the exception being participants between 18-24 years old, who attend only one. The data on congress attendance per age group is summarised on Figure 2.

Figure 2. number of congresses attended per year by age group

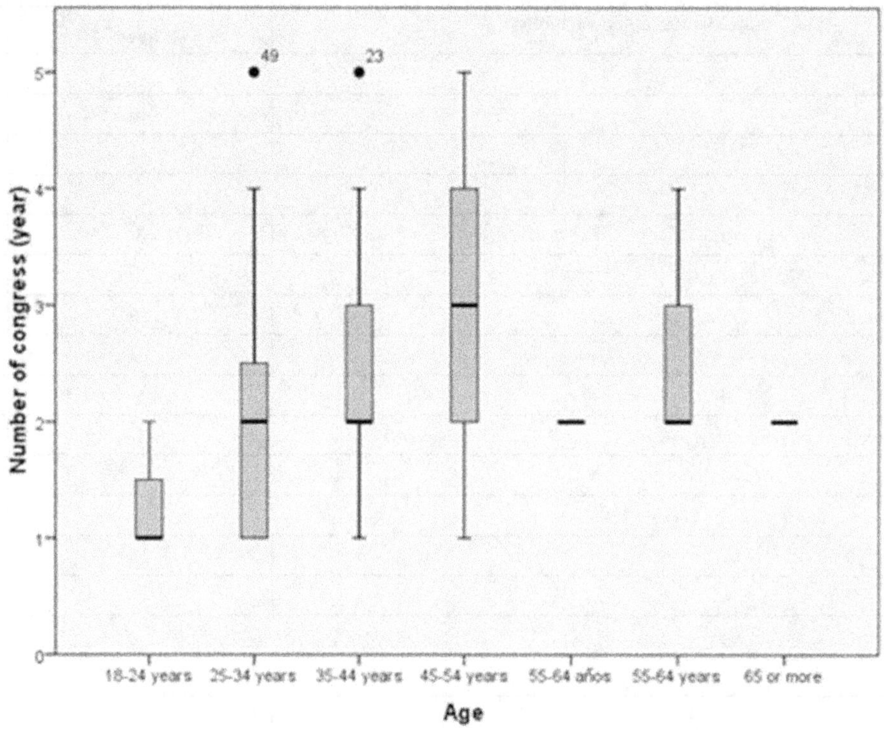

The great majority of respondents, 90%, informed that they have submitted communications to the congresses they attended. Considering that, typically, not all the co-authors attend the congress, this number is quite high. Regarding the people with whom respondents attended congresses, there were three options: by oneself (35.5%), with work colleagues (48.4%) and with partner (16.1%). The Pearson chi-square test indicated that this variable shows statistically significance relationship only with gender. In this context, men are more likely to go to congresses by themselves, while women more often go with their partners or work colleagues, as summarised in Figure 3.

Figure 3. Type of group differences between genders

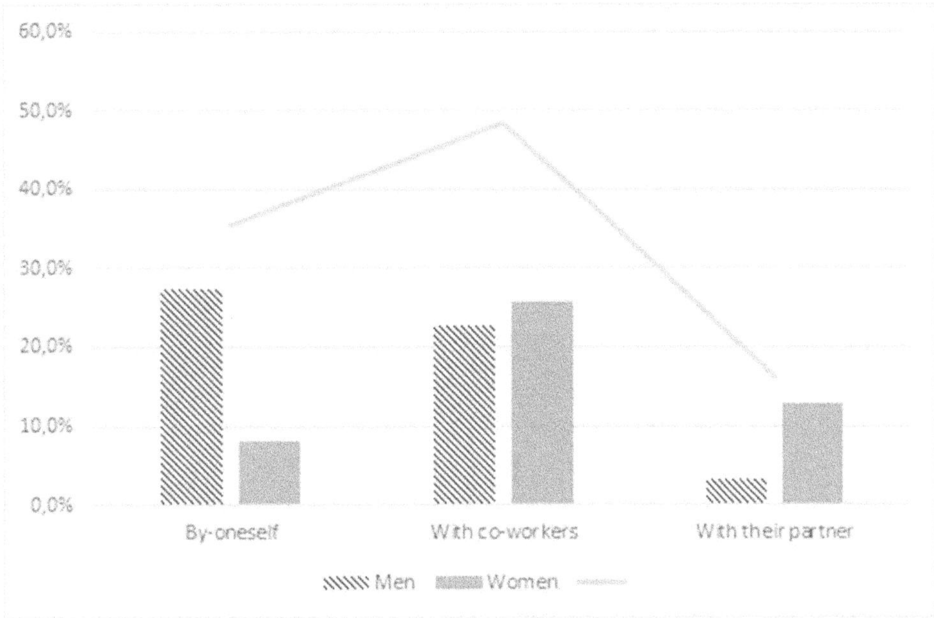

Transport and Accommodation

Regarding the means of transportation used by participants to get to the event's host destination, most drove their own car (60.8%). A significantly smaller portion of respondents travelled by train (13.5%) or rented car (8.1%). Respondents who travelled by plane had to use a secondary means of transportation from the airport, namely train (6.8% or the total sample) and rented car (5.4% of the total sample). A very small portion of participants also used a combination of train and bus (2.7%), train and taxi (2.7%) and plane and bus (1.4%), or shared a car (1.4%).

The Pearson chi-square test showed that three sociodemographic variables have a statistically significant relationship ($p < .05$) with the means of transportation used: age ($p = .000$), occupation ($p = .001$) and nationality ($p = .000$). Regarding age, the private car dominates in all groups. For respondents between 18 and 24 years old, shared car is the second most popular option; for respondents between 35 and 44 years old, the train is the second most popular means of transportation; for those between 45 and 54 it is the rented car; and for those between 55 and 65 it is the plane plus rented car. Regarding main occupations, the private car is again the most popular amongst all groups. The second place for students is the train, while for professors, it is plane plus train. Regarding nationality, after the private car, Spanish respondents mostly use the train, while the Portuguese mostly drive rented cars.

Regarding accommodations, 4-star hotels are the most popular option (55.9%), followed by 3-star hotels (28.8%), hostels (8.5%), 5-star hotels (3.4%) and friends' and relatives' homes (3.4%). The Pearson chi-square test did not point to any statistically significant relationship between the means of accommodation chosen and the sociodemographic variables.

Figure 4. Complementary tourism activities carried out by congress attendees.

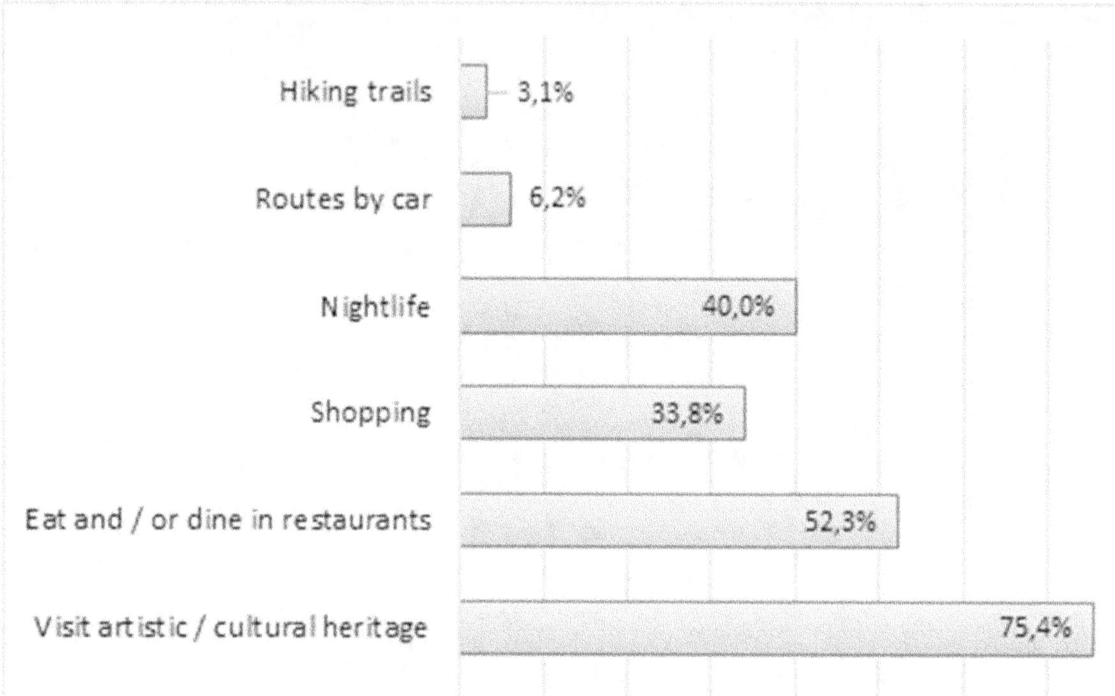

Figure 5. Means through which participants learned about the congresses

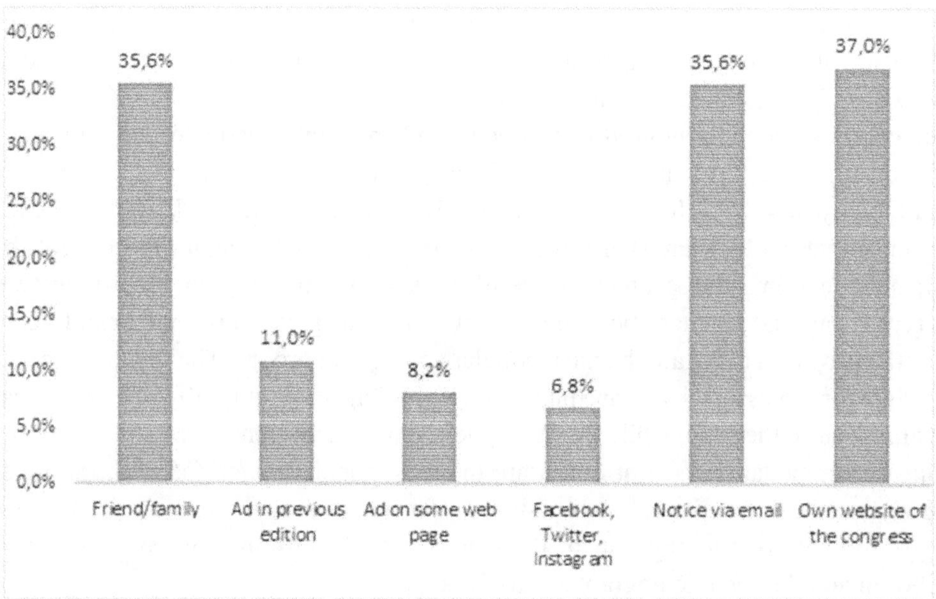

Figure 6. Participants daily expenditure variations among different age groups

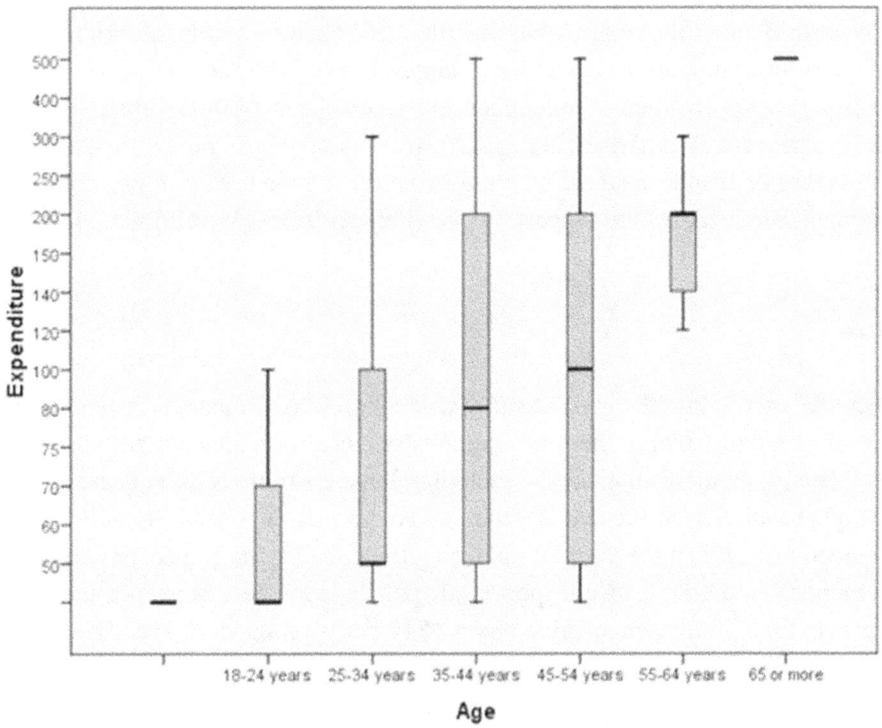

Complementary Tourism Within the Destination

As mentioned earlier, people who participate in congresses take advantage of being in a different place to do some sightseeing and other leisure activities. This has also been corroborated within the present study. As showed in Figure 4, walking around town is the activity carried out by most respondents (80%), followed by visiting cultural/artistic heritage sites (75.4%) and having lunch or dinner in restaurants (52.3%) – in some congresses, meals are included in the event, therefore, in these cases, eating in restaurants is considered a complementary tourism activity. The chi-square test pointed to a statistically significant relationship between the complementary tourism activities engaged in by participants and their level of formal education ($p = .001$). In this context, respondents who hold a Ph.D. tittle tend to combine the following activities: Taking a walk, experiencing local nightlife, visiting cultural/artistic heritage sites and having lunch or dinner in restaurants (10.6% of doctors). Still within doctors, others combine walking around town with shopping and visiting cultural/artistic heritage sites (8.5% of doctors). Meanwhile, master's degree holders tend to just take a walk and visit cultural/artistic heritage sites (11.5% of masters).

Promotional Methods' Effectiveness

When organising a congress, it is essential to promote it to the greatest possible amount of people who might be interested. In this context, the information provided at the end of previous editions is quite relevant, since 11% of participants stated that they mark it in their calendars right away. The most effective promotional method, however, it the event's own webpage, as it was the means through which 37% of participants learned about the congresses they attended. Therefore, it is important to keep it updated with new and interesting contents, as it is an important source of information. Next, e-mail alerts (35.6%) is a close second, along with friends, colleagues and relatives (35.6%). Results regarding promotional methods' effectiveness are summarised in Figure 5. It is interesting to notice that, unlike in most other contexts, for this type of event, social media platforms are not main means of promotion and interaction between participants. Indeed, results suggest they are the least relevant promotional method, among the alternatives.

Expenditure

The last aspect analysed within the present study was congress participants' expenditure patterns. This topic arises a lot of interest within the tourism industry, since previous studies show that the average expenditure of meetings tourists in general – including business tourists as well as congress tourists – is significantly higher than that of the average tourist. To assess this topic, respondents were asked how much they planned to spend on the destination during their stay, including accommodation, food, shopping, complementary activities and transportation (participation fees were not included). Responses varied from 50€ to 500€, the average value being 154€ per person. More specifically, 50€ is the most repeated value (14,9%), followed by 100€ (12%). Results also pointed to a statistically significant relationship between participants daily expenditure and their age ($p = .000$). As shown in Figure 6, younger participants tend to spend less money during the trip, while older participants spend significantly more.

CONCLUSION

Business tourism, or meetings tourism, has grown exponentially within the last decades, and is expected to grow even further. Meeting tourists are considered a high-quality segment, as although their average stay is shorter than that of leisure tourists, their average daily expenditure is considerably higher. In many cases, especially in conventions or business meetings, employers cover travellers' basic expenses (transport, accommodation and meals). Therefore, they become more willing to spend their own money in leisure activities, instead of limiting their trip to attending a specific event. In the case of congresses, these expenses are often covered by the participant's university or research centre, which also turns them more likely to spend more whining the destination.

The characterisation of Spanish and Portuguese congress tourists carried out within the present study corroborates all these trends, showing that they also apply to this specific research universe. For example, the results corroborate that participants do carry out complementary tourist activities whilst in the destination. More specifically, 80% of respondents said they walked around town while in the host destination, 75% visit cultural/artistic heritage sites and 52% has lunch or dinner in restaurants. These activities, combined with the necessary expenses associated with the travel, such as transports within the city and accommodation, add to the total economic impact of congress tourists in host destinations, making it a very profitable segment and an attractive alternative for destination managers willing to increase and stabilise tourism incomes.

The findings also corroborate an above average willingness to pay premium prices on high quality services. Regarding accommodation, for instance 4-stars hotels are an absolute majority (55.9%). Such result is expectable, considering the segment's previously addressed higher average expenditure. More specifically, respondents' average daily expenditure is 154€ per person. Meanwhile, the average daily expenditure of foreign leisure tourists visiting Spain and Portugal is 146€ (Instituto Nacional de Estadística, INE, 2019), and 79,20€ (Instituto Nacional de Estatística, 2019), respectively. Additionally, findings show that congress attendees' average daily expenditure increases with age. In this context, people between 55 and 64 spend on average 200€ daily, while those above 65 spend 500€.

These findings corroborate and illustrate the importance of MICE for the tourism industries in Portugal and Spain. Both countries are currently recovering from a recent economic crisis, and in both cases, tourism has been a key industry for such recovery. In the case of Spain, tourism has always been a major source of income, as the country leads the European ranking of destinations in terms of international tourists' expenses and is a close second to the United States in the world ranking. Moreover, when it comes to tourist arrivals, Spain is second only to France (UNTWO, 2019). In Portugal, on the other hand, the tourism industry has boomed within the last decade, as especially the last five years. The industry not only helped the country's economy recover from the crisis, but actually thrived, partially, as a result of such crisis, which stimulated domestic tourism (as many Portuguese could no longer afford international travel), attracted the more price sensitive European leisure tourist, and became an attractive alternative for entrepreneurs coming from other industries that were negatively affected.

Tourism in both countries, however, suffer with seasonality, as the offer is mostly based on the leisure market and sun and sea tourism accounts for a significant part of the demand. This particularly affects Portugal, whose seaside occupies a relatively greater portion of the territory, which also lacks well developed winter tourism offers. As inferred from the present study's findings, MICE tourism can be used as a tool for counter-balancing such effects, increasing the added value of destinations and the stability of tourism economy, especially on what concerns to the jobs they generate. In this context, cities and regions whose tourism offer is particularly prone to seasonality should attempt to bring or organise events to balance tourism demand and income throughout the year. Accordingly, places with an incipient tourism offer, but with some potential, should invest in events in order to capitalise on a new tourism demand.

Moreover, seasonality is not the only characteristic of tourist consumption that threatens the sustainability of the industry's economy. Concentration of tourists, especially in areas which are already dense, such the countries' biggest cities, also pose challenges to destination managers and other stakeholders, as it creates problems to residents and detracts the experience of the tourists themselves. This issue is more evident in Barcelona and the Balearic Islands, where protests against over tourism became a common sight. In the case of Portugal, Lisbon and Porto have been facing similar problems, although with less international notoriety. In this context, given the benefits of MICE tourism illustrated in the present study's findings, developing a MICE industry in periphery towns and regions should also help manage tourists' concentration in Portugal and Spain, spreading the activity's benefits more evenly and contributing to a better experience and improved attitude of locals.

Despite its contributions, the present study is not without its limitations. Some of the results could not be properly addressed, as they were beyond the study's scope, and required more detailed information to be fully comprehended. For instance, two diametrically distinct expenditure patterns of the congress tourists have been identified: 1. the cases in which the universities or research centres cover the travellers' basic costs, which include attendance fees, travel and accommodation; and 2. the cases in which individual attendees fully pay their own expenses. In other situations, the costs are shared between congress attendees and their home institutions. Future research on this subject could delve into the relationship between the patterns of expenditures and the specific situations regarding who covers the travel expenses.

REFERENCES

Bigné, J. E., Alcañiz, J. E. B., Font, X., & Andreu, L. (2000). *Marketing de destinos turísticos: análisis y estrategias de desarrollo* [Marketing of Tourist Destinations; analysis and development strategies]. Madrid: Esic editorial.

Binimelis, A., & Ordinas, A. (2003). El turismo de reuniones y negocios en Mallorca [Marketing and business tourism in Mallorca]. *Cuadernos de Turismo, 12*, 35–52.

Centre for the Promotion of Imports from Developing Countries. (2016). *CBI product factsheet: MICE from the EU.* Author.

Chiang, C. C., King, B., & Nguyen, T. H. (2012). Taiwan's MICE visitors: Business, leisure and education dimensions. *International Journal of Culture, Tourism and Hospitality Research, 6*(1), 21–33. doi:10.1108/17506181211206225

Clark, G. (2008). *Local development benefits from staging global events.* Publications de l'OCDE.

CWT Meetings & Events (2019). *Informe 2019 de CWT Meetings & Events.* Available from https://www.cwt-meetings-events.com/es/es/insights/meetings-and-events-trends/

Davidson, R., & Cope, B. (2003). *Business travel: Conferences, incentive travel, exhibitions, corporate hospitality and corporate travel.* Pearson Education.

Díaz, C. G. Q. (1995). El mercado de reuniones, congresos e incentivo en España [The market for meetings, conferences, and incentives in Spain]. *Estudios Turísticos, 126*, 179–190.

Espejo, C., & Ponce, M. D. (1999). Turismo de congresos y reuniones en la ciudad de Murcia [Congress and meeting tourism in the city of Murcia]. *Cuadernos de Turismo, 3*, 45–60.

Eurostat. (2000). *Methodological Manual for Satistics on Congresses and Conferences.* Brussels: Author.

Fenich, G. (1992). Convention center development: Pros, cons and unanswered questions. *International Journal of Hospitality Management, 11*(3), 183–196. doi:10.1016/0278-4319(92)90017-P

Flamarich, M., & Duro, J. A. (2011). Business (events) tourism in Spain: Analysis and prospects. *Papers de Turisme, 49/50*, 59–75.

Getz, D., & Page, S. J. (2016). Progress and prospects for event tourism research. *Tourism Management, 52*, 593–631. doi:10.1016/j.tourman.2015.03.007

Global Business Travel Association. (2013). *Global Business Travel Association (GBTA).* Retrieved from: http://www.gbta.org.

Global Business Travel Association. (2019). Available from https://www.gbta.org/

Gouvêa, M., & dos Santos, R. M. (2005). Marketing de servicios en el contexto del turismo de negocios [Marketing services in the context of business tourism]. *Estudios y Perspectivas en Turismo, 14*(2).

Hosteltur. (2019). *Las tres tendencias que mueven el turismo MICE en 2018* [The three trends that move MICE tourism in 2018]. Retrieved from https://www.hosteltur.com/126296_tres-tendencias-mueven-turismo-mice-2018.html

IBTM. (2019). *Trends Watch Report 2019.* Retrieved from https://www.ibtmworld.com/PageFiles/397712/IBTM%20World%202018%20Trends%20Watch%20Report.pdf?v=636789096955805051

IBTM Americas. (2019). *MICE Travel Report.* Available from https://puntomice.com/tag/ibtm-americas/

Instituto Nacional de Estadística de España. (2019). www.ine.es

Instituto Nacional de Estatística. (2019). https://www.ine.pt/xportal/xmain?xpid=INE&xpgid=ine_publicacoes&PUBLICACOEStipo=ea&PUBLICACOEScoleccao=107668&selTab=tab0&xlang=pt

International Congress and Convention Association. (2019). https://www.iccaworld.org/

Javalgi, R. G., & Steven White, D. (2002). Strategic challenges for the marketing of services internationally. *International Marketing Review, 19*(6), 563–581. doi:10.1108/02651330210451926

Lau, C. (2009). *Manual on elective. I, Meetings, incentives, conventions and exhibitions (MICE).* Hong Kong: Education Bureau.

Lee, M. J., & Back, K. J. (2005). A review of convention and meeting management research 1990–2003: Identification of statistical methods and subject areas. *Journal of Convention & Event Tourism, 7*(2), 1–20. doi:10.1300/J452v07n02_01

Leiper, N. (1990). Tourist attraction systems. *Annals of Tourism Research, 17*(3), 367–384. doi:10.1016/0160-7383(90)90004-B

Lickorish, L. J., Jefferson, A., Bodlender, J., & Jenkins, C. L. (1991). *Developing tourism destinations: policies and perspectives.* London: Financial Times Prentice Hall.

Petersen, J., & Belchambers, K. (1990). Business Travel – a Boom Market. In M. Quest (Ed.), *Horwath Book of Tourism* (pp. 56–65). London: Palgrave Macmillan. doi:10.1007/978-1-349-11687-4_6

Ponce, M. D. (2006). Las administraciones ante el turismo de reuniones y congresos en Murcia [The administrations before the tourism of meetings and congresses in Murcia]. *Papeles de Geografía, 44*, 95–118.

Richards, B. (1992). Tourism blues. *Leisure Management, 12*(11), 30–32.

Rogers, T. (2008). Forewords. In T. Rogers (Ed.), *Conferences and Conventions: A global industry* (pp. 13–17). Oxford: Butterworth Heinemann.

Statista. (2019). *Global Business Travel Industry – Statistics & facts.* Retrieved from https://www.statista.com/topics/2439/global-business-travel-industry/

Swarbrooke, J., & Horner, S. (2007). Main concepts in consumer behaviour, including models of consumer behaviour adapted for tourism. *Consumer Behaviour in Tourism*, 40-49.

UNTWO. (2019). *UNTWO Tourism Highlights, 2018 edition*. Retrieved from https://www.e-unwto.org/doi/pdf/10.18111/9789284419876

UNWTO. (2006). *Informe global sobre la industria de reuniones* [Global report on the meetings industry]. Retrieved from http://cf.cdn.unwto.org/sites/all/files/pdf/am_8_meetings_esp-red.pdf

UNWTO. (2014). *Global Report on the Meetings Industry*. Madrid: World Tourism Organization (UNWTO).

Villena Navarro, M. (2014). *Turismo de Negocios: Análisis del gasto diario de los turistas de negocios* [Business Tourism: Analysis of the daily expenditure of business tourists]. Cartagena: Universidad Politécnica de Cartagena.

Wan, Y. K. P. (2011, April). Assessing the strengths and weaknesses of Macao as an attractive meeting and convention destination: Perspectives of key informants. *Journal of Convention & Event Tourism, 12*(2), 129–151. doi:10.1080/15470148.2011.565704

Wootton, G., & Stevens, T. (1995). Business tourism: A study of the market for hotel-based meetings and its contribution to Wales's tourism. *Tourism Management, 16*(4), 305–313. doi:10.1016/0261-5177(95)00020-O

ADDITIONAL READING

Abulibdeh, A., & Zaidan, E. (2017). Empirical analysis of the cross-cultural information searching and travel behavior of business travelers: A case study of MICE travellers to Qatar in the Middle East. *Applied Geography (Sevenoaks, England), 85*, 152–162. doi:10.1016/j.apgeog.2017.06.001

Altareri, K. (2016). *Development of multicultural MICE tourism in the Middle East: the case of Saudi Arabia* (Doctoral dissertation). Retrieved from Victoria University Bibliography Database.

Becker, E. (2016). *Overbooked: The exploding business of travel and tourism*. New York: Simon and Schuster.

Getz, D., & Page, S. J. (2016). Progress and prospects for event tourism research. *Tourism Management, 52*, 593–631. doi:10.1016/j.tourman.2015.03.007

Rocha, E. D. S., Petry, T. R. E., & Riciardi, L. D. P. (2017). Business and events tourism in Balneário Camboriú, SC: Prospects of expansion and consolidation with the establishment of the events centre of the municipality. *Applied Tourism, 2*(2), 161–179. doi:10.14210/at.v2n2.p161-179

Rogers, T. (2008). Conferences and Conventions – A global industry, 2nd edition. Ljublja, Elsevier Ltd.

Rogerson, C. M. (2015). The uneven geography of business tourism in South Africa. *The South African Geographical Journal, 97*(2), 183–202. doi:10.1080/03736245.2015.1028984

Spenceley, A. (2015). Tourism and the IUCN World Parks Congress 2014. *Journal of Sustainable Tourism, 23*(7), 1114–1116. doi:10.1080/09669582.2015.1046704

Tsui, W. H. K., Balli, F., Tan, D. T. W., Lau, O., & Hasan, M. (2018). New Zealand business tourism: Exploring the impact of economic policy uncertainties. *Tourism Economics*, *24*(4), 386–417. doi:10.1177/1354816617731387

Uansa-ard, S., & Binprathan, A. (2018). Creating the awareness of halal MICE tourism business in Chiang Mai, Thailand. *International Journal of Tourism Policy*, *8*(3), 203–213. doi:10.1504/IJTP.2018.10015683

Vaid, J., Kesharwani, S., & Dubey, A. K. (2016). A Study of MICE Tourism Dynamics in Auto Expo 2016. *Global Journal of Enterprise Information System*, *8*(4).

KEY TERMS AND DEFINITIONS

Commercial Fair or Exposition: An event in which products, services, or even tourist destinations are exhibited, promoted, and traded.

Congress: A type of conference that typically takes place with a regular frequency and gathers members of a professional or academic area with the goal of discussing previously stablished issues.

Convention: An association meeting (of a political party, for example) aiming to deliberate programs, make decisions, or solve problems.

Corporate Meeting: A gathering of a certain number of people with common interests regarding an organisation, which might include employees, clients, suppliers, partners, or other stakeholders.

Incentive Trips: Trips, with all expenses covered, offered to employees as a form of acknowledging and rewarding his/her work performance. It is a commonly used management tool that aims to encourage dedication among an organisation's members.

Leisure Trip: A trip where the main goal is to rest or engage in leisure activities.

MICE: Meetings, incentives, conferencing, exhibitions. It includes trips where the primary objective is to attend an event.

ENDNOTE

1 According to data from the first journey of the Congress Cities network's and Spain Conventions Bureau's (SCB) annual meeting, which took place during the Spanish Federation of Municipalities and Provinces (FEMP) meeting, held in Benidorm.

Chapter 3
Work-Based Learning Strategies and Innovative Work Behavior:
Business Tourism Perspective

Sadia Shaheen
Government College University, Faisalabad, Pakistan

Muhammad Waseem Bari
Government College University, Faisalabad, Pakistan

ABSTRACT

This study investigates the relationship between learning strategies and innovative work behavior. The author chooses business organizations (e.g., textile industry, sports industry, and pharmaceutical companies) for the purpose of data collection. Particularly, those employees are targeted who avail business tours and conferences on a regular basis. Hence, the convenience sampling technique is used to collect data. A total of 800 questionnaires were distributed, but 535 proper filled questionnaires having response rate 66.87% were utilized. Data were collected in two-time waves. At T1 (Time lag 1), data were collected regarding employee demographics (e.g., gender, age, and education and work experience) as well as independent variables (e.g., cognitive learning strategies as well as behavioral learning strategies). At T2 (Time lag 2), data were collected on the dependent variable (innovative work behavior). The results of the study indicate that CLS and BLS have a positive impact on IWB. Managerial and theoretical implications are also discussed along with limitations and future directions.

INTRODUCTION

Learning is an essential part of professional life for sustainable competitive advantage and growth of individuals, groups as well as organizations (Kim & Lee, 2013; Widmann, Messmann, & Mulder, 2016). To obtain relevant knowledge, expertise and capabilities through formal ways (e.g., training, workshops, and conferences, diplomas, certificates and degrees) and informal ways (e.g., interaction with colleagues, learning that happens in the daily life, in the workplace, in the family, and in the communities) is es-

DOI: 10.4018/978-1-7998-3142-6.ch003

sential part of working life (Kortsch, Schulte, & Kauffeld, 2019). Both types of learning are essential for the growth and development of individuals and organizations (Nikolova, Van Ruysseveldt, De Witte, & Syroit, 2014).

Previously, the concerned of management was on formal learning and enhancing knowledge and expertise through formal ways of learning, but the trend has been changing with the changing demands of the working environment (Bunderson & Boumgarden, 2010). With the changing requirements of professional life, the practitioners focused on work-based learning (WBL) and thinking on enhancing their knowledge, skills, and expertise through interaction with other colleagues and communities as well (Malmberg et al., 2019). Enhancing employee's learning has become the necessity of the management to survive in the dynamic work settings (Kwakman, 2003). Despite having all necessary formal expertise and skills, informal learning has grabbed the attention of management to enrich employee's expertise as well as to face the changing demands of their stakeholders (Graen, Hui, & Taylor, 2006). Despite having the ardent importance of WBL in organizations, there is a few research evidence on how to enhance WBL as well as little is known about the underlying mechanism of enhancing employee's learning at the workplace, particularly contextual factors understudied. Holman, Epitropaki, & Fernie (2001) defined dimensions of WBL as Cognitive learning Strategies CLS and Behavioral Learning Strategies BLS. Organizations are motivating their employees to polish their expertise by utilizing both types of learning CLS as well as BLS so that they could be more creative, innovative and they could complete their professional assignments in a unique way (Penfold, 2008). Innovation at the workplace is essential for the individual as well organization to cope with the changing workplace demands (Widmann et al., 2016). The key ingredients of innovation and creativity are learning and WBL is understudied in relation to innovative work behavior (IWB). IWB is defined as the generation as well as the execution of new ideas (Wynen, Boon, Kleizen, & Verhoest, 2019).

Figure 1. Theoretical framework

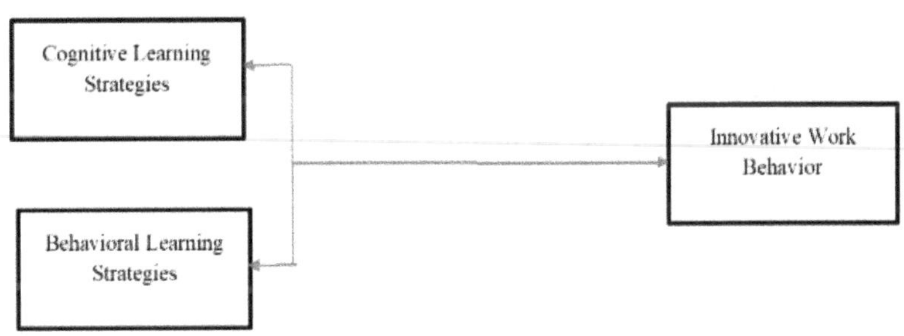

WBL is directly associated with employee IWB but this theoretical relationship gained scant attention from academic researchers (Graen et al., 2006; Warr & Downing, 2000). According to componential theory of creativity social and psychological components are essential for enhancing an individual's creativity at the workplace (Pandey, Gupta, & Gupta, 2019). We argue that employees interaction with colleagues and supervisors in business tourism (e.g. in conferences, seminars, business galas) and self-motivation such as learning through reflection and experimentation is positively related to IWB. On the basis of learning literature, we argue that in business tourism CLS and BLS result in employees IWB.

In business, tourism employees interact with new talent, workforce, innovative ideas of colleagues and enhance learning through interaction with new technology at a new place (Thomas, 2012). In business tourism both types of learning e.g. CLS and BLS could be boosted up which could have a positive impact on employees IWB. Based on available learning literature we propose a positive relationship between WBL and employee IWB. In the existing literature, a number of antecedents of IWB have been discussed but the theoretical relationship between learning strategies and IWB has remained unexplored in the existing literature.

Moreover, we tried to test the WBL in business tourism, for the stated purpose we choose the export sector, the textile sector and pharmaceutical organizations to investigate the impact of business tourism learning on IWB.

Most of the previous studies have been conducted in education sectors (Warr & Downing, 2000; Entwistle & Ramsden, 1983), but the focus of the current study is to test the role of learning strategies in business organizations.

There are a few key objectives of this chapter. First, to highlight the importance of WBL in business tourism. Second, a detail investigation of dimensions of WBL (e.g. CLS and BLS) in business tourism. Third, to investigate the impact of WBL on IWB we choose those employees who attend business conferences, business galas, and seminars. The purpose of this research evidence is to lighten the importance of learning in business tourism.

LITERATURE REVIEW

IWB

According to a number of researchers, there is a significant difference between creativity and innovation. Creativity is known as the generation of novel ideas while innovation is known as generation as well as promotion and implementation of new ideas (De Jong & Den Hartog, 2010; Williamson, Battisti, Leatherbee, & Gish, 2019). Hence innovation is a broader concept as a contrast to creativity (Janssen, 2000; Scott, Bruce, Scott, & Bruce, 2013; Amabile, 1996; Pieterse et al, 2010; Shalley Zhou & Oldham, 2004). IWB is essential for the positive change in individual behavior as well as organizational structure (Miron, Erez, & Naveh, 2004). Moreover, research evidence suggests that creativity and innovation foster outstanding performance, the adaptation of new ways of working and help in maintaining a sustainable competitive advantage (Mertins Heisig & Vorbeck, 2003; Palacios & José Garrigós 2006; Teece, 2007; Gottfridsson, 2012; Sundbo, 2009). There are numerous factors which influence on IWB such as leadership support, organizational structure, knowledge sharing practices, and individual personality traits (Acemoglu, Akcigit, & Kerr, 2016; Borgatti & Cross, 2003; Manuti, Pastore, Scardigno, Giancaspro, & Morciano, 2015; Sigala & Chalkiti, 2015) In order to enhance IWB employees need to engage in various activities such as a regular review of existing work methods and interaction with others (Kolb, International, Hempstead, & Lewin, 1987; Matsuo, 2019). Besides all other essential factors of motivating innovative behavior, we particularly focus on WBL. We argue that essential factors of enhancing employee IWB are CLS as well BLS. To test the significance of these learning strategies we establish a positive relationship between both learning strategies and employee IWB. In the following section of this study, we established the positive relationship between CLS and BLS on employee IWB based on the componential theory of creativity and available literature on learning strategies.

CLS AND IWB

Organizations need to be innovative so that they could remain alive as well as compete with the changing needs of their customers (Williamson et al., 2019). Recently, practitioners and scientists are also emphasizing that creativity and innovation are essential for the survival, growth and competitive advantage of any organization (Bos-Nehles & Veenendaal, 2017). For the stated reason organizations need to have competent individuals who can help them in the generation and implementation of innovative ideas (Nguyen, Nguyen, Do, & Nguyen, 2019). IWB is defined as generation and implementation of unique procedures regarding products as well as services and these unique procedures could be advantageous for the organization (Janssen, 2005; Scott et al., 2013). Researchers suggest employees cognitive and emotional resources are required to enhance IWB at the workplace (Carmeli, Meitar, & Weisberg, 2006). Other than emotional cognitive resources there are certain factors which play a significant role in enhancing IWB e.g. interaction with colleagues, supervisory support and guidance as well as conducive work environment (Manuti et al., 2015).

We argue that employee IWB could be enhanced through learning and learning is defined as one's devotion to, apprehension for, evolving to an individual's competencies, knowledge, skills, and abilities by gaining and sharing information about changing customers' needs, changing market conditions as well as opponents' actions (Atitumpong & Badir, 2018). Individual learning is a fundamental factor for IWB as well as organizational development (Margana, Widodo, & Mukhtar, 2019). According to Pandey, Gupta, & Gupta (2019) learning is connected with innovation because in learning phase individuals seek new ideas as well as get opportunities to implement those ideas. Hence, individual learning could be enhanced through learning strategies and learning strategies are defined as a behavioral and cognitive process that individuals use to gain and consolidate knowledge at the workplace (Holman et al., 2001). Learning strategies have been divided into two border categories e.g., cognitive learning strategies (CLS) and behavioral learning strategies (BLS) (Nikolova et al., 2014).

CLS is described as learning through reflection as well as experimentation. In CLS employees seek new information in the light of existing knowledge and enhance their existing knowledge with the help of different experimentation at the workplace (Holman et al., 2012; Kortsch et al., 2019; Nikolova et al., 2014). CLS is defined as a rational and intentional way of thinking which demands time and effort to learn new things (Matsuo, 2019). Researchers defined the concept of CLS with the help of two-dimension such as learning through experimentation and learning through reflection (Wielenga-Meijer, Taris, Kompier, & Wigboldus, 2010).

Experimentation and reflection are essentials for WBS as Kolb defined cognitive learning process in four sequential steps which are interrelated and have a strong impact on individuals' WBL (Kolb et al., 1987). According to Kolb different experiences at the workplace leads towards observation which ultimately results in observers' abstract conceptualization and investigation (Kolb et al., 1987; Sugarman, 1987). Wielenga-Meijer et al. (2010) defined learning through reflection and experimentation is an ongoing process of learning at the workplace where reflection demands knowledge acquisition and expansion in existing knowledge with the help of observation and experimentation. Particularly, we learn from everyday experiences and those everyday experiences direct us to make innovative experiences (Widmann et al., 2016). To take advantage of WBL and everyday experiences it is essential to reflect on everyday experiences to seek lessons from them. According to Matsuo (2019) reflection is described as the degree to which an individual regularly examines his work methods and procedures (Seggelen - Damen & Dam, 2016). According to this definition 'reflection is a temperament and predisposition to reflect on daily activities and work practices (Kolb et al., 1987). As well as, social cognition theory suggests reflection plays a motivational role in self-regulation (Matsuo, 2019). Employees who review their work activities and methods they seek numerous new ideas to learn how they could make their job better than ever (Wielenga-Meijer et al., 2010). In CLS employees review existing working procedures on regular basis and actively seek how they could improve their existing work procedures (Park, Song, Yoon, & Kim, 2014). Based on these arguments we hypothesize a positive relationship between CLS and IWB

H1: Cognitive learning strategies is positively related to innovative work behavior

BLS AND IWB

BLS is defined as the attainment of knowledge from colleagues as well as supervisor (e.g. peers, subordinates, partners, boss, leader, and manager) (Borgatti & Cross, 2003; Burt, 2004; Holman et al., 2012). BLS includes enrichment in existing knowledge through written as well as the practical presentation of ideas (Holman et al., 2001). By keeping on dimensionality and extensive literature review of learning strategies researchers introduced two dimensions of BLS which are known as learning from colleagues and learning from supervisor (Nikolova et al., 2014; Sugarman, 1987; Wielenga-Meijer et al., 2010). Learning from colleagues as well as supervisors is also known as learning through social networking, social networks and ties provide a strong ground to address work-related problems and solve those problems with the help of constructive exchange of ideas (Fritsch & Kauffeld-Monz, 2009; Luke, Maxine, Kamal, David, & Andy, 2004). Therefore, Fritsch & Kauffeld-Monz (2009) suggest networks and workplace relationships enhance individual's learning which leads towards IWB. (-) suggest support and constructive feedback from colleagues strengthen workplace ties which motivate employees to exchange ideas and knowledge (Afsar, 2016). Thus, this exchange of ideas help in the generation of the better solution of the problem and having multiple solutions to a problem leads towards IWB (Widmann et al., 2016). Through the mutual exchange of ideas with coworkers and supervisor employees seek constructive feedback and assistance which ultimately influence on employees innovation and creative outcomes. Getting useful ideas from coworkers is positively associated with the constructive solution and new idea generation which ultimately leads towards IWB (Amabile, Schatzel, Moneta, & Kramer, 2004; Matsuo, 2019). Hence, the overview of both types of learning strategies suggests that WBL could be enhanced through reflection and experimentation as well as interaction with colleagues and supervisor (Matsuo, 2019; Nikolova et al., 2014). Employees reflect their job activities by reviewing and comparing it with other colleagues (Matsuo, 2019).

Xerri & Brunetto (2011) examined the impact of social networks on employee IWB in engineering organizations and conclude that employees workplace relationships, the association among different groups and social networks are positively related with employee IWB. Regardless of advancement in decision-making techniques and tool still employees need social support to share and exchange information as well as they also need to gather information from their colleagues so they could make the right and innovative decision (Carolina, 2006). Bani-Melhem, Zeffane, & Albaity (2018) hypothesized that peers support, workplace happiness and job stress are positively associated with IWB. This study has been conducted in the hotel industry of the United Arab Emirates but according to the statistical result of the study, only job stress in not positively related with IWB but workplace happiness and peers support are positively related with IWB.

Qammach (2016) Suggest that knowledge sharing through different ways such as discussion with colleagues, exchange of information and learning through observation result in IWB. As well as Darroch & Mcnaughton (2002) suggest employees involvement in knowledge sharing through social networks result in employee IWB.

As well as, interaction with colleagues and supervisor influence on employees learning which motivate them in improving their WBL (Nikolova et al., 2014). Both types of learning could be enhanced through interaction with colleagues and peers and these types of informative meetings happen in business tourism where people meet with other colleagues and share or exchange ideas (Li & Liu, 2019; Widmann et al., 2016). Business tourism is also known as the knowledge-based activity where individuals learn by direct or indirect interaction with others (Richards, 2011; Shanker, 2008). This direct and indirect interaction has been discussed under cognitive and behavioral learning strategies. Therefore, we suggest that cognitive learning and behavioral learning is linked with business tourism which helps individuals in polishing their knowledge, skills, and abilities.

H2: Behavioral learning strategies are positively related to innovative work behavior

METHODS

Population and Sample

The data are collected under a research plan which aimed to inspect the impact of learning strategies on employee IWB in export, textile and pharmaceutical organizations of Pakistan. The key reason behind the selection of this population is that the management of these organizations actively participate in business tours and exchange their knowledge in business tours, conferences and seminars (McAdam & McClelland, 2002; Montani, Battistelli, & Odoardi, 2017). Therefore, the focus of this study is particularly on those employees who actively attend business tours. Due to the stated reason conveniences sampling technique is used to approach targeted employees of these organizations.

The respondents were recruited by following a step by step procedure. The principal author contacted the head of the human resource department of ten textile, ten export, and ten pharmaceutical organizations and explained the reason behind data collection. The questionnaire was distributed along with a detail cover letter and in a cover letter, a detail description was provided that the author did not know any participant personally. No answer is right and wrong only your opinion is required. Moreover, all responses will be kept confidential and will have no individual identity.

Data were collected in two-time waves to solve the issue of common method bias. In the first wave (T1), participants were requested to fill the demographic information e.g. gender, age, education, and work experience as well as theoretical variables CLS and BLS. In wave two (T2), data are collected regarding employee IWB. Total one month gap was given between wave 1 and wave 2 to separate independent variables from dependent variables.

Total of 800 questionnaires was distributed in all organizations. After discarding incomplete information the data were analyzed on 535 observations (a 66.87% response rate). The sample characteristics are presented and Table 1.

Table 1. Demographic characteristics of respondent

Respondent's Demographic characteristics	Percentage
Gender	
Male	64.1
Female	45.9
Age in Years	
20-30	5.4
30-40	49
40-50	44.1
50-60	1.5
Education	
Metric	0.2
Intermediate	25.4
Graduation	29.9
Masters	29.7
Any other	14.8
Work Experience	
1-5	13.8
6-10	26.2
11-15	30.5
16-20	29.5

MEASURES

CLS

Cognitive Learning Strategies were measured by using 8-items scale developed Holman et al. (2012) by and further investigated by (Holman et al., 2001). The sample items of the scale are ''I think about how my work ðts into the ''bigger picture'' at this organization. ''I try to think how the different parts of this ðt together. I try to think how my work relates to that of others at this organization. The α reliability of this scale was 0.87.

BLS

Behavioral Learning Strategies were measured by using 7-items scale developed by Holman et al. (2012) and further investigated by (Holman et al., 2001). The sample items of the scale are ''I try out new things by applying them in practice'' ''I do practical things to help myself to learn'' ''I ask others questions when I am uncertain about something''. The α reliability of this scale was 0.77.

IWB

Innovative Work Behavior was measured by using 9-items scale developed by (Janssen, 2000) based on (Scott et al., 2013). Sample items of the scale were ''Creating new ideas for difficult issues'' ''Searching out new working methods, techniques, or instruments'' ''Generating original solutions for problems''. The α reliability of this scale was 0.93.

Table 2. The α reliabilities of all theoretical variables.

α reliabilities

Variables	α reliabilities
Cognitive Learning Strategies	0.87
Behavioral Learning Strategies	0.77
Innovative Work Behavior	0.93

RESULTS

Correlation Analysis

Table 3 represents the mean, standard deviation and Correlation analysis of all the variables. According to statistical results CLS are positively and significantly related to BLS (r=.022. P<0.05), BLS are positively and significantly related to IWB (r=0.104, p<0.05).

Table 3. Correlation analysis

Table 4 (a) Original Model

Model	?2	DF	RMSEA	IFI	TLI	CFI
Original Model	5.550	249	092	0.87	0.83	0.846

?2= Chi-square, DF=Degree of freedom, RMSEA= Root mean square of approximation, IFI=Incremental fit index, TLI= Truker-Lewis fit,CFI= Comparative fit index

Table 4 (b) Revised Model

Model	?2	DF	RMSEA	IFI	TLI	CFI
Revised Model	1.59	230	0.70	0.92	0.903	0.919

?2= Chi-square, DF=Degree of freedom, RMSEA= Root mean square of approximation, IFI=Incremental fit index, TLI= Truker-Lewis fit,CFI= Comparative fit index

Measurement Model

Structural equation modeling SEM was used through AMOS to test the hypothesized relationships of the theoretical variables. Confirmatory factor analysis CFA was performed before SEM to ratify measurement model. The measurement model comprised of a total of three theoretical variables, two independent variables e.g. cognitive learning strategies, behavioral learning strategies, and one dependent variable innovative work behavior. To confirm the measurement model we used model $\chi 2$, comparative fit index (CFI), Truker-Lewis fit (TLI), Comparative fit index (CFI) and root mean square of approximation (RMSEA).

A Hu & Bentler (1999) and Kline (2005) Insignificant $\chi 2$ indicates a good fit if the model and the threshold values of CFI, TLI, and IFI are 0.95 and above are acceptable The values of RMSEA below 0.05 are considered the good fit of the model. Table 4 represents the values of $\chi 2$, CFI, TLI, and IFI of the original model. However, in the original model, we could not reach the goodness of fit values

Table 4. (a) original model (b) revised model

Correlation Analysis

	Mean	Mean	S.D	1	2	3	4	5	6	7
1	Gender	1.35	0.48	-						
2	Age	2.41	0.61	.006	-					
3	Education	3.33	1.01	.001	-.04	-				
4	Tenure	3.75	1.02	-.032	-.106	.115*	-			
5	CLS	2.34	0.79	.030	-.047	-.074	.022	-		
6	BLS	2.37	0.72	-.016	-.062	.030	.006	.104	-	
7	IWB	2.13	0.88	-.015	.008	-.006	.040	.614**	.174**	-

CLS=Cognitive Learning Strategies, BLS=Behavioral Learning Strategies, IWB=Innovative Work Behavior

Figure 2. Original model

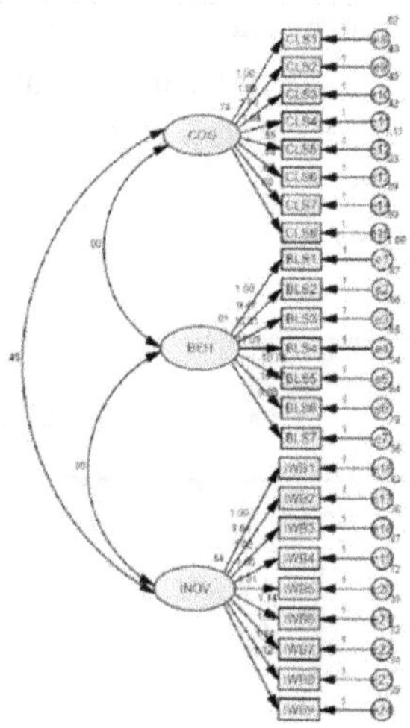

Figure 3. Revised model

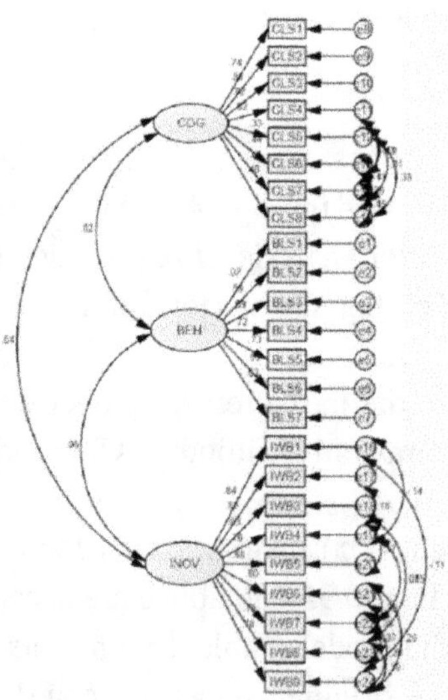

therefore, we revised the analysis with few modifications and both results are presented in Table 4 (A) and Table 4 (B). After performing modification analysis we got the goodness of fit of CFI, TLI, IFI as well as RMSEA, table 4 (B) revised model represents and all values are within an acceptable range. Hence, proving a good model fit.

HYPOTHESIS TESTING

Figure 4 represents the results of path analysis. Hypothesis 1 states there is a positive relationship between CLS and IWB which is supported with statistical results (β=.667, $P<0.05$), therefore, hypothesis 1 is accepted. Hypothesis 2 states there is a positive relationship between BLS and IWB and this is also supported by statistical results (β=.134, $P<0.01$). Hence, hypothesis 2 is also accepted.

Figure 4. Path Analysis

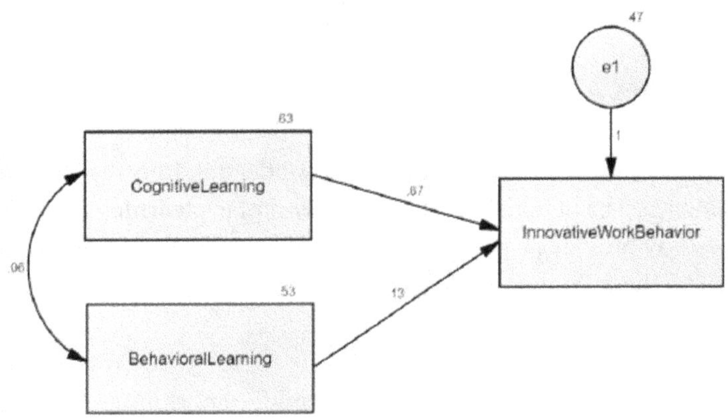

DISCUSSION

The results of the study prove that there is a positive relationship between CLS and IWB and BLS and IWB. On the basis of the componential theory of creativity, we propose a positive relationship between learning strategies and innovative work behavior. There are a number of research evidence which states that employees WBL result in an enhancement in their existing knowledge, skills, and expertise. Based on this argument we investigate the impact of WBL on employee IWB.

The results of the study are consistent with the existing studies. WBL strategies such as CLS and BLS results in employee development, enhancement in knowledge as well as in employee creativity (Holman et al., 2012; Pintrich & Garcia, 1991). IWB is influenced by a number of sources such as psychological and physical resources, relationships with colleagues and leadership support (Janssen, 2004; Scott et al., 2013; Warr & Downing, 2000; Young, 2012). Examined that learning and development are positively associated with IWB. CLS and BLS result in engagement in daily work activities as well as interaction with peers, colleagues, and supervisor which leads them in improving their work activities and enrichment in existing knowledge (Warr & Downing, 2000). Employee WBL is positively related to employees' problem-solving skills and practical knowledge (Kolb & Kolb, 1999; Warr & Bunce, 1995).

The impact of learning strategies on employee knowledge has been relatedly examined in existing literature based on this literature and componential theory of creativity we establish a positive relationship between WBL and IWB. We got good support for our hypothesis as the statistical also support for our hypothesized relationships. We collect data from textile, exports and pharmaceutical organizations. We found that the individuals who regularly reflect and experiment their daily work activities, as well as employees who interact with other colleagues and supervisors, are more innovative, anxious towards learning. Particularly, business tourism and business meetings enhance employees' knowledge and they learn a number of unique ways to fulfill their work assignments. In interaction with colleagues, employees exchange their ideas, knowledge, and ways of working and in this interaction employees intentionally and un-intentionally learn new ways of performing their job. Therefore, we argue that business tourism plays an important role in enhancing employee IWB.

THEORETICAL AND MANAGERIAL IMPLICATIONS

Managerial Implications

The current study helps managers and practitioners in understanding the importance of workplace learning. Moreover, it also assists in understanding how workplace learning could be enhanced through business tourism. This study provides a detailed analysis of both WBL strategies such as CLS and BLS on employee IWB. Hence, the following are a few recommendations on how they could enhance WBL through business tourism:

- Employees should be provided with workplace learning opportunities, such as reflection and experimentation on daily work activities should be appreciated by leader and leaders should provide their followers/subordinates opportunities to review their daily work activities and learn from them.
- Interaction with colleagues is necessary for enhancement in one's knowledge and expertise. It is the responsibility of the management to provide a platform to their employees where they could interact with their colleagues and exchange their ideas.
- Business meetings, conferences, and business galas are the best platforms to exchange ideas and knowledge. Hence, management should take proactive actions in regularly conducting these events in the workplace.
- Management should support and encourage their employees to participate in business conferences and seminars through tangible and non-tangible rewards.

Practical Implications

WBL learning strategies have been repeatedly studied and empirically tested in education sectors. But there are a few studies in the existing literature, where the importance of WBL has been tested in business organizations. This study has been conducted in business organizations such as the textile industry, sports industry, and pharmaceutical organizations. Moreover, the relationship of WBL has been tested previously in relation to employees' development and enhancement in knowledge. In this study, we have tested the impact of WBL on employee IWB. Moreover, based on the componential theory of creativity we infer that WBL results in IWB.

FUTURE DIRECTIONS AND LIMITATIONS

Despite having many positive points there are a few limitations of this study. First, we use a direct impact on WBL on IWB but this relationship could be investigated with meditating and moderating mechanism. Because, there are a number of factors which could explain this direct relationship such as supervisory support, leader-member exchange, and individual indicators such as employee motivation and personality traits. Second, we collect data only from those organizations who support business tourism, a comparative study (between organizations who support business tourism and organizations who do not support business tourism) can present a clear picture of WPL. Third, although we tried to overcome the common method bias problem still it is a single source study and data are collected only from employees. For better results, data should be collected from multiple sources (e.g. employee as well as their respective supervisors).

REFERENCES

Acemoglu, D., Akcigit, U., & Kerr, W. R. (2016). Innovation network. *Proceedings of the National Academy of Sciences of the United States of America*, *113*(41), 11483–11488. doi:10.1073/pnas.1613559113 PMID:27681628

Afsar, B. (2016). The impact of person-organization fit on innovative work behavior: The mediating effect of knowledge sharing behavior. *International Journal of Health Care Quality Assurance*, *29*(2), 104–122. doi:10.1108/IJHCQA-01-2015-0017 PMID:26959893

Amabile, T. M., Schatzel, E. A., Moneta, G. B., & Kramer, S. J. (2004). Leader behaviors and the work environment for creativity: Perceived leader support. *The Leadership Quarterly*, *15*(1), 5–32. doi:10.1016/j.leaqua.2003.12.003

Atitumpong, A., & Badir, Y. F. (2018). Leader-member exchange, learning orientation and innovative work behavior. *Journal of Workplace Learning*, *30*(1), 32–47. doi:10.1108/JWL-01-2017-0005

Bani-Melhem, S., Zeffane, R., & Albaity, M. (2018). Determinants of employees' innovative behavior. *International Journal of Contemporary Hospitality Management*, *30*(3), 1601–1620. doi:10.1108/IJCHM-02-2017-0079

Borgatti, S. P., & Cross, R. (2003). A relational view of information seeking and learning in social networks. *Management Science*, *49*(4), 432–445. doi:10.1287/mnsc.49.4.432.14428

Bos-Nehles, A. C., & Veenendaal, A. A. R. (2017). Perceptions of HR practices and innovative work behavior: The moderating effect of an innovative climate. *International Journal of Human Resource Management*, *5192*, 1–23. doi:10.1080/09585192.2017.1380680

Bunderson, J. S., & Boumgarden, P. (2010). Structure and learning in self-managed teams: Why "bureaucratic" teams can be better learners. *Organization Science*, *21*(3), 609–624. doi:10.1287/orsc.1090.0483

Burt, R. S. (2004). Structural holes and good ideas. *American Journal of Sociology*, *110*(2), 349–399. doi:10.1086/421787

Carmeli, A., Meitar, R., & Weisberg, J. (2006). Self-leadership skills and innovative behavior at work. *International Journal of Manpower*, *27*(1), 75–90. doi:10.1108/01437720610652853

Carolina, N. (2006). Building Collegial Information Networks : A Theory of Advice Network Generation. *Academy of Management Review*, *31*(3), 615–637. doi:10.5465/amr.2006.21318921

Darroch, J., & Mcnaughton, R. (2002). Examining the link between knowledge management practices and types of innovation. *Journal of Intellectual Capital*, *3*(3), 210–222. doi:10.1108/14691930210435570

De Jong, J., & Den Hartog, D. (2010). Measuring innovative work behaviour. *Creativity and Innovation Management*, *19*(1), 23–36. doi:10.1111/j.1467-8691.2010.00547.x

Fritsch, M., & Kauffeld-Monz, M. (2009). The impact of network structure on knowledge transfer: An application of social network analysis in the context of regional innovation networks. *The Annals of Regional Science*, *44*(1), 21–38. doi:10.100700168-008-0245-8

Graen, G. B., Hui, C., & Taylor, E. A. (2006). Experience-based learning about LMX leadership and fairness in project teams: A dyadic directional approach. *Academy of Management Learning & Education*, *5*(4), 448–460. doi:10.5465/amle.2006.23473205

Holman, D., Epitropaki, O., & Fernie, S. (2001). Understanding learning strategies in the workplace: A factor analytic investigation. *Journal of Occupational and Organizational Psychology*, *74*(5), 675–681. doi:10.1348/096317901167587

Holman, D., Totterdell, P., Axtell, C., Stride, C., Port, R., Svensson, R., & Zibarras, L. (2012). Job Design and the Employee Innovation Process: The Mediating Role of Learning Strategies. *Journal of Business and Psychology*, *27*(2), 177–191. doi:10.100710869-011-9242-5

Hu, L. T., & Bentler, P. M. (1999). Cutoff criteria for fit indexes in covariance structure analysis: Conventional criteria versus new alternatives. *Structural Equation Modeling*, *6*(1), 1–55. doi:10.1080/10705519909540118

Janssen, O. (2000). Job demands, perceptions of effort-reward fairness and innovative work behaviour. *Journal of Occupational and Organizational Psychology*, *73*(3), 287–302. doi:10.1348/096317900167038

Janssen, O. (2004). How fairness perceptions make innovative behavior more or less stressful. *Journal of Organizational Behavior*, *25*(2), 201–215. doi:10.1002/job.238

Janssen, O. (2005). The joint impact of perceived influence and supervisor supportiveness on employee innovative behaviour. *Journal of Occupational and Organizational Psychology*, *78*(4), 573–579. doi:10.1348/096317905X25823

Kim, T. T., & Lee, G. (2013). Hospitality employee knowledge-sharing behaviors in the relationship between goal orientations and service innovative behavior. *International Journal of Hospitality Management, 34*(1), 324–337. doi:10.1016/j.ijhm.2013.04.009

Kortsch, T., Schulte, E. M., & Kauffeld, S. (2019). Learning @ work: Informal learning strategies of German craft workers. *European Journal of Training and Development, 43*(5–6), 418–434. doi:10.1108/EJTD-06-2018-0052

Kwakman, K. (2003). Factors affecting teachers' participation in professional learning activities. *Teaching and Teacher Education, 19*(2), 149–170. doi:10.1016/S0742-051X(02)00101-4

Li, Y. Q., & Liu, C. H. (2019). The power of coworkers in service innovation: The moderating role of social interaction. *International Journal of Human Resource Management, 30*(12), 1956–1976. doi:10.1080/09585192.2017.1314310

Luke, P., Maxine, R., Kamal, M., David, D., & Andy, N. (2004). Networking and innovation: A systematic review of the evidence. *International Journal of Management Reviews, 5–6*(3–4), 137–168. doi:10.1111/j.1460-8545.2004.00101.x

Malmberg, J., Järvelä, S., Holappa, J., Haataja, E., Huang, X., & Siipo, A. (2019). Going beyond what is visible: What multichannel data can reveal about interaction in the context of collaborative learning? *Computers in Human Behavior, 96*, 235–245. doi:10.1016/j.chb.2018.06.030

Manuti, A., Pastore, S., Scardigno, A. F., Giancaspro, M. L., & Morciano, D. (2015). Formal and informal learning in the workplace: A research review. *International Journal of Training and Development, 19*(1), 1–17. doi:10.1111/ijtd.12044

Margana, G. S., Widodo, S. E., & Mukhtar, M. (2019). *A Study of the Influence of Transformational Leadership, Organizational Learning on Employees Innovative Behavior and Work Engagement at Sekolah Bahasa Polri*. Academic Press.

Matsuo, M. (2019). Effect of learning goal orientation on work engagement through job crafting: A moderated mediation approach. *Personnel Review, 48*(1), 220–233. doi:10.1108/PR-11-2017-0346

McAdam, R., & McClelland, J. (2002). Sources of new product ideas and creativity practices in the UK textile industry. *Technovation, 22*(2), 113–121. doi:10.1016/S0166-4972(01)00002-5

Miron, E., Erez, M., & Naveh, E. (2004). Do personal characteristics and cultural values that promote innovation, quality, and efficiency compete or complement each other? *Journal of Organizational Behavior, 25*(2), 175–199. doi:10.1002/job.237

Montani, F., Battistelli, A., & Odoardi, C. (2017). Proactive Goal Generation and Innovative Work Behavior: The Moderating Role of Affective Commitment, Production Ownership and Leader Support for Innovation. *The Journal of Creative Behavior, 51*(2), 107–127. doi:10.1002/jocb.89

Nguyen, T. P. L., Nguyen, K. N., Do, T. D., & Nguyen, T. T. M. (2019). Knowledge sharing and innovative work behavior: The case of Vietnam. *Uncertain Supply Chain Management, 7*, 619–634. doi:10.5267/j.uscm.2019.5.001

Nikolova, I., Van Ruysseveldt, J., De Witte, H., & Syroit, J. (2014). Work-based learning: Development and validation of a scale measuring the learning potential of the workplace (LPW). *Journal of Vocational Behavior, 84*(1), 1–10. doi:10.1016/j.jvb.2013.09.004

Pandey, A., Gupta, V., & Gupta, R. K. (2019). Spirituality and innovative behaviour in teams: Examining the mediating role of team learning. *IIMB Management Review, 31*(2), 116–126. doi:10.1016/j.iimb.2019.03.013

Park, Y. K., Song, J. H., Yoon, S. W., & Kim, J. (2014). Learning organization and innovative behavior: The mediating effect of work engagement. *European Journal of Training and Development, 38*(1–2), 75–94. doi:10.1108/EJTD-04-2013-0040

Penfold, P. (2008). Learning through the world of second life-a hospitality and tourism experience. *Journal of Teaching in Travel & Tourism, 8*(2–3), 139–160. doi:10.1080/15313220802634224

Pillai, R., Schriesheim, C. A., & Williams, E. S. (1999). Fairness perceptions and trust as mediators for transformational and transactional leadership: A two-sample study. Journal of Management, 25. doi:10.1177/014920639902500606

Qammach, N. I. J. (2016). The Mediating Role of Knowledge Sharing on Relationship between IT Capability and IT Support as Predictors of Innovation Performance: An Empirical Study on Mobile Companies in Iraq. *Procedia Economics and Finance, 39*, 562–570. doi:10.10162212-5671(16)30300-8

Richards, G. (2011). Creativity and tourism. The state of the art. *Annals of Tourism Research, 38*(4), 1225–1253. doi:10.1016/j.annals.2011.07.008

Scott, S. G., Bruce, R. A., Scott, S. G., & Bruce, R. A. (2013). Determinants of innovative behavior: a path model of individual innovation in the workplace. *Academy of Management Journal, 37*(3), 580–607. doi:10.2307/256701

Seggelen - Damen, I., & Dam, K. (2016). Self-reflection as a mediator between self-efficacy and well-being. *Journal of Managerial Psychology, 31*(1), 18–33. doi:10.1108/JMP-01-2013-0022

Shanker, D. (2008). ICT and Tourism : Challenges and Opportunities. *Proceedings Ohe International Conference on Tourism in India-Challenges Ahead*, 51–58.

Sigala, M., & Chalkiti, K. (2015). Knowledge management, social media and employee creativity. *International Journal of Hospitality Management, 45*, 44–58. doi:10.1016/j.ijhm.2014.11.003

Sugarman, L. (1987). Experiential learning: Experience as the source of learning and development, David A. Kolb, Prentice-Hall International, Hemel Hempstead, Herts., 1984. No. of pages: xiii + 256. *Journal of Organizational Behavior, 8*(4), 359–360. doi:10.1002/job.4030080408

Thomas, R. (2012). Business elites, universities and knowledge transfer in tourism. *Tourism Management, 33*(3), 553–561. doi:10.1016/j.tourman.2011.06.009

Warr, P., & Downing, J. (2000). Learning strategies, learning anxiety and knowledge acquisition. *British Journal of Psychology, 91*(3), 311–333. doi:10.1348/000712600161853 PMID:10958577

Widmann, A., Messmann, G., & Mulder, R. H. (2016). The Impact of Team Learning Behaviors on Team Innovative Work Behavior: A Systematic Review. *Human Resource Development Review*, *15*(4), 429–458. doi:10.1177/1534484316673713

Wielenga-Meijer, E. G. A., Taris, T. W., Kompier, M. A. J., & Wigboldus, D. H. J. (2010). From task characteristics to learning: A systematic review. *Scandinavian Journal of Psychology*, *51*(5), 363–375. doi:10.1111/j.1467-9450.2009.00768.x PMID:20180922

Williamson, A. J., Battisti, M., Leatherbee, M., & Gish, J. J. (2019). Rest, Zest, and My Innovative Best: Sleep and Mood as Drivers of Entrepreneurs' Innovative Behavior. *Entrepreneurship Theory and Practice*, *43*(3), 582–610. doi:10.1177/1042258718798630

Wynen, J., Boon, J., Kleizen, B., & Verhoest, K. (2019). How Multiple Organizational Changes Shape Managerial Support for Innovative Work Behavior: Evidence From the Australian Public Service. *Review of Public Personnel Administration*, *§§§*, 1–25. doi:10.1177/0734371X18824388

Xerri, M., & Brunetto, Y. (2011). Fostering the innovative behaviour of SME employees: A social capital perspective. *Research and Practice in Human Resource Management*, *19*(2), 43–59.

Young, L. D. (2012). How to promote innovative behavior at work? The role of justice and support within organizations. *The Journal of Creative Behavior*, *46*(3), 220–243. doi:10.1002/jocb.15

KEY TERMS AND DEFINITIONS

Behavioral Work-Based Learning Strategies: Learning new skills, expertise, and knowledge through peers, subordinates, and supervisors. Hence, learning through interaction with coworkers and supervisors is known as behavioral learning strategies. The subtypes of behavioral work-based learning strategies are known as (1) learning through interaction with the supervisor and (2) learning through interaction with colleagues.

Cognitive Work-Based Learning Strategies: Examination and organization of new data in the light of current knowledge e.g. reflection and experimentation. Therefore, the subtypes of cognitive learning strategies are described as (1) learning through reflection and (2) learning through experimentation.

Learning Through Experimentation: Experimentation is described as the way of obtaining new knowledge and addition in the existing knowledge through the experimentation of the existing work procedures and practices.

Learning Through Interaction With Colleagues: When employees exchange ideas and learn new ways of doing things through the discussion with peers and coworkers this is known as learning through interaction with colleagues.

Learning Through Interaction With Supervisor: When employees exchange ideas and learn new ways of doing things through the discussion with the supervisor/leader this is known as learning through interaction with the supervisor.

Learning Through Reflection: Reflection is described as the way of obtaining new knowledge and addition in the existing knowledge through reflection on the current work practices and approaches.

Chapter 4
Authentic Leaders and Business Tourism:
An Exploration of Authentic Relationships

Matthew Knox
https://orcid.org/0000-0002-6092-071X
University of Tasmania, Australia

Joseph Crawford
https://orcid.org/0000-0002-2191-6216
University of Tasmania, Australia

ABSTRACT

With increasing global accessibility, business travel has become a key mechanism to share knowledge across modernist nation borders. Such business travel typically includes conferences, training events, meetings, pitches, and working retreats. Yet, at the end of each of these activities, the rate of continued knowledge sharing, and network establishment likely diminishes. This chapter explores the role of authentic leaders and authentic relationships as a catalyst to create meaningful exchanges of knowledge and a genuine desire to maintain relationships after the business tourism activity ends. The implications, if founded, include a need to train and develop authentic leader behaviors for employees who are likely to engage in business tourism activities. The effect? Higher creativity and innovation, more knowledge sharing, and better cross-cultural connectivity.

INTRODUCTION

The world is changing, globalization has created the conditions of rapid evolution in traditional brick-and-mortar business operations. The digital era is characterized by higher accountability and more innovation, and each of these has challenged the contemporary business. Knowledge workers, or creative class members as Florida (2006) highlights, are not restricted like their ancestors. Instead, the localized geographic boundaries offer only a restriction when Wi-Fi, cellular service, and forms of transport are

DOI: 10.4018/978-1-7998-3142-6.ch004

not available. For first world countries, among many others, access to reliable internet and international plane flights have created a global equalizer for business networking.

Tourism itself is one of the world's largest economic industries. Boasting a growth rate faster than the entire global economy (approx. 1.7 trillion USD), the sector is a catalyst for jobs, innovation and prosperity (UNWTO, 2019). Australia's domestic market alone accounts for a significant portion of their GDP, with overnight spending reaching $44.3 billion during 2018 (Tourism Research Australia, 2019a). Business travel continues to be a main proponent of the industry, with business tourists making up a significant proportion of the market (UNWTO, 2019). International travel has experienced the most significant and consistent growth the last 7 years, with the Pacific, especially Australia boasting the highest rate (UNWTO, 2019). This is reflected in both Australia's domestic and international visitors, with business tourists making up more than 20 percent, and their individual average spend rivaling the highest spenders in the market (Tourism Research Australia, 2019a; 2019b). Likewise, business travel showed the largest rate of growth when considering tourists reasons for travelling in 2018 (Tourism Research Australia, 2019b).

Business tourism offers a unique proposition to enable boundary spanning behavior from those tourists engaging collectively in knowledge sharing activities. The value proposition of such knowledge sharing stems from an ability to create solutions to problems by applying a creative lens. Underpinning the knowledge exchange are authentic relationships, yet we know little of how such relationships enable the transfer of knowledge during business tourism. The key to knowledge sharing is meaningful connections that transcend the short tourism period. As such, we focus on the authentic relationship establishment, development, and sustainability in the business tourism context. Ongoing maintenance of relationships is possible, we posit, through effectively networked communication. This chapter proposes that, to ensure the shortcomings of networked communications do not impact the relationships developed in the business tourism context, authentic leadership behaviors may be applied.

To undertake this, we discuss the context of business tourism. We continue with a critical review on some literature to inform the reader of some antecedent knowledge this Chapter draws upon. First, is a presentation of how business tourism is defined and some of the limitations of current conceptualizations, before articulating a working definition. Next, we explore how communications in the 'digital era' is divergence from the way in which modernist business has operated. Following, is an explication of knowledge sharing behaviors and how the business tourist is affected by current communication technology and theory. Innovation and creativity are defined within the current organizational scholarship literature, and subsequently interpreted through the lens of the business tourist.

In the remainder of the literature, authentic leaders and authentic relations are defined and explored. The authentic leader (Crawford et al., 2019), like many styles of leader behaviors, has not been applied to the tourism setting; but it should be. The synergies between the five behaviors of the authentic leader and other components of the literature (such as knowledge sharing behaviors) are considered. The literature concludes by seeking to understand a contemporary understanding of how authentic relations are intertwined with effective business tourist activities. The discussion section goes on to propose a series of hypotheses surrounding our broader question of inquiry: how can authentic leader behaviors and authentic relations accelerate knowledge sharing, innovation, and creativity in business tourism? The Chapter concludes with exploring the practical implications of our theoretical work and provides opportunities for future research.

CONTEXT

Tourism is a key proponent of the global economy. Business activity now spans boundaries once unheard of, going beyond the traditional geographic boundaries with rapid evolutions in the technology of movement. Before the Wright brothers created the first plane in 1903, and for many years after while plane technology was scaling to affordable global levels, travel beyond the immediate community or country was not common practice. Since then, tourism has become a tool for seeing family, enjoying leisure, and conducting business (Swarbrooke & Horner, 2012). For the past seven years, the tourism sector has outpaced the combined global economy and does not seem to be slowing down (UNWTO, 2019). The following will explore the historic roots of business tourisms emergence, its prominence in the global economy, and what it all means for the contemporary businessperson.

The travelling businessperson is not a new phenomenon. From traders travelling from tribe to tribe, journeymen touring villages, to Marco Polo's Silk Road travels in the thirteen century, diverse and profitable opportunities away from home have been a cornerstone of business for millennia. Contemporary understandings can trace their comparatively short lineage back to the deterritorialized society globalization has molded. While not the only antecedent to rapid growth in tourism, globalization acts as the foundations for interconnectivity in both travel and ongoing relationships of business travel, especially those of international origins (Hirst et al., 2015). The perpetuation of global trade, as well as international production are both linked to the phenomena of globalization (Dunning, 2014). In social sciences, globalization is understood as the intensification of global socialization, linking local happenings around the world and effecting events beyond traditional spatial restraints (Giddens, 1990). This understanding reflects the success business has received as consequence of globalizations effect (Levitt, 1993). As cross-cultural interactions become increasingly common between businesspeople, so too does their ability to communicate long-term. Contemporary platforms such as LinkedIn, Skype, Zoom, and similar allow businesspeople to continue relationships that would otherwise have been prevented by distance. The ease of communication one can now benefit from perpetuates much of business tourisms appeal, providing reason to form long-distance relationships (Swarbrooke & Horner, 2012).

Various forms of business tourism exist, and this variety offers much beyond creating potential for trade. In their seminal work on business travel, Swarbrooke and Horner (2012) discuss the emergence of business tourism, its perpetuation, and the forms it comes in. Swarbrooke and Horner (2012) explicate 15 different types of business tourism, from general business trips, to the more abstract examples, such as "military service at places other than one's normal base" (p. 4). Between these groups, it is important to recognize the individual differences, distinctions, and subtypes of the categories which have been generally adopted in the literature (Rogers, 2013). Trends in business tourism have emerged consequently, with discourses on the topic commonly discussing conventions, conferences, meetings, training courses, incentive trips, and product launches (Swarbrooke & Horner, 2012; Hankinson, 2005; Rogers, 2013). The significance of these trends is not the events themselves, rather their multidimensional opportunities for business tourists to interact.

To elaborate with an example, the Workplace Mental Health Conference, held in Hobart, Australia boasted full days of collaborative conference proceedings and workshops (see Sanderson et al., 2017). It also facilitated a trip by museum ferry to a nearby internationally acclaimed museum, the Museum of New and Old Art (MONA) with guided tours and a conference dinner. Following, the conference attendees were invited to collaborate on the authoring of a White Paper (Martin et al., 2017). While many conferences create and innovate new ways to enable attendees to develop lasting relationships, that conference enabled knowledge sharing, informal connections through shared experiences, and subsequent opportunities to continue to collaborate digitally. Other events such as training courses that boast a high volume of businesspeople offers vast opportunities to form relationships and share knowledge. Likewise, one-on-one meetings allow the business tourist to form meaningful connections, facilitating ongoing business and promote creativity and innovation (Thakur et al., 2012). From an organizational perspective, incentive trips and company retreats have been shown to increase job satisfaction as well as productivity (Robinson, 2012). The various benefits business tourism offers provides much to its perpetuation and its prominence is reflected in the global economy.

With the saturation of business tourists in the market, and potential for its various benefits, we suggest that the onus of responsibility is on the business tourist to implement strategies to go beyond the standard attend, listen, return, and forget, but to ensure each relationship developed has the capacity to go beyond the finite business tourist activity to filter to the digital, or subsequent tourist activities. As such we posit authentic leader behaviors could be a toolkit the business tourist can embody to accelerate their knowledge-sharing, creativity and innovation. A proposition, we return to in the discussion.

CRITICAL REVIEW

This Chapter adopts a critical literature review stance, with a focus on the research question *how can authentic leader behaviors and authentic relationships be a catalyst for more effective business tourism?* The aim of this Chapter is to develop a critical perspective of authentic relations in the business tourism setting. The focus for this Chapter is on the role of authentic leader behaviors in interactions of the business traveler. In doing so, we explore the role of authentic leader behavior in business tourism and its effect on developing and sustaining relationships. We also explore the role of authentic relationships in networked communications and knowledge sharing.

Our rationale for engaging in a critical review over a potentially more rigorous review (e.g. systematic review) is due to the disparate theories nested within our literature review. The first, business tourism has reasonably limited information on it. A search on EBSCOhost for the phrase "business tourism" in the abstract or title yields only 233 results. The sheer lack of available information surrounding business tourism requires far more critical and exploratory approaches to the construct. Authentic leadership and authentic relations have not, to the authors' knowledge, been applied to the business tourism context and requires a critical lens to examine the potential synergies. To this end, we will begin by presenting a brief critical review of each of the underlying constructs that we intend to draw upon in the discussion.

LITERATURE

This literature review begins with the business tourist. While there is limited information surrounding the practice (business tourism) or the individual (business tourist), it does represent a significant subsector of tourism. This review explores the business tourist from a conceptual level, seeking clarity on the terms before progressing. We also acknowledge that some business tourism research is set before the digital era, and any knowledge added should consider the role that communication technology has in enabling or challenging the longevity of business tourist relationships. Knowledge sharing behaviors, creativity, and innovation are explored next with a view to understand how they may be both enablers and outcomes of effective business tourism relationships.

The literature review continues to define the authentic leader and how the theory of authentic leadership has evolved from a response of unethical leadership conduct to a theory of leadership effectiveness in a wide range of contexts. Given the empirical links established between authentic leadership and some enablers of business tourism (knowledge sharing behaviors, creativity, and innovation), there is an opportunity to understand how the business tourism relationships may be better supported by authentic leaders. Finally, the literature review concludes with an exploration of authentic relations and how these may turbocharge the business tourist relationships. Business Tourist Business tourism is a term pertained to a wide array of activities. From conferences to product launches, business encapsulates the ambiguous, as well as the obvious. These events are however regulated by one key component. The individual business person involved in in the various types of business tourism lies at the core of business tourism, and as such exists a key variable to research. The business tourist is not strictly denied an accepted definition, however definitions that exist are more descriptive, rather then conceptual. Initially, the business tourist lacked consideration. Business for travel purposes had been discussed in literature and recognized statistically, however up until recent times the terms had not be given strict conceptualization. Davidson (1994) first discussed the pair in his work on travelling for business. Davidson described business tourism as the earliest form of tourism, noting the business tourist as any man who travels for work. Asides from the obvious, Davidsons definitions does not provide much in the way of theory, merely stating the business tourists as someone who travels for trading purposes. Swarbrooke and Horner (2012) attempt to rectify this, though as previously discussed, their work relates to business tourisms various forms. Given the lack of clarity, the United Nations World Tourism Organization provide their own definition to allow for targeted research and data, highlighting the importance of purpose of the business visitors trip (UNWTO, 2008). Like those before it however, this definition lacks specifics, opting instead for a broad, far-reaching description. Rogers (2013) problematizes business tourisms conceptualization due to the failing commonalities, suggesting the concept is trying to encapsulate too much. While some argument presents itself for this stance, we posit that the prominence of the industry, as well as its clear distinction from other forms of tourism requires clarity to utilize its potential benefits. As such, we look to explore the proponents and uniqueness that separates business tourism from other forms. While past definitions opt for sweeping conceptualizations, Swarbrooke & Horner (2012) do however note the importance of the travel component. Destinations are a result of the business travel, meaning one travels specifically for their event, meeting, or retreat. Destination is therefore crucial to business tourisms distinction. Likewise, the types of business travel highlights the importance of intention in travelling for business (Swarbrooke & Horner, 2012; Rogers, 2013). While one may send a few work emails on an overseas trip, if their trip is not to reach a business-related event, their travel can not be distinguished as for reasons of business. This distinction has been adopted in contemporary statistics (cf. UNTWO 2019; Research Tourism Australia, 2019a; 2019b), and is required to maintain consistency in research. Business related events is the last area of contention in past definitions. This distinction is the most ambiguous, as while one can claim to be travelling for business, what defines this conduct is inherently subjective (Beaverstock et al., 2009). This is where the significance of distinguishing the business tourist presents itself. While one may travel from place to place, the business tourist does so with the main purposes of forming connections, continuing on-going relationships, and facilitating knowledge sharing, creativity and innovation. The manner in which the business tourist does this falls into the explication of business travel events of above. This is to say that the business tourist maximizes their potential to achieve their goals through attending conferences, conducting meetings with new relations, undergoing training, and utilizing the potential of business travel. In doing so the business tourist is active in their pursuit of success, adopting appropriate behaviors to utilize the potential of business travel.

To clarify the current state of the business tourism field, as well as providing for the significance of intention, destination, events, and the business tourist, we provide the following definition of business tourism: Business tourism is the utilization of travel with the intent to achieve business objectives, form connections, and share knowledge. Communication in the Digital Era In the globalized world, communication through online profiles has allowed for a nonstop business trade. Networked communications have become a contemporary outlet for interactions, whether social, business or pleasure. In essence, the networking of humanity can be seen through the lens of technology (Tapscott, 1996). The benefits of networked communications in business can be seen to range from negotiations over email, too interviews through video calling, transcending time zones and relative business hours (Tapscott, 1996). Networked communications, despite their limitations, are driving contemporary business, providing outlet for maintaining business relations regardless of contextual restrictions. Initial contact, negotiations, and ongoing relations have all been shown to suffer in considerations of networked communications, resulting in a significant demand for face to face meetings and personalized trust development (Gustafson, 2012).

The barriers of international borders now only exist in the abstract of time-zones. Globalized technology however provides the businessperson with the opportunity to transcend time, continuing relationships and business across borders (Dunning, 2014). With the advent of the internet considerations of business grew beyond the locale. New forms of marketing had to be created, brands could now be advertised across the world, reaching those who were previously far beyond reach (Hammil, 1997). The growth of online communication allows the conduct of business without physical presence. Business can essentially be done at any time. One can finalize a contract at 3am in their time zone, while another receives it instantaneously at 2pm in theirs. Social media arguably pertains the most significance in this area. The digital revolution of the contemporary world provides much to the realm of business tourism. From an events management perspective there is new concentration on utilizing the potential for connectivity through technology to promote pro-social initiatives like minimizing waste and navigate the evolving industry (Hawkins & Bonney, 2019). Further research into the tourism sector suggests the need to combat burnout in business tourists. Due to the rigors of travel, as well as brief and intense interactions, organizational psychologists raise concerns over the well-being of business travelers (Niessen et al., 2018). Communicating in the digital era provides a combatant to this issue in that the business tourist can relegate their communications to the online world, forgoing the need to travel for every meeting. Likewise, the pressure of forming on-going relationships need not be influenced by difficulties in their continuation, with instant online connections available to all. The ability to communicate across boarders has not only presented new opportunities, but changed the way business is created, developed, and maintained (Aral et al., 2013). Like the humble traders of antiquity, the business tourist travels to the now not so far-away places in record time. Nowadays initiating relationships is not dependent on locale. The non-stop trade of international business and ongoing relationships has resulted in a whole new desire to travel for business. Businesspeople meet to negotiate, cooperate, manage, enable productivity, brainstorm, innovate, and develop professional networks and interpersonal trust (Beaverstock et al., 2009). Networked communications can be an ineffective way of conducting complex business. This is not to say that networked communications do not hold merit, rather, their respective value stems from effective personal relations (Lyons, 2013). This chapter proposes that authentic leader behaviors can allow for this foundational development, which we posit as crucial in sustaining business beyond the tourism period through networked communications. In summary, due to the short-term nature of the tourism period, the business tourist must develop authentic relationships to ensure their continued existence beyond the short-term tourism setting. Developing authentic relations can be an expensive venture, when considering the trans-national nature of business tourism, as such the development of foundations must be effective. To ensure this, the business tourist must demonstrate their authenticity through trust and the ability to reciprocate relationships, consequently ensure the exchange of knowledge. This chapter proposes that authentic leader behaviors provide an appropriate framework for the business tourist. Knowledge Sharing Behaviors Despite the ease of communicating through technology, evidence maintains the need for physical travel to ensure distance does not affect authentic relations (Larsen, Axhuasen, & Urry, 2006). Business tourism involves, in general, a higher level of per head spending than any other type of tourism and exists as the core market for most airlines and hotel chains (Swarbrooke & Horner 2012). The business tourist must consequently justify their expensive travelling through ensuring that the knowledge created exceeds the value of the trip, or the relations they are establishing are sustainable and beneficial. Business tourism can be encapsulated through three distinctive activities: meetings, conferences, and incentive trips (Weber & Chon, 2002) in which all rely on forming relations and sharing knowledge. We believe that business tourism theory should also encapsulate the informal activities also: ad hoc interactions in nearby cafes and out-of-conference interactions as two examples. Conceptually, knowledge sharing is seen as the exchange of information through willing participants (McDermott & O'Dell, 2001). Knowledge sharing involves various considerations to the nature of the exchange, with specific attention to the actors involved (Snowden, 2002). An individual's attitude toward knowledge sharing is driven primarily by anticipated reciprocal relationships (Bock et al., 2005). The formation, development, and sustainability of reciprocal relations are required to ensure knowledge sharing in cases of brief and infrequent interactions (Cohen, 1988), such as business tourism. Said interactions have been shown to require a far higher rate of motivational fulfilment. The nature of the tourism period heightens the requirement of developing sustainable relations, as much as it fuels the exchange of knowledge (Larsen, Axhuasen, & Urry, 2006). Trust is likewise an integral component of knowledge sharing, individuals have been shown to demonstrate a higher exchange rate with more authentic individuals, especially in climates wherein knowledge sharing is inherent to the context (Hinds & Pfeffer 2003).

CREATIVITY AND INNOVATION

The conceptualization of creativity and innovation in the organizational setting exists as a wide area of intrigue in organizational research. From their impacts on profit (Davila et al., 2012), too their relationship with productivity and well-being (Ekvall, 1996), their capabilities are debated through a range of scholarly lenses. The insight creativity and innovation provide into the factors of conducting successful business offers vast and varied conceptualization (Anderson et al., 2004). For example, Schumpeter (1934) initially posits innovation as a key to profit, adopting a definitional stance in attempting to understand its role in business. With more recent definitions acknowledge the cognitive processes behind innovation, though it's conceptual existence has remained definitional (Gumusluoglu & Ilsev, 2009; Quintane et al., 2011). In contrast, several scholars conceptualize creativity as a process too complex for a singular definition (Amabile, 1988; Kahn, 1990; Drazin et al., 1999). Favor instead is given to understanding creativity through its embodiment, a distinction not shared with innovation. In a rigorous review of literature, Amabile (1988) provides distinct, and universal criteria for creativity, positing intrinsic motivation, and a relevant skillset as crucial to its emergence. Creativity in this sense is not one single notion, rather an accumulation of virtues that are applied procedurally (Drazin et al., 1999). Further to this is creativity's conceptualization as a social phenomenon (Ford, 1996). One applies their relevant skills and motivations with others of a like mind to form a useful new product, service, idea, procedure, or process (Woodman et al., 1993). Creativity like this requires the balancing of resources and management skills (Kahn 1990; Andriopoulos, 2001). The organizational culture as such acts as premise to creativity's social emergence, with leadership in management and resource distribution as a mediator (Andriopoulos, 2001). Creativity is a vehicle leader's drive to promote success. Leading creativity in the organizational setting involves the management of both resources and ideas. This premise to knowledge creation and utilization is not dissimilar to how innovation is understood in organizational research. In a review on innovation's conceptualization in research Crawford (et al., in press) distinguishes innovation as a tool leaders can use to apply creativity to create solutions. Like creativity, innovation is posited as a process one applies (Gumusluoglu & Ilsev, 2009; Quintane et a., 2011; Crawford et al., in press), an understanding reflected in Australian governmental policy. In an effort to more effectively respond to the Third Sector, the government report, 'Australia 2030: Prosperity through Innovation' marks innovation as a process, distinguishing its effectiveness in science, education, industry, government and culture. Like leaderships effect on creativity, creativity acts as a moderator to innovations success (Thakur et al., 2012). As such, some scholars suggest that the two are dependent on one another, going as far to argue that the two should be synthesized under one notion (Ekvall, 1996). We problematize this due to the conceptual significance of both creativity and innovation. While the two both hold merit too each-others success, both maintain separate cognitive processes, as well as separate social motivations and resource requirements (Shalley et al., 2004). Anderson (et al., 2014) does integrate the two under one definition, although creativity and innovation are distinguished as two separate stages. This understanding incorporates their separate complexity while acknowledging both concepts independent variables and processes.

We acknowledge the broad implications of conceptualizing creativity and innovation as an organizational process with separate stages, motivations and considerations. Doing so allows for application across a wide array of organizational situations. Creativity and innovation above all require leadership in fulfilling these various processes (Amabile, 1988; Anderson et al., 2014; Crawford et al., 2019). As such we posit the need to consider this leader-follower interaction in facilitating the business tourist in leading creativity and innovation in their relationships. Authentic Leaders

The development of authentic leadership is the product of a scholarly response to the continued concerns of areas of the business sector acting unethically (Luthans & Avolio, 2002). In more recent times, this has included examinations of current leaders (like Corbyn, Obama, and Trump: Iszatt-White et al., 2018; Sidani & Rowe, 2018; Williams et al., 2012). Originating in the field of positive organizational scholarship (Cameron & Dutton, 2003) and positive organizational behavior (Luthans & Avolio, 2002), the authentic leader was formed to articulate what it means to be a good leader. More recently, a good leader relates normatively to a person who can be both ethical and effective; motivated by the 'right' reasons (Newstead et al., 2019). Authentic leaders, in many ways, align to such a foundation; the ability to be ethical with a strong positive moral perspective, and the ability to be effective through the synergy of informal influence and a strong self-concept (Crawford et al., 2019). Authentic leader behaviors (see Figure 1) explicates five behaviors of a leader who can consider their context and act appropriately.

Figure 1.

Authentic leader	influences and motivates followers to achieve goals through their sincerity and positive moral perspective, enabled through heightened awareness and balanced processing.
Awareness	Having insight into the behaviors of one's self, other individuals, and groups.
Sincerity	Presenting one's true self to others, honestly and openly in all relationships and with consideration for the context.
Balanced processing	The tendency to consider all relevant information available and using this to make decisions that benefit the collective (e.g. followers).
Positive moral perspectives	The commitment to one's intrinsic ethical framework, and a willingness to subdue personal interests and ego to facilitate collective interests.
Informal influence	The ability to inspire and motivate individuals to accomplish goals of their own volition, regardless of rank or position.

Definitions quoted from multiple pages in Crawford et al., (2019).

Much of the literature on authentic leadership has focused on organizational and individual outcomes. Many of these have useful parallels to the business tourism context. Authentic leaders have a positive influence on the creativity (Rego et al., 2012) and innovation of others (Černe et al., 2013). Importantly, however, authentic leader behaviors have a theorized relationship with higher levels of social innovation (Crawford et al., in press) and facilitating positive behavioral development in organizations (Knox et al., 2019). In a corporate landscape characterized with continual ethical failures, the ability to support social value creation (through social innovation) is important for societal progression. Authentic leaders also create conditions where their employees are willing to share knowledge, and exhibit knowledge sharing behaviors (Edú-Valsania et al., 2016). In business tourism, the ability to support knowledge sharing is critical. Some business tourists can attend conferences without building any new lasting connections, instead listening to the content, making small talk, and then returning home to never hear from those individuals again. Authentic leaders in these contexts are different. In the tourism setting, authentic leadership has been explored with less frequency; leadership theory in tourism has a similar narrative. Despite the lack of scholarly research into the behaviors of effective tourism leaders, there are synergies that emerge. Local leadership capability was established as a critical success factor in rural tourism development (Jackson & Murphy, 2006; Wilson et al., 2001). Leadership, in the context of regional tourism, has been discussed with recognition that many theories of leadership emerge within organizations of clear boundaries and hierarchies (e.g. Valente et al., 2015). Authentic leaders are different, their ability to engage and develop informal influence transcends the formal power structures within an organization. The nature of business tourism is such that the formal constraints of the modernist organization are typically non-existent. Corporate credit cards, and management preapproval of spending limits creates hurdles for engagement. These hurdles are not usually insurmountable, however. Instead, an effective business tourist embodying authentic leader behaviors navigates their environment effectively to achieve the desired objectives of such travel.

This Chapter explores the application of the authentic leader to business tourism, a field of emerging research previously described. One-on-one interactions between business travelers are reliant on the development of authentic relations, for without which knowledge sharing and future communications have been found to go wanting (Weber & Chon, 2002). Through their ability to act transparently and effectively communicate ideas, the authentic leader can foster these meaningful and sustainable relationships (Bird et al., 2009). The notion of authentic relations is explored in more depth below.

AUTHENTIC RELATIONSHIPS

Businesspeople meet and engage with one another daily. While much of business is conducted in the abstract realm of corporations, significant portions of business is created, maintained, and deconstructed through relational conflict (or otherwise) between individuals within (or related to) the entity. Professional networks and interpersonal trust are foundations created by corporate citizens to enable the sustainable development of such relationships (Beaverstock et al., 2009). These interactions can provide repeated opportunities to collaborate, discuss, and create. In the short-term dosage-like nature of business tourism engagements, the ability to quickly develop the foundations of potential long-term corporate relationships is necessary.

Authenticity has a positive role in interpersonal interactions (Avolio & Gardner, 2005). Authenticity can be understood as the moral significance of being true to ones' self, and likewise having the knowledge to act morally (Taylor, 1992). While authenticity has vast historical roots, tracing its conceptualization to Aristotle, the notion has found resurgence in organizational literature. Not dissimilar is the 'sister' concept of sincerity. Often being understood in relation to authenticity, sincerity is the perception of the how authentic one person is (Trilling, 2009). While one can appear to be sincere when they represent their authentic self, they may be acting or being false in their presentation (Ferris et al., 2007). Goffman (1978) highlights that all humans have a degree of multiplicity in the way they present. Such multiplicity could be considered a divergence from ones' true self and may affect another's interpretation of their sincerity. One's authentic performance, however, fosters positive relationships through building trust (Avolio & Gardner, 2005). Authentic relationship, we posit, are the intersection of the interpretation of another's sincerity between individuals seeking to be authentic in their actions and their words.

Authentic relationships are built on inclusivity, respect, collaboration, transparency, and care (Crippen, 2012). This is not to say that those performing in authentic relationships are free of accountability, rather it is perpetuated (Fries-Britt & Snider, 2015). Although herein lies the problem. Apparent sincerity is the false presentation of one's authentic self (Ferris et al., 2007); employed in political science as a skill of manipulation, and deception, political skill can be utilized to engage in ulterior motives. The 'Watergate' scandal exemplifies how unethical behavior can occur behind since pretenses. Due to the competitive nature of business, such performances are a tool for unethical behavior and risk (Crawford et al., 2017). The significance of this in the business tourism context lies in the formation and perpetuation of relationships. Due to the short-term nature of interactions, the business tourist must rely on strong foundations to ensure ongoing relations. Such foundations can be found in the inherent trust developed in authentic interactions (Avolio & Gardner 2005; Bird et al., 2009). A great challenge in these settings is that an authentic person may be engaging sincerely with another person. If the latter person is adept at a display of apparent sincerity, it may not be obvious. A disjoint foundation of sincerity is unlikely to create sustainable business relationships in geographically proximate setting.

In a business tourism setting for example, be attending a multiday conference with the expectation that their knowledge will increase as a result. We posit that the authentic relationships established, by authentic leaders, during this finite period will have a profound impact on the knowledge that the tourist walks away with, or the new relationships formed. Meetings between conference delegates are inherently fueled from context, knowledge shared, and furtherance of business. We believe that authenticity underpinning this relationship will further enable the knowledge dispersion among tourists.

DISCUSSION

In organizational settings, empirical research demonstrates strong relationships between authentic leadership and knowledge sharing behaviors, creativity, and innovation. Within the tourism setting, the relationships between leadership and those engaging in tourism are lacking. This Chapter begins the conversation to explicate the role of the authentic leader in the business tourism context, by drawing on the existing literature of the outcomes (both theoretical and empirical) that authentic leadership can produce. For this, we present a model as follows (Figure 2.

Figure 2.

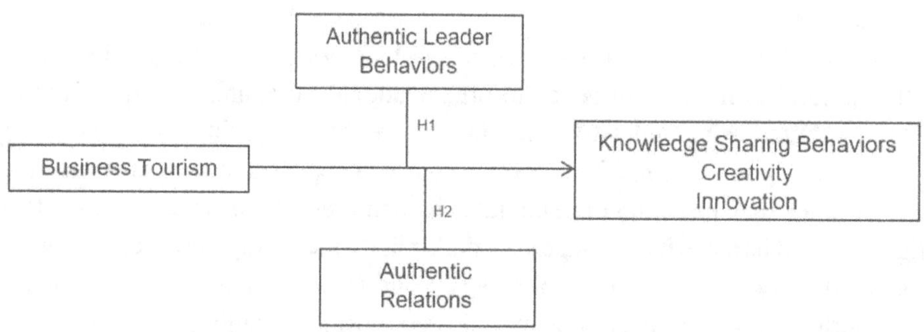

Authentic leadership behaviors provide much in the way of rectifying obstacles for business tourists. Likewise, authentic leaders are as versatile as they are specific, allowing the business tourist to be flexible in the face of business travels various types. As such we posit that the embodiment of authentic leadership behaviors facilitates the business tourist in encapsulating an array of opportunities.

Hypothesis 1. *Authentic leader behaviors moderates the relationship between business tourism and knowledge sharing, creativity, and innovation.*

Business tourism encapsulates a diverse and opportunistic variety of experiences. Events such as conferences and product launches, as well as one-on-one meetings and trade discussions provide business tourists with opportunities to interact with potential partners. Targeted education enables the authentic leadership behavior of awareness, aiding business tourists in clarifying their beliefs and values (Eriksen, 2009). The authentic leadership behavior, sincerity works coherently with awareness. Through the embodiment of sincerity, the business tourist can present their beliefs and values authentically, facilitating relationships and providing a positive climate for knowledge sharing (Ilies et al., 2005). Authentic leaders are theorized to have a heightened presentation of authenticity tourism experiences that are plagued with expectations and apparent sincerity (Knox et al., *in-press).* Awareness in professional interactions likewise promotes positive relationships through the conservation of an authentic climate (Luthans, 2002). Given the multitude of interactions the business tourist experiences, the ability to communicate one's intentions authentically, and promote positive interactions through awareness is crucial in capitalizing on potential opportunities.

Balanced processing benefits authentic leaders with the ability to consider all available information and benefit the collective. The development of new partnerships and continuing those on-going, is moderated by the benefit each side experiences (Mohr & Spekman, 1994). Developing relationships built on trust assures mutual benefits in partnerships (Das & Ten, 1998). Authentic leaders facilitate trust through the presentation of an authentic self and can make decisions that are mutually beneficial (Luthans & Avolio, 2009). Authentic leaders can also understand their own strengths and weaknesses (Avolio & Gardner, 2005), and have practical reflexivity (Eriksen, 2009). In the competitive and fast paced world of business tourism, adapting to a variety of situations provides much for embracing all available opportunities, as well as assuring collective interests are maintained. While increased success and opportunities encourage ongoing business, frequent interactions and heightened levels of trust have the potential to breed unethical behavior that results in one sided payoffs (Brass et al., 1998). Much research exists exploring the role of unethical conduct in professional settings, commonly shown to result in unproductive climates and business failure (Paine, 1994). As such, managing one's intrinsic moral framework is crucial in ensuring one does not succumb to deceitful temptations. A positive moral perspective encourages authentic leaders to manage their ego and personal desires with the collective interest (Yammarino et al., 2008) Authentic leaders also engage and develop informal influence, subduing traditional roles of power in favor of encouraging ethical behavior and success (Crawford et al., 2019).

While the nature of business tourism surpasses much of the traditional roles of contemporary organizations, the actors within those roles act within the business tourist space also. Obstacles of manipulation and unclear motivations as such remain. By embodying authentic leader behaviors, the business can manage navigate their environment without suffering and falling into the pitfalls of the inauthentic, therefore capitalizing on the multitude of opportunities available.

Hypothesis 2. Authentic relationships moderate the relationship between business tourism and knowledge sharing, creativity, and innovation.

FUTURE RESEARCH DIRECTIONS

This research has taken a largely unexplored area of contemporary business life (the business tourist) and sought to better understand how the behaviors of the authentic leader, and the relationships they engage in, could enhance the sustainability and longevity of the business tourism relationship. Yet, as the field of business tourism is still quite infant, there is need for significantly more research. To begin with, we examined one style of leadership; others may be more effective. We examined one foundation of relationships; others may be more effective. Future research should seek to empirically examine the relationship between the authentic leader and the outcomes espoused by a business tourist.

In addition, there is more research needed to understand the outcomes of effective business tourism. Indeed, there is a need to adequately consider what the value of business tourism should be, beyond the initial dosage or hit of new knowledge or new networks associated with conferences and training days. If the field of business tourism is to hold form, it is necessary to therefore seek to understand if conferences with a dinner have greater network establishment, or training days with more collaborative pedagogies applies have greater knowledge sharing. There are still significant unknowns in the theory of business tourism.

CONCLUSION

We have sought to enter a behavioral lens of business tourism, and we see immense value in other scholars taking on such a lens. We began this Chapter with an exploration of the emergence of business tourism, and how globalization has enhanced the ability for businesses to send their employees on knowledge and network gathering activities. We sought to understand how the business tourism landscape had evolved with new communication technology. We continued to explore some of the possible benefits associated with business tourism but acknowledge our limited capacity to comprehensively explore these given the limited information available. We found knowledge sharing behaviors, creativity, and innovation could all be both enablers and outcomes of effective business tourism.

We explored the authentic leader and theorized that business tourists with higher authentic leader behaviors are likely to be more effective at engaging in sustainable business tourist activities. Authentic relations were also considered to be of importance in this setting. The practical implications we identified were that authentic leadership development for employees who engage in business tourism activities would have a positive influence on their effectiveness abroad.

REFERENCES

Amabile, T. (1988). A model of creativity and innovation in organizations. *Research in Organizational Behavior*, *10*(1), 123–167.

Anderson, N., De Dreu, C., & Nijstad, B. (2004). The routinization of innovation research: A constructively critical review of the state-of-the-science. *Journal of Organizational Behavior*, *25*(2), 147–173. doi:10.1002/job.236

Anderson, N., Potočnik, K., & Zhou, J. (2014). Innovation and creativity in organizations: A state-of-the-science review, prospective commentary, and guiding framework. *Journal of Management*, *40*(5), 1297–1333. doi:10.1177/0149206314527128

Aral, S., Dellarocas, C., & Godes, D. (2013). Introduction to the special issue—social media and business transformation: A framework for research. *Information Systems Research, 24*(1), 3–13. doi:10.1287/isre.1120.0470

Avolio, B., & Gardner, W. (2005). Authentic leadership development: Getting to the root of positive forms of leadership. *The Leadership Quarterly, 16*(3), 315–338. doi:10.1016/j.leaqua.2005.03.001

Beaverstock, J., Derudder, B., Faulconbridge, J., & Witlox, F. (2009). International business travel: Some explorations. *Geografiska Annaler. Series B, Human Geography, 91*(3), 193–202. doi:10.1111/j.1468-0467.2009.00314.x

Bird, J., Wang, C., Watson, J., & Murray, L. (2009). Relationships among principal authentic leadership and teacher trust and engagement levels. *Journal of School Leadership, 19*(2), 153–171. doi:10.1177/105268460901900202

Bock, G., Zmud, R., Kim, Y., & Lee, J. (2005). Behavioral intention formation in knowledge sharing: Examining the roles of extrinsic motivators, social-psychological factors, and organizational climate. *Management Information Systems Quarterly, 29*(1), 87–111. doi:10.2307/25148669

Brass, D., Butterfield, K., & Skaggs, B. (1998). Relationships and unethical behavior: A social network perspective. *Academy of Management Review, 23*(1), 14–31. doi:10.5465/amr.1998.192955

Cameron, K., & Dutton, J. (2003). *Positive organizational scholarship: Foundations of a new discipline.* San Francisco, CA: Berrett-Koehler Publishers.

Černe, M., Jaklič, M., & Škerlavaj, M. (2013). Authentic leadership, creativity, and innovation: A multilevel perspective. *Leadership, 9*(1), 63–85. doi:10.1177/1742715012455130

Cohen, E. (1988). Authenticity and commoditization in tourism. *Annals of Tourism Research, 15*(3), 371–386. doi:10.1016/0160-7383(88)90028-X

Crawford, J., Dawkins, S., Martin, A., & Lewis, G. (2017). Understanding the organizational climate of unethical leadership in the Australian Football League. *The Journal of Leadership Studies, 11*(2), 52–54. doi:10.1002/jls.21525

Crawford, J., Dawkins, S., Martin, A., & Lewis, G. (2019). Putting the leader back into authentic leadership: Reconceptualising and rethinking leaders. *Australian Journal of Management*; Advanced Online Publication. doi:10.1177/0312896219836460

Crawford, J., Kelder, J.-A., & Knox, M. (in-press). What does it take to be a social entrepreneur? Authentic leaders and their effect on innovation. In Ç. Dogru (Ed.), *Leadership styles, innovation, and social entrepreneurship in the era of digitalization.* Hershey, PA: IGI Global.

Crippen, C. (2012). Enhancing authentic leadership– followership: Strengthening school relationships. *Management in Education, 26*(4), 192–198. doi:10.1177/0892020612439084

Das, T., & Teng, B. (1998). Between trust and control: Developing confidence in partner cooperation in alliances. *Academy of Management Review, 23*(3), 491–512. doi:10.5465/amr.1998.926623

Davila, T., Epstein, M., & Shelton, R. (2012). Making innovation work: How to manage it, measure it, and profit from it. Upper Saddle River, NJ: FT Press.

Drazin, R., Glynn, M., & Kazanjian, R. (1999). Multilevel theorizing about creativity in organizations: A sensemaking perspective. *Academy of Management Review*, *24*(2), 286–307. doi:10.5465/amr.1999.1893937

Dunning, J. (2014). *The globalization of business (routledge revivals): The challenge of the 1990s*. Abingdon, UK: Routledge. doi:10.4324/9781315743691

Edú-Valsania, S., Moriano, J., & Molero, F. (2016). Authentic leadership and employee knowledge sharing behavior: Mediation of the innovation climate and workgroup identification. *Leadership and Organization Development Journal*, *37*(4), 487–506. doi:10.1108/LODJ-08-2014-0149

Ekvall, G. (1996). Organizational climate for creativity and innovation. *European Journal of Work and Organizational Psychology*, *5*(1), 105–123. doi:10.1080/13594329608414845

Eriksen, M. (2009). Authentic leadership: Practical reflexivity, self-awareness, and self-authorship. *Journal of Management Education*, *33*(6), 747–771. doi:10.1177/1052562909339307

Ferris, G., Treadway, D., Perrewe, P., Brouer, R., Douglas, C., & Luxm, S. (2007). Political skills in organizations. *Journal of Management*, *33*(3), 290–320. doi:10.1177/0149206307300813

Florida, R. (2006). The flight of the creative class: The new global competition for talent. *Liberal Education*, *92*(3), 22–29.

Ford, C. (1996). A theory of individual creative action in multiple social domains. *Academy of Management Review*, *21*(4), 1112–1142. doi:10.5465/amr.1996.9704071865

Fries-Britt, S., & Snider, J. (2015). Mentoring outside the line: The importance of authenticity, transparency, and vulnerability in effective mentoring relationships. *New Directions for Higher Education*, *2015*(171), 3–11. doi:10.1002/he.20137

Giddens, A. (1990). Structuration theory and sociological analysis. *Anthony Giddens: Consensus and controversy*.

Goffman, E. (1978). The presentation of self in everyday life. London, UK: Harmondsworth.

Gumusluoglu, L., & Ilsev, A. (2009). Transformational leadership, creativity, and organizational innovation. *Journal of Business Research*, *62*(4), 461–473. doi:10.1016/j.jbusres.2007.07.032

Gustafson, P. (2012). Managing business travel: Developments and dilemmas in corporate travel management. *Tourism Management*, *33*(2), 276–284. doi:10.1016/j.tourman.2011.03.006

Hamill, J. (1997). The Internet and international marketing. *International Marketing Review*, *14*(5), 300–323. doi:10.1108/02651339710184280

Hankinson, G. (2005). Destination brand images: A business tourism perspective. *Journal of Services Marketing*, *19*(1), 24–32. doi:10.1108/08876040510579361

Hawkins, C., & Bonney, M. (2019). Lean thinking in leisure: Continuously improving event volunteering and management. *Annals of Leisure Research, 22*(3), 362–372. doi:10.1080/11745398.2019.1568267

Hinds, P., & Pfeffer, J. (2003). Why organizations don't "know what they know": Cognitive and motivational factors affecting the transfer of expertise. In M. Volker (Ed.), *Sharing expertise: Beyond knowledge management* (pp. 3–26). Boston, MA: MIT Press.

Hirst, P., Thompson, G., & Bromley, S. (2015). *Globalization in question.* Hoboken, NJ: John Wiley & Sons.

Ilies, R., Morgeson, F., & Nahrgang, J. (2005). Authentic leadership and eudaemonic well-being: Understanding leader–follower outcomes. *The Leadership Quarterly, 16*(3), 373–394. doi:10.1016/j.leaqua.2005.03.002

Iszatt-White, M., Whittle, A., Gadelshina, G., & Mueller, F. (2018). The 'Corbyn phenomenon': Media representations of authentic leadership and the discourse of ethics versus effectiveness. *Journal of Business Ethics*; Advanced Online Publication.

Jackson, J., & Murphy, P. (2006). Clusters in regional tourism An Australian case. *Annals of Tourism Research, 33*(4), 1018–1035. doi:10.1016/j.annals.2006.04.005

Kahn, W. (1990). Psychological conditions of personal engagement and disengagement at work. *Academy of Management Journal, 33*(4), 692–724.

Knox, M., Crawford, J., & Kelder, J.-A. (2019). Developing authentic leadership and fostering social entrepreneurial innovation. In A. Ayandibu (Ed.), *Reshaping entrepreneurship education with strategy and innovation.* Hershey, PA: IGI Global.

Knox, M., Crawford, J., & Young, S. (in press). Ecotourism Emergence in Tasmania through Social Entrepreneurs and Authentic Leaders. In D. Batabyal (Ed.), *Global Entrepreneurial Trends in the Tourism and Hospitality Industry*. Hershey, PA: IGI Global.

Larsen, J., Axhausen, K., & Urry, J. (2006). Geographies of social networks: Meetings, travel and communications. *Mobilities, 1*(2), 261–283. doi:10.1080/17450100600726654

Levitt, T. (1993). The globalization of markets. In R. Aliber & R. Click (Eds.), *Readings in international business: a decision approach* (pp. 249–267). Boston, MA: MIT Press.

Luthans, F. (2002). The need for and meaning of positive organizational behavior. *Journal of Organizational Behavior: The International Journal of Industrial, Occupational and Organizational Psychology and Behavior, 23*(6), 695–706. doi:10.1002/job.165

Luthans, F., & Avolio, B. (2009). Inquiry unplugged: Building on Hackman's potential perils of POB. *Journal of Organizational Behavior, 30*(2), 323–328. doi:10.1002/job.590

Lyons, G. (2013). Business travel—The social practices surrounding meetings. *Research in Transportation Business & Management, 9*, 50–57. doi:10.1016/j.rtbm.2013.03.001

Martin, A., Sanderson, K., Warnecke, E., Dawkins, S., Bartlett, L., Memish, K., & Peebles, D. (2017). *An integrated approach to workplace mental health.* Hobart, Australia: University of Tasmania.

McDermott, R., & O'Dell, C. (2001). Overcoming cultural barriers to sharing knowledge. *Journal of Knowledge Management, 5*(1), 76–85. doi:10.1108/13673270110384428

Mohr, J., & Spekman, R. (1994). Characteristics of partnership success: Partnership attributes, communication behavior, and conflict resolution techniques. *Strategic Management Journal, 15*(2), 135–152. doi:10.1002mj.4250150205

Newstead, T., Dawkins, S., Macklin, R., & Martin, A. (2019). *We don't need more leaders–We need more good leaders. Advancing a virtues-based approach to leader (ship) development. Leadership Quarterly*; Advanced Online Publication.

Niessen, C., Müller, T., Hommelhoff, S., & Westman, M. (2018). The impact of preventive coping on business travelers' work and private life. *Journal of Organizational Behavior, 39*(1), 113–127. doi:10.1002/job.2215

Paine, L. (1994). Managing for organizational integrity. *Harvard Business Review, 72*(2), 106–117.

Quintane, E., Mitch Casselman, R., Sebastian Reiche, B., & Nylund, P. (2011). Innovation as a knowledge-based outcome. *Journal of Knowledge Management, 15*(6), 928–947. doi:10.1108/13673271111179299

Rego, A., Sousa, F., Marques, C., & Cunha, M. (2012). Authentic leadership promoting employees' psychological capital and creativity. *Journal of Business Research, 65*(3), 429–437. doi:10.1016/j.jbusres.2011.10.003

Robinson, P. (2012). *Tourism: The key concepts*. Abingdon, UK: Routledge. doi:10.4324/9780203104910

Rogers, T. (2013). Conferences and conventions: A global industry (3rd ed.). Abingdon, UK: Routledge. doi:10.4324/9780203119402

Sanderson, K., Martin, A., Warnecke, E., Dawkins, S., Peebles, D., Bartlett, L., . . . LaMontagne, A. (2017). Workplace mental health: Co-production of an action plan for Australia. In Work, Stress, and Health Conference: Contemporary Challenges and Opportunities (p. 2). Academic Press.

Schumpeter, J. (1934). The theory of economic development, translated by Redvers Opie. *Harvard: Economic Studies, 46*.

Shalley, C., Zhou, J., & Oldham, G. (2004). The effects of personal and contextual characteristics on creativity: Where should we go from here? *Journal of Management, 30*(6), 933–958. doi:10.1016/j.jm.2004.06.007

Sidani, Y., & Rowe, W. (2018). A reconceptualization of authentic leadership: Leader legitimation via follower-centered assessment of the moral dimension. *The Leadership Quarterly, 29*(6), 623–636. doi:10.1016/j.leaqua.2018.04.005

Snowden, D. (2002). Complex acts of knowing: Paradox and descriptive self-awareness. *Journal of Knowledge Management, 6*(2), 100–111. doi:10.1108/13673270210424639

Swarbrooke, J., & Horner, S. (2012). *Business travel and tourism*. Abingdon, UK: Routledge. doi:10.4324/9780080490601

Tapscott, D. (1996). *The digital economy: Promise and peril in the age of networked intelligence.* New York, NY: McGraw-Hill.

Taylor, C. (1992). *The ethics of authenticity.* Boston, MA: Harvard University Press.

Thakur, R., Hsu, S., & Fontenot, G. (2012). Innovation in healthcare: Issues and future trends. *Journal of Business Research, 65*(4), 562–569. doi:10.1016/j.jbusres.2011.02.022

Tourism Research Australia. (2019a). *State tourism satellite accounts: 2017-2018.* Barton, Australia: Austrade.

Tourism Research Australia. (2019b). *International visitors in Australia: Year ending March 2019.* Barton, Australia: Austrade.

Trilling, L. (2009). *Sincerity and authenticity.* Boston, MA: Harvard University Press. doi:10.2307/j.ctvjhzrdp

UNWTO. (2008). *International recommendations for tourism statistics 2008.* United Nations World Tourism Organization.

UNWTO. (2019). *International tourism highlights* (2019 edition). United Nations World Tourism Organization.

Valente, F., Dredge, D., & Lohmann, G. (2015). Leadership and governance in regional tourism. *Journal of Destination Marketing & Management, 4*(2), 127–136. doi:10.1016/j.jdmm.2015.03.005

Weber, K., & Chon, K. (2002). *Convention tourism: International research and industry perspectives.* Hove, UK: Psychology Press.

Williams, E., Pillai, R., Deptula, B., & Lowe, K. B. (2012). The effects of crisis, cynicism about change, and value congruence on perceptions of authentic leadership and attributed charisma in the 2008 presidential election. *The Leadership Quarterly, 23*(3), 324–341. doi:10.1016/j.leaqua.2011.07.003

Wilson, S., Fesenmaier, D., Fesenmaier, J., & Van Es, J. C. (2001). Factors for success in rural tourism development. *Journal of Travel Research, 40*(2), 132–138. doi:10.1177/004728750104000203

Woodman, R., Sawyer, J., & Griffin, R. (1993). Toward a theory of organizational creativity. *Academy of Management Review, 18*(2), 293–321. doi:10.5465/amr.1993.3997517

Yammarino, F., Dionne, S., Schriesheim, C., & Dansereau, F. (2008). Authentic leadership and positive organizational behavior: A meso, multi-level perspective. *The Leadership Quarterly, 19*(6), 693–707. doi:10.1016/j.leaqua.2008.09.004

ADDITIONAL READING

Crawford, J., Dawkins, S., Martin, A., & Lewis, G. (2019). Putting the leader back into authentic leadership: Reconceptualising and rethinking leaders. *Australian Journal of Management*; Advanced Online Publication. doi:10.1177/0312896219836460

Crawford, J., Kelder, J.-A., & Knox, M. (*in-press*). What does it take to be a social entrepreneur? Authentic leaders and their effect on innovation. In Ç. Dogru (Ed.), *Leadership styles, innovation, and social entrepreneurship in the era of digitalization (n.p.)*. Hershey, PA: IGI Global.

Rego, A., Sousa, F., Marques, C., & Cunha, M. (2012). Authentic leadership promoting employees' psychological capital and creativity. *Journal of Business Research*, *65*(3), 429–437. doi:10.1016/j.jbusres.2011.10.003

Swarbrooke, J., & Horner, S. (2012). *Business travel and tourism*. Abingdon, UK: Routledge. doi:10.4324/9780080490601

KEY TERMS AND DEFINITIONS

Authentic Follower: An authentic follower is an individual who, through their capacity for authenticity and positive organizational engagement, are self-managing and follow leaders with whom they share values (Crawford et al., 2018, p. 274).

Authentic Leader: An individual who "influences and motivates followers to achieve goals through their sincerity and positive moral perspective, enabled through heightened awareness and balanced processing" (Crawford et al., 2019).

Authentic Relationships: The intersection of the interpretation of another's sincerity between individuals seeking to be authentic in their actions and their words.

Business Tourism: The utilization of travel with the intent to achieve business objectives, form connections, and share knowledge.

Change Success: The degree to which a change initiative is perceived to be effective.

Creativity: The process managing ideas and resources to promote creative products, services, ideas, procedures, or processes.

Knowledge Sharing: The exchange of information through willing participants involving specific attention to the actors involved.

Positive Organizational Behavior: A field that emerged as a result of organizational behavior and positive psychology to focus on strength-based approaches to human behavior at work.

Chapter 5
Role of SOR Theory in Business Tourism Activities for Effective Knowledge Transfer:
Pluralistic Learning Theory for Project Knowledge Transfer

Sadia Hanif
https://orcid.org/0000-0003-0931-8608
Foundation University, Pakistan

Ali Ahsan
https://orcid.org/0000-0002-2023-0511
Chifley Business School, Torrens University, Australia

ABSTRACT

This study takes on a course of "phenomenon before theory" approach. For that purpose, firstly, the phenomenon regarding knowledge transfer barriers is developed, and then it is confirmed through expert suggestions. Knowledge transfer theories from project management literature are then applied and systematically reviewed against observed phenomenon to obtain the desired solutions. The systematic review of knowledge transfer theories is conducted through content analysis. The observed gaps are then filled with pluralism with components of required theories. The findings were that the discussed theories are not fully mitigating the knowledge transfer hitches. The most important factor for competence development is effective learning during the knowledge transfer activities which has not been addressed by the knowledge transfer theories. The chapter proposes a "Pluralistic Learning Theory for Project Knowledge Transfer" as a practical solution for all sorts of business tourism activities to transfer knowledge effectively.

DOI: 10.4018/978-1-7998-3142-6.ch005

INTRODUCTION

In Pakistan after the arrival of China Pakistan Economic Corridor (CPEC), the business tourism opportunities developed and has increased day by day, both at Public Sector institutes and even in universities. There are lots of opportunities for a developing country Pakistan to learn technological, management and administrative skills from a developed country China. In this regard, the various centres of excellence developed at public sector institutes and universities but still due to lack of awareness those institutes were not working in a way to transfer knowledge effectively as mostly done in excellence centres of developed countries. Besides, there are various knowledge transfer problems on ground regarding absorptive capacities, motivation to learn, willingness and absence of effective training/instructing methodologies. These knowledge transfer barriers were hitching the flow of knowledge from various channels of CPEC in the form of trainings, workshops, seminars, conferences, discussions, working together opportunities, sharing of knowledge, and thinking about change.

After the inauguration of CPEC, the business tourism activities increased at public sector organizations. The website of the 'Islamabad Policy Research Institute' provides a detailed list of seminars, conferences, workshops, pieces of training and discussions related to CPEC. The universities in Pakistan also organized various seminars conferences and symposiums to introduce the benefits of CPEC. The public sector organizations send their employees for pieces of training and workshops to China to learn updated skills.

The public organizations and CPEC work process has been implemented in the form of projects. Pakistan public sector organizations were already at project management maturity level 2, as according to Ali (2010), there were various problems in project management at all phases of projects in all knowledge areas as shown in table 1.

Table 1. Issues faced by Public sector Project Management in Pakistan

Phases	PM Issues	Cultural Issues	Issues related to LDC_s	PSO_s issues
Initiation	Lack of proper costing, design, and activity duration estimation	Unrealistic claims to please the boss	Political involvement/ pressure in selecting the project, Lack of funds	Lengthy project approval
Planning	Lack of stakeholder, risk,quality, communication management and planning	Contractor not following the plan, lack of involvement of middle managers	Lack of feasibility study on medium to small-scale projects, Lack of skilled staff, No written policy in selecting project team members	
Execution	Scope changes, project managers have limited decision power	Fear of change, executives not willing to delegate powers,	Lack of capacity in the organization to implement the project, Capacity issues with contractors, equipment, resources, and data management	
Monitoring	Attempt to provide wrong information, no electronic data management system	Lack of following the SOPs	Lack of accountability corruption	Lack of attention by the authorities, biases
Closing	No lessons learned and benefit achieved report			

Source: Ali, A. (2010). Investigating project management practices in public sector organizations of a less developed country

Considering the discussed issues in table 1, the present study discusses the knowledge transfer barriers with respect to project management. Because these barriers influence the learning process during business tourism activities. Due to which there is no observed effect on project management applications in the public sector of Pakistan, besides the number of training, and other business tourism activities.

This study takes on a course of "*Phenomenon before theory*" approach. For that purpose, firstly, the phenomenon is developed, and then it is confirmed through expert's suggestions. Knowledge transfer theories from project management literature are then applied and systematically reviewed against observed phenomenon to obtain the desired solutions. The systematic review of knowledge transfer theories is conducted through content analysis. The observed gaps are then filled with pluralism with components of required theories. The findings were that the discussed theories are not fully mitigating the knowledge transfer hitches. The most important factor for competence development is effective learning during the knowledge transfer activities which has not been addressed by the knowledge transfer theories. The paper proposed a "*Pluralistic Learning Theory for Project Knowledge Transfer*" as a practical solution for all sorts of business tourism activities to transfer knowledge effectively.

RESEARCH PROBLEM

In the context of Pakistan, it is observed that despite the demonstration of many improvement programs such as training, certifications, courses, seminars, conferences, and various business tourism activities, project management of Pakistan still encounters the failures both in public and in the private sector organizations. It is assumed that it is so because of the lack of effective learning in order to develop absorptive capacities, willingness, and motivation through relevant training and other business tourism activities. Also, the organizational barriers exist because of the nonexistence of absorptive capacities.

In the current study, the observed phenomenon of knowledge transfer barriers goes through expert's opinion with the help of unstructured interviews and a solution is proposed with the pluralism of components of relevant learning theories in SOR theory.

Background

According to Sazulanski, there are various elements that influence the transfer of knowledge including source, recipient, channel, and context. The organizations encounter the barriers associated with each of these elements. These barriers can either be any of the characteristics of the knowledge transferor and the receiver or their relationship. Knowledge transfer is a dyadic exchange process in which a source ensures the availability of knowledge to be acquired by the recipient (O'Sullivan, 2007, p7). Therefore, knowledge transfer depends upon the way in which knowledge is exchanged between the transferor and the transferee. Sazulanski defines this relationship as an *arduous relationship* i.e. Emotionally exhausting and distant between the source and recipient (Sazulanski, 1996). The quality of an *arduous relationship* influences the way of transferring the required knowledge and the ability of the recipient to learn and apply the acquired knowledge (Ko et al, 2005, p64). This means that knowledge transfer occurs through a bridge of learning processes. In this context, Ko et al, (2005) defined knowledge transfer as, "*The communication of knowledge from a source that is learned and applied by a receipt*" p62. The receiver should learn first and then apply the acquired knowledge. Therefore, knowledge transfer needs an effective learning process. Knowledge transfer among organizational members enhances their individual competencies (Sveiby & Simons, 2002). Effective knowledge transfer ensures that the employee receives and understands (learn) the knowledge he needs to perform at his best. Knowledge means what is known, but competence refers to the ability to do something. Competence demonstrates the ability to apply learned knowledge (Epstein, & Hundert, 2002). Competencies can be developed through the process and the transfer of learning. Therefore, it is important to understand the learning processes and barriers in one context that leads to the transfer of knowledge for its application in another context (National Research Council. 2000, p51).

Kim (1993), defined individual learning as '*increasing one's capacity to take effective action*'p38. The transfer of learning commences with the learner becoming able enough to articulate his abilities i.e. applying physical ability for innovation (Kim, 1993, p2 and 3).

The present study is centered on a phenomenon influencing the learning processes in the context of developing countries that are hindering the effective knowledge transfer. The paper focuses on the most applicable knowledge transfer theories in the field of project management only that typically highlight the knowledge transfer process to convert knowledge into action. The research questions of the study are;

Research Question 1: What are the phenomena influencing the learning processes during knowledge transfer through training and business tourism mechanisms for the competency development of Project Managers?

Research Question 2: What is the role of SOR theory for effective learning processes during knowledge transfer?

Research Question 3: What are the gaps in Knowledge Transfer theories against the phenomenon influencing the learning process?

Research Question 4: How pluralism helps to mitigate the solution by combining components of different learning theories in SOR theory ?

This chapter is organized according to research questions as in the literature review section the phenomenon that influences learning processes discussed along with the role of SOR theory with reference to the discussed phenomenon. After discussing the methodology and expert's opinion, in the analysis and results section, there is discussion of results by identifying gaps in the knowledge transfer theories about learning barriers in Pakistan, the recommendation of solutions through pluralism and future research directions and limitations.

LITERATURE REVIEW

Phenomenon Influencing Learning Processes in Developing Countries

The phenomena of Knowledge Transfer inherently involve interdependencies and asymmetries between the entities involved such as providers and recipients (Grant, 1996). The effective knowledge transfer through learning mechanisms highlights the importance of the interaction of 'Transferor' and 'Transferee' as 'Stimulus' and 'Response' following the SOR (Stimulus-Organism-Response) Theory. Effective learning depends on understanding the behaviors of the learners according to the situation. Behaviours are shaped by reasons, in case of learning the need and reason to learn, in order to shape the behaviour of the learner. The capabilities of learners to understand the context of knowledge also shapes the behaviour of the learner. The understandability of behaviours by trainers/instructors makes the processes of learning productive (Wilson, & Peterson, 2006). Although there is a difference in training and learning, they are inextricably linked. The quality of learning depends on the quality of training, because of which, the role of the trainer remains very critical. Effective learning enables a learner/trainee for non-specific and unexpected situations and to face future challenges (Epignosis, 2014, p23/24). In the case of project management, unexpected situations and future challenges are linked with sponsoring organizations and project management situations in the field. The project managers must develop such skills which will be helpful in tackling the problems that may arise due to the sponsoring organizations. In such cases, the trainers should be aware of the systems and structures of their sponsoring organizations and about the practical situations of project management (Yanwen, 2012).

As mentioned above, it is observed that the trainers, capacities of the trainees and the support from the sponsoring organizations play an interconnected role to enable the trainees to learn the needed skills and develop necessary competencies. Furthermore, each role is aligned with associated problems to define the research problem.

Role of Trainers Capabilities

Mcleod (2017), suggested that educators/trainers design and carry out their activities according to ways offered by each learner/trainee to engage them in a manner that suits them. This type of approach aids in filling the learning gaps effectively. In this regard, it is important to know the teaching methodologies and to define whether they are aligned with the existing behaviours? As per SOR theory (Buxbaum, 2016, ch:2), the behavior of the trainee alters, after getting the training from the trainer. Thus, the learning mechanisms used by the trainer should effectively capture the learning gaps and strategies that lead to a change in the knowledge behaviour of the trainee in a real situation (Bedilescu, 2013).

Prior research confirms that human behaviours are shaped by psychological factors of the organism. However, the environmental/social changes alter the psychological factors (Fekete et al., 2014, (Dretske, 2004). This reverse influence of the social actors on the psychological factors alters the responses of trainees as soon as their stimulus changes. This is a common practice in the training sessions of international project management conducted by expert trainers in Pakistan. The trainees participate with a positive psychological mindset, they learn through trainers effectively and exhibit good assessment results. But, when they come out of training sessions and face the real project contextual constraints, they are unable to apply the knowledge they obtained during the training session to show their competencies. This is because of the absence of learning to apply the knowledge in changing and challenging situations (Epignosis, 2014, p23/24)

With respect to training, if the trainer effectively learns and understands the organismic components prior to the commencement of the training, then he can make a conscious effort to deliver the required knowledge accordingly. Whereas, learning organismic factors is quite difficult for social scientists and especially for trainers, because of their biological nature. Trainers should design and carry out their activities, catering to the needs of the trainees needs as they vary from person to persona and organization to organization. This approach will be able to fill the knowledge transfer gaps more effectively. (Mcleod, 2017). For that purpose, trainers should take guidelines from the learning theories while designing their training content and presentations. Upgrading knowledge and skills and changing attitudes and behaviours is not a one-time activity, it requires constant stepwise reinforcement and up-gradation with the guideline of the existing theories (Filipatali, 2013).

In practice, training must follow the scientific methodologies that consist of a set of approaches and processes focusing on the situations in which specific gaps need to be filled or where performance improvement is required (Milhem, Abushamsieh, & Pérez, 2014). The scientific methodologies are based on the theories of situated learning or cognition, constructivism, experiential learning, transformative learning theory, and action theory (Milhem et al, 2014). Following the situational learning and cognition theories, the training material should be prepared for the training session according to the issues and problems faced by the trainees. In this way, trainees will be able to apply the skills learned during the training sessions in other similar situations. The training activities should be designed in a way that could be shared during the ongoing training for the active participation of the trainees in order to resolve the identified issues (Anderson et al., 1996). The process, content, and methodology of training can influence the learning processes and helps to improve absorptive capacities to improve knowledge transfer activities across the training programs.

Role of Absorptive Capacity of Trainees

Competencies can be developed through knowledge transfer if individuals have the absorptive capacity so that they can recognize, assimilate, and utilize external knowledge (Cohen & Levinthal, 1990, p.128, Sazulanski, 1996). Acquiring and absorbing project knowledge is not possible without relevant and updated absorptive capacity (Maninga, 2017). In 2007, Parent, Roy and St-Jacques followed systematic thinking in knowledge management suggested by Rubenstein-Montano (2001). They proposed a dynamic knowledge transfer model (DKTM) that focuses on the components of the social system required for knowledge transfer via two pre-existing conditions (need and prior knowledge) and the four capacities (adaptive and responsive capacity, generative capacity, disseminative capacity, and absorptive capacity). This model assumes that appreciation of knowledge depends on the need of the users and the level of existing relevant knowledge. Generative capacity depends on the system's intellectual creative capital that defines the member's ability to discover and improve the knowledge. Disseminative capacity depends on social ties and other intermediaries. Absorptive capacity typically exists in systems where there is the prior related knowledge that enables the members to recognize the value of knowledge acquired and apply it to relevant issues within the system. Finally, adaptive and responsive capacity depends on prior learning experiences and processes of systems, critical thinking approach, the vision of the members, distributed leadership, feedback loops, and monitoring mechanisms. It ensures the ability of the members to continuously learn and update the knowledge transfer systems in use for continuous development. The model assumes that all four capacities should be possessed by the system for a successful transfer and application of knowledge. The absence of any capacity will create a need to recover through knowledge transfer mechanisms first, and then further improvements could be made possible (Parent, Roy, & St-Jacques, 2007). According to the theory of absorptive capacity, individuals should possess a knowledge base to acquire knowledge (Miles, 2012). Knowledgebase refers to the adaptive and responsive capacity in DKTM as in the absence of it, the individuals will not be able to adapt and respond to the new knowledge.

The characteristics of the absorptive capacity, the knowledge transfer mechanisms, and its usability are different in the developed-developed and developed-developing interfirm collaborations. In the case of developed-developed interfirm collaborations, the learning process could be two way that leads to developing the competitiveness of both the entities. Whereas, in the case of developed-developing collaboration, the learning process becomes one-way for each entity with different intentions. Developing countries with low skill, technological and other resource constraints make knowledge development difficult. In the relationship between transferor and transferee, the transferees should have some prior knowledge in order to absorb the knowledge acquired (Narteh, 2008).

Hence, first, there is a need to develop absorptive capacities through major shifts in the learning processes in developing countries. Moreover, the learning processes should also maintain and evolve over time for continuous learning. In this way, the knowledge transfer process can be made effective for adequate competency development (Soderlund, 2008).

The absorptive capacity develops employees' basic skills and knowledge, and enhance the employees' motivation to learn from the knowledge given at the training. However, for the effective utilization of the knowledge transfer, besides absorptive capacity, the ability and willingness of employees to apply knowledge should also be taken into consideration (Minbaeva et al., 2007). The employees can learn if they are able as well as willing to learn (Anatoliivna, 2013).

Role of Organizations in Learning and Knowledge Transfer Processes

The formal systems and structures of the organizations, the coordination among organizational members, their expertise, and their behaviours affect the knowledge transfer. The organizations that facilitate knowledge transfer are deemed as fertile. Conversely, the organizations that create hindrances in the knowledge transfer gestures and practices are said to be barren (Sazulanski, 1996, p32).

These hindrances are also called organizational barriers which can adversely affect the learning of the individuals. Schilling and Kluge (2009) discussed various organizational barriers to learning and distinguished them in three broad categories (Action-personal, structural-organizational, and societal-environmental). *Action-personal barriers* evolve from individuals, including the biases and deficiencies of the employees in receiving and assimilating the organizational knowledge as well as the lack of general knowledge, technical knowledge and stereotyping. It refers to a lack of knowledge base and absorptive capacity as discussed previously. This leads to high-level stress and fear of disadvantages. *Structural-organizational barriers* arise from organizations in which those employees work. These are lack of clear and measurable goals, poor or inactive performance feedback, strict rules and regulations, lack of opportunities to learn and apply the new knowledge as well as the lack of experience in applying the new knowledge. Developing countries often suffer a great deal because of poor organizational structures (Yanwen, 2012). The third type of barrier i.e. *societal-environmental* originates from a society in which these employees live. These barriers include complex, dynamic and competitive market environments, cultural distances and low level of experience as well as complicated, ambiguous and difficult knowledge (Schilling & Kluge, 2009). All these barriers exist in the context of the present study.

THE ROLE OF SOR THEORY IN MITIGATING THE RESEARCH PROBLEM

In 2008, Hollan reviewed the SOR theory in contrast to cognitive theories. He also discussed Spence's (1948, 1950) criticisms. Spence said that in SOR learning capacity, responses are generated according to the stimulus and it becomes difficult for the respondents to show performance later when the usual paths are blocked, and shortcuts of spatial environments are opened. SOR theorists consider only receptor-effector connections of the brain, whereas, cognitive theorists also consider the re-organization of 'brain currents' and 'neurophysiological' trace systems' to specify the hypothetical relations of behaviours tied with the learning process. Cognitive theorists conceive the associations among the sensory-perceptual processes involved in learning, however, SOR theorists take into consideration, only the biological factors. During the 1950s, most of the cognitive theorists posited that the SOR theorists did not consider the influence of organizational variables. However, Spence addressed that SOR theorists had an interest in the orientation and attention variable; for example, reinforcement contingencies and changes in learning rate parameters. According to Hollan (2008), the outcome of Spence's study pendulum decidedly swung towards the cognitive side and this is because of Spence's little knowledge of the nervous system. The neurophysiological characterization of learning has become very advanced now. However, to understand behaviour, it is vital to study it as neural analysis can provide valuable clues and connect the abstract phenomenon of learning with the pure scientific community (Hollan, 2008).

Therefore, certain measures are required to enhance the quality of the stimulations by the trainers during the knowledge transfer activities so that trainees are able to learn and apply them in various interventions. The trainers must keep track of the learning experiences of trainees and consider the context of knowledge application. Likewise, the trainees must be able to recognize their need to learn.

Organizational Context and SOR Theory

The organizational and contextual perspective is missing in SOR theory because the organismic components are only individualistic. The organizations tend to influence the responses of trainees, hence, as soon as the stimulations change, the response for applying that knowledge will alter. The public sector organizations of Pakistan are characterized by bureaucratic structures and traditional project management systems and hence, encounter knowledge transfer problems associated with such structures and systems. Hence, even if the respondents learn new knowledge, because of the existing knowledge transfer problems, they are unable to fully exploit it.

Therefore, considering only the psychological factors for the effective transfer of knowledge during the training sessions is not enough. In order to ensure a successful knowledge transfer and its successful application, there is a need to all such factors that control the responses of the trainees according to real project contextual constraints.

Knowledge Retention in SOR Theory

Knowledge retention refers to capturing knowledge in a way so that it can be used later (Aggestam et al., 2014). The transferred knowledge is useless if the trainees are unable to retain over a considerable period of time for its application (Maskell & Malmberg, 1999). This happens when learners cannot actualize the value of the learned knowledge (Burleson, 2005). If the transferred knowledge is not able to resolve the problems, is not easy to adopt and exploit, then the learners will not actualize its importance (Reilly, 1998). The discussion on training approaches shows that there should be a purposeful learning environment considering the real situation. The consideration of organizational context shows that learning through real-life settings can make the learners able to exploit the learned knowledge. But, there is also a need for motivation to accept challenges and risks in applying new knowledge to have a greater good. Such motivations can be developed through organizational commitment i.e. 'sense of oneness' which drives the employees' commitment to stay in the organization. The associated factors like loyalty, trust, work-life policies, procedural justice, and career development also influence employee motivation and commitment (Coombs, 2009). The open communication channels policy, rewards, and incentives also aid the employees in applying the knowledge and to retain it for continuous wellbeing (Bessick & Naicker, 2013). Hence, the contextual factors like human resource management, supportive leadership, learning culture, autonomy and systems of IT influences the knowledge exploitation and exploration (Gonzalez & De Melo, 2018).

As compared to the above discussion, the scenario is totally different in Public Sector organizations of Pakistan. The bureaucratic structures hinder the open door communication and motivation to apply new knowledge through rewards, incentives, and other above-discussed factors. The knowledge retention is an organizational attribute (Kumar, 2017), the employers and employees need to adopt it through learning because learning can change and improve behaviours. By ignoring human cognitive factors and organizational context, SOR theory alternatively can't work alone for effective knowledge transfer for the competency development. It needs to associate itself with the theories of learning to solve the research problem discussed.

METHODOLOGY

To develop the theories against the social phenomenon, Tellis (2017) recommended begininig with the study of the phenomena, instead of firstly considering already developed theories. He argued that by studying the assumptions of theories first, the researcher can become a believer of those assumptions and it makes it difficult for him to refute. The theories enable biased interpretation of his fresh insights of realism.

So keeping that in mind, the observed phenomenon of knowledge transfer barriers in developing countries will be first reviewed from the literature. It will then be confirmed through the expert's opinions in the context of Pakistan. The phenomena will then be compared against the knowledge transfer theories to find a way to resolve the issues of knowledge transfer for effective learning processes.

In this regard, the search engines used were HEC digital library Pakistan, Aiou E-Library Pakistan, Emerald, Sciencedirect, Research gate, Academia, and Google scholar. The study will modify *'concept-centric matrix augmented with a unit of analysis'*, by Webster (2002) to *'phenomenon-centric matrix augmented with the content of theories'*. The review of theories will be conducted through the content analysis of relevant articles shown in table 3. The study used MAXQDA for content analysis.

Expert's Opinion on Discussed Research Problem

The existence of the knowledge transfer barriers discussed in the literature was confirmed through expert interviews using the window model. A snowball sampling technique used to search the experts in the field through personal contacts and through professional networks like Research Gate, LinkedIn, and Google Scholar. Ilya (2011), suggested an *'Eight Window Model'* for sampling in numerous qualitative researches. The window model is used to mitigate the expected risk of homogeneity of the experts and to frame the sample of experts.

Figure 1. Expert Interview eight window model
Source: Steinberg (2011)

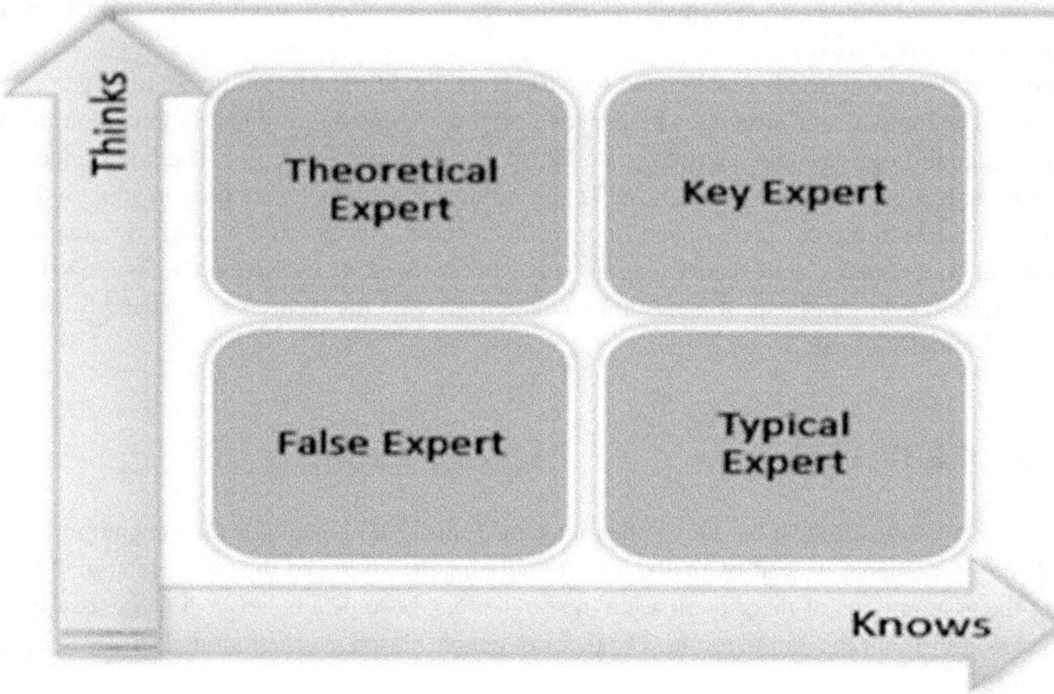

The purpose of the expert's opinion was to clarify the issues relevant to the observed phenomenon. Individual experts can be consulted but a group of experts is usually recommended for a wide range of opinions. The results of the expert's opinion can be a list of perceived problems as well as a list of recommendations and improvements. A good rule of thumb is to consult 4 to 6 experts (Rauterberg, March 2019). In the present study, a total of 14 experts were contacted and 10 out of them corresponded in the interviews. According to the window model, the first three windows are most important and should be fulfilled. In the present study, the ten respondents fulfilled all four windows. The experts available within the country were interviewed in person and those present in foreign countries were interviewed through emails. The unstructured interviews from experts were then analysed through content analysis in MAXQDA.

Analysis and Results

The first step in the research process was to gather expert opinion. The expert's opinions were transcribed in a document form and analysed using MAXQDA. The codes were developed according to the phenomenon of knowledge transfer barriers in the developing countries and Pakistan, which shows a few of the unique suggestions enlisted in table 2.

Table 2. Experts opinions

Suggestion	Frequency	%	Memo
Develop Model	7	70.00	Experts have suggested that there should be a model or structure based on the phenomenon discussed as a theoretical basis for resolving learning issues
Absorptive capacity	6	60.00	There was a lack of understanding of the concept of 'absorptive capacity', the expert's link it with 'Knowledgebase' and 'capacity building'. However, the knowledge base is the prerequisite of effective absorptive capacity
Training Problems	6	60.00	Experts have suggested focusing more on training, the process of need identification and how much these training enable PMs to apply the knowledge
Organization's support	2	20.00	Experts have suggested looking towards the organization's support as well, that why PM's cannot practice their competencies, why learned knowledge cannot be retained and how learned knowledge could use on a sustainable basis. In practice, PM's have a problem regarding an organization's environment in our PM system
culture's comparison	2	20.00	Compare cultures to see whether the problem really exists in Pakistan only? Also, reduce the high contextuality to enhance the generalizability of the knowledge gap
DOCUMENTS with code(s)	10	100.00	
DOCUMENTS without code(s)	0	0.00	
ANALYSED DOCUMENTS	10	100.00	

Table 3. The list of observed phenomena of knowledge transfer barriers in Public sector project management of Pakistan

Source	Phenomena	
	Barrier	Sub-Barrier
Learner	Absorptive capacity	1. Knowledge Base 2. Motivation 3. Willingness 4. Knowledge retention
Trainer	Training	1. Philosophy 2. Methodology 3. Agenda
Organizations	Organizational Barriers	1. Stickiness 2. Context 3. Political influence 4. Societal-environment 5. Structural organizational barriers 6. Actionable-personal barriers

Table 4. List of theories considered with source articles

Theory	References
Analogical transfer	Gick, M. L., & Holyoak, K. J. (1983). Schema induction and analogical transfer. Cognitive psychology, 15(1), 1-38. Novick, L. R. (1988). Analogical transfer, problem similarity, and expertise. Journal of Experimental Psychology: Learning, Memory, and Cognition, 14(3), 510. Catrambone, R., & Holyoak, K. J. (1989). Overcoming contextual limitations on problem-solving transfer. Journal of Experimental Psychology: Learning, Memory, and Cognition, 15(6), 1147. Mair, C., Martincova, M., & Shepperd, M. (2009). A literature review of expert problem solving using analogy. Nokes, T. J. (2004, January). Testing three theories of knowledge transfer. In Proceedings of the Cognitive Science Society (Vol. 26, No. 26). Chen, Z. (2002). Analogical problem solving: A hierarchical analysis of procedural similarity. Journal of Experimental Psychology: Learning, Memory, and Cognition, 28(1), 81. Gentner, D., Loewenstein, J., & Thompson, L. (2003). Learning and transfer: A general role for analogical encoding. Journal of Educational Psychology, 95(2), 393.
Nonaka's four modes of knowledge	Nonaka, I. (1994). A dynamic theory of organizational knowledge creation. Organization Science, 5(1), 14-37. Nonaka, I., Byosiere, P., Borucki, C. C., & Konno, N. (1994). Organizational knowledge creation theory: a first comprehensive test. International Business Review, 3(4), 337-351. Spencer, B. (1997). Organizational Knowledge Creation-Ikujiro Nonaka. In Knowledge Advantage Conference (pp. 11-12). Nissen, M. E. (2006). Harnessing Knowledge Dynamics Principled Organizational Knowing and Learning. Bratianu, C. (2010, March). A critical analysis of Nonaka's model of knowledge dynamics. In Proceedings of the 2nd European Conference on Intellectual Capital (pp. 115-120). Academic Publishing Limited, Reading.
Process Model of Knowledge Transfer	Szulanski, G. (1999). The process of knowledge transfer: A diachronic analysis of stickiness. Organizational behaviour and human decision processes, 82(1), 9-27.
(DKTC)	Parent, R., Roy, M., & St-Jacques, D. (2007). A systems-based dynamic knowledge transfer capacity model. Journal of Knowledge Management, 11(6), 81-93.
(CBKSM)	Lodhi, S. A. (2005). Culture based knowledge sharing model (Doctoral dissertation, National College of Business Administration & Economics, Lahore).
CRAI model	Oluikpe, P. (2007). Knowledge management in development projects (Doctoral dissertation, © Paul Oluikpe).
Knowledge transfer model of interfirm collaboration	Narteh, B. (2008). Knowledge transfer in developed-developing country interfirm collaborations: a conceptual framework. Journal of knowledge management, 12(1), 78-91.
Knowledge transfer-A Process model	Liyanage, C., Elhag, T., Ballal, T., & Li, Q. (2009). Knowledge communication and translation–a knowledge transfer model. Journal of Knowledge management, 13(3), 118-131.
Knowledge governance mechanism- fit barrier matrix	Fang, S. C., Yang, C. W., & Hsu, W. Y. (2013). Inter-organizational knowledge transfer: the perspective of knowledge governance. Journal of Knowledge Management, 17(6), 943-957.

The expert's review confirms the existence of the observed phenomenon and explains the nature of their existence. The phenomenon of training problems and absorptive capacity was considered as the most severe of all the problems by the reviewers. Effective learning during training develops absorptive capacities that can further make the employees competent. Organizational support is also important because unless there are no processes and routines to aid the application of the learned knowledge stepwise, the trainees will not be able to retain that knowledge. The experts also suggested developing a model based on the existing phenomenon to give a practical solution. Therefore, existing knowledge transfer theories in project management were systematically reviewed using content analysis and pluralism approach to manipulate the phenomenon of knowledge transfer barriers in public sector project management of Pakistan. The observed phenomenon is listed in a tabulated form in table 3.

Table 5. Code coverage of phenomena

#	Code System	Sub codes	Code coverage (%)
1	Stickiness[1]		7%
2	Training		0%
3		Training philosophy	6%
4		Training methodologies	12%
5		Training Agenda	10%
6	Political influence		0%
7	Barriers to organizational learning		1%
8		Organizational Context	10%
9		Societal-environment	3%
10		Structural organizational barriers	3%
11		Actional-personal barriers	4%
12	Absorptive capacity		20%
13		Motivation	9%
14		Willingness	5%
15		Knowledge retention	1%
16		Knowledge Base	13%
	NOT CODED		2,428,027
	CODED		100% (60,300)
	WHOLE TEXT		2,488,327

Reviewed Theories

The study systematically reviewed the knowledge transfer theories using content analysis against the discussed phenomena of learning barriers for competency development. Knowledge transfer theories were reviewed to assess their applicability in this situation. The theories that were reviewed were enlisted along with relevant source articles in table 4.

The articles enlisted went through a systematic review by critically discussing and relating the phenomenon with the theories. For this purpose, the assumptions, components, and limitations of the theories were reviewed thoroughly. Using MAXQDA, the theories were coded against phenomena of knowledge transfer barriers (shown in table 2) to develope '*phenomenon-centric matrix augmented with the content of theories*. The results were developed using code configurations, code frequencies, code coverage and interactive code matrix from MAXQDA analysis.

The discussed observed phenomenon entered in MAXQDA as codes and subcodes. The coded segments were then searched through the MAXQDA document search using codes as keywords. The selected segments for codes from documents (enlisted in table 4) are those that matched with the occurrence of the phenomenon only and remaining (in headers, footer, footnotes, headings, and references) were not considered. This practice was adopted to ensure the consideration of literature and theories of the observed phenomena. The code coverage table 5 shows the percentage of entire documents coded with their respective codes.

Table 6. Code configuration in documents

#	Code combinations	Frequency	%	% (valid)
1	Absorptive capacity	46	12.81	12.81
2	Actional-personal barriers	36	10.03	10.03
3	Barriers to organizational learning	4	1.11	1.11
4	Knowledge Base	30	8.36	8.36
5	Knowledge Base + Absorptive capacity	7	1.95	1.95
6	Knowledge retention	1	0.28	0.28
7	Motivation	35	9.75	9.75
8	Motivation + Knowledge Base	1	0.28	0.28
9	Motivation + Willingness	3	0.84	0.84
10	Organizational context	24	6.69	6.69
11	Organizational context + Absorptive capacity	1	0.28	0.28
12	Training	1	0.28	0.28
13	Political influence	1	0.28	0.28
14	Societal-environment	18	5.01	5.01
15	Stickiness	18	5.01	5.01
16	Stickiness + Absorptive capacity	1	0.28	0.28
17	Stickiness + Motivation	2	0.56	0.56
18	Stickiness + Organizational context	1	0.28	0.28
19	Structural organizational barriers	31	8.64	8.64
20	Training Agenda	32	8.91	8.91
21	Training methodologies	32	8.91	8.91
22	Training philosophy	18	5.01	5.01
23	Willingness	15	4.18	4.18
24	Willingness + Organizational context	1	0.28	0.28
	TOTAL (valid)	359	100.00	100.00
	Missing	0	0.00	-
	TOTAL	359	100.00	-

Figure 2. Gap analysis through phenomenon-centric matrix augmented with the content of theories

Theories / Phenomenon	Analogical transfer	Nonaka's four modes of knowledge	Process Model of Knowledge Transfer	Dynamic Knowledge transfer capacity model	Cultural Based Knowledge Sharing Model (CBKSM)	CRAI model	Knowledge transfer model of interfirm collaboration	Knowledge transfer-A Process model	Knowledge governance mechanism- fit barrier matrix
Stickiness	×	×	√	×	×	×	×	×	√
Training	×	×	√	×	×	×	×	×	×
Training philosophy	×	×	×	×	×	×	×	×	×
Training methodologies	×	×	×	×	×	×	√	√	×
Training Agenda	×	×	×	×	×	×	×	×	×
Political influence	×	×	×	×	×	×	×	×	×
Organizational context	×	√	√	×	×	×	√	√	√
Barriers to organizational learning	×	×	×	×	×	×	×	×	√
Societal-environment	×	×	×	×	×	×	√	×	×
Structural organizational barriers	×	×	×	×	×	×	×	×	×
Actional-personal barriers	×	×	×	×	×	×	×	×	×
Absorptive capacity	×	×	√	√	×	×	√	√	√
Motivation	×	×	√	√	√	√	×	×	√
Willingness	×	×	√	×	√	√	×	√	×
Knowledge retention	×	×	×	×	×	×	×	×	×
Knowledge Base	×	×	√	√	√	√	√	×	√

It is observed in Table 5 that the phenomenon '*Absorptive capacity*', has the highest coverage followed by the '*knowledge base*'. According to the theory of '*Absorptive capacity*', the '*knowledge base*' is the prerequisite of absorptive capacity. However, for the development of both '*knowledge base*' and '*absorptive capacity*', there is a need for knowledge transfer. Further, table 5 shows that *organizational context, training methodologies*, and *agenda* also have higher coverage. According to the literature discussed, effective knowledge transfer can occur with 'organizational support' and 'trainer's capabilities' to engage the trainees in a way that suits them by resolving their problems and make them competent according to their needs.

The code configuration analysis in table 6 gives codding patterns by displaying how frequently the same set of codes appeared in the activated documents. The results provided a combination of codes that helped in understanding various attributes of codes and confirms the support of the discussed phenomena for an effective knowledge transfer.

In table 6, the highlighted text shows some combinations that occurred in various documents. Among these combinations, the '*absorptive capacity + knowledge base*' had the highest percentage. It is observed from these combinations that most of the phenomena are related to each other. Hence, by focusing on only one or a few phenomena, the problems will not be addressed. It is impossible to resolve base problems regarding the *knowledge base* and *absorptive capacity* without the support of motivation, willingness of individuals and the favourable organizational context.

To view whether the knowledge transfer theories support all the discussed phenomena of the knowledge transfer barriers in Pakistan, a '*phenomenon-centric matrix augmented with the content of theories*' was developed using '*interactive quote matrix*' and '*valid percentage*'. The occurrence of phenomena against each set of theories was analysed in MAXQDA

The phenomenon-centric matrix shows the percentage of valid occurrence of the phenomena against each theory. The (Ö) shows that the theory has coded the segments related to the phenomenon. Whereas, the (×) shows that theory has not coded the segments related to the phenomenon, and hence, shows a gap against the selected phenomena. To ensure a successful knowledge transfer considering all phenomena, a thorough framework is required in the context of Pakistani project management practices. The table shows that *knowledge base, absorptive capacity, organizational context, and* motivation are covered more comprehensively in KT theories, but sources of knowledge transfer i.e. *training followed by training philosophy, methodology, and agenda*. The political influence on the knowledge transfer and the role of societies is also ignored by knowledge transfer theories. Likewise, the concept of *knowledge retention* also overlooked as it can sustain continuous learning long after the knowledge transfer process.

Role of Learning in Gaps of Knowledge Transfer Theories

The identified gaps in the knowledge transfer theories are exhibited in table 6 against the code configuration. Table 6 shows that the transfer of knowledge can be effective if there are an adequate knowledge base and absorptive capacity that reinforces the motivation and willingness to learn. The development of absorptive capacity prerequisites effective learning in the knowledge transfer processes. Effective learning requires the transferor to understand the knowledge gaps and follow the most appropriate methodology of transferring the project management knowledge with respect to the context of the organization.

Project management training limitations show that the journey from knowledge transfer to competence development cannot be completed effectively without an effective learning process. During the training, the relationship between transferor and transferee i.e. arduous relationship can be strengthened through the effective process of learning (Sazulanski, 1996).

The barriers regarding organizational context are internal barriers but they are shaped by some external factors also. These external factors can be either political or societal. These factors have never considered by theories as shown in phenomena coverage percentage in the augmented matrix (table 7). Although these situations are beyond the organizational control, the flow of knowledge with effective strategies taking the inferences and intuitions into consideration can be helpful to manipulate the organizational, societal and political barriers.

The transferred knowledge if not retained, can vanish as it occurs in Pakistan quite often. The lessons learned to go in vain because of ineffective knowledge retention management in terms of its usability and application according to the situations. Thus, it has become inevitable to consider the knowledge retention processes to sustain the transferred knowledge for further development

There remains an avid requirement of developing effective learning processes so that knowledgebase the absorptive capacity issues can be resolved in order to develop project management competencies in the managers which can be used whenever needed. There are different theories to explain the discussed phenomena and the identified gaps. However, a pluralistic approach can be helpful to integrate all the relevant and applicable components from different theories as suggested by Soderlund (2011) to ensure that the knowledge transfer journey successfully leads to competence development in the project managers considering the nature of the project organizations. The present study incorporates SOR theory as a basic leaning theory to address the issue of the learning barriers.

PLURALISM

Soderlund (2011) described project management as a 'fragmented adhocracy', like many other sub-disciplines in management. This refers to the scientific field having no common principles and ideas and are based on several theoretical traditions. The projects are intentionally designed to disappear. Hence, it exists for a very short period of time. It has various perspectives, typologies and multi-dynamics problems due to which it is quite difficult to apply a single theory to its problem. Hence, it is effective to view the problem from different elements of theories and integrate those elements into the theories to solve the problems.

Inferences

As illustrated in inferential reasoning learning theory by Boddez, Houwer, and Becker. (2017), the commencement of the learning process should be cause or reason specific and aimed at a certain objective. It should start from premises and end on conclusions. Thus, inferential reasoning is basically an effortful cognitive process that is typically based on proportional premises and conclusions. In case of the absence of inferences, these premises start influencing the behaviours in people to develop inconsistency in them as all the knowledge exists in the form of propositions that can be true or false.

Hence, in project management, the learner should come up with some inferential reasoning of learning that can be in the form of 'Goals of learning' i.e. The reason for the learning session. With other questions that should be addressed such as, what are its shortcomings? How can he apply it? On the 'Stimulus' side, the trainers should be aware of the 'basic inferential reasoning' that motivate the trainees to attend the training session. Boddez, Houwer, and Becker also argued in the summary of their work that inferential reasoning theory is not enough to explain the cognitive account of causal learning. Learning is also influenced by sub-conscience besides experience, knowledge, and consciences (Underwood, 1982). This sub-conscience is basically linked with intuitive thinking.

Intuitions

The intuitive thinking is developed by contemplation without the use of analytical reasoning. Hence, it is important to understand how people discern and comprehend something new, for which there are no existing proportional premises. Crossan, Lane, and White (1999) suggested a framework of organizational learning that starts with an individual's intuition. Intuition refers to the past pattern of recognition based on highly sophisticated mind maps and expert knowledge that novices cannot. This means that experts do not think consciously, they know spontaneously using their subconscious and that's why unable to share the knowledge verbally (Crossan et al., 1999). Such a piece of knowledge becomes implicit knowledge and in project management, most of the knowledge is implicit. Because of the typological nature of the projects, such an implicit knowledge differs according to sectors and organizations.

Crossan, Lane, and White (1999) discussed some of the issues in his framework. Intuitive knowledge is non-verbal and hence difficult to fathom because of which it becomes difficult to transfer into the minds. This issue has been resolved by various intuitive trainers of the current era in the field of project management. Secondly, only experts in the field can have intuitive knowledge. However, extensive and additional knowledge through the workshops, training, certifications and advance research in the area across the borders can also develop intuition. Intuitive knowledge is usually limited to the experts which limits the promulgation of new knowledge. In developing countries, this is one of the main factors that obstruct the development of new methods of learning and knowledge transfer. Also, he said in conclusion, that once the intuitive knowledge is institutionalized, it is embedded in the collective mindset of the organization that becomes a cognitive barrier to the change. The experts suggest resolving this problem through the suggestion of conducting cascade training sessions within the organizations as well as adopting ethical management styles that reduce depersonalization and helps in adopting continuous change.

Actualization

Rosanas and Velilla (2005) argued that the basic purpose of any management control system is to attain set individual actions that are congruent with the organizational goals. Therefore, there is a need to influence human actions that are directly linked with ethical dimensions that relate to ultimate goals of the human beings. In this regard, managers should respect the characteristics of human actions and make it a personal action i.e. link with oneself. They further argued that control systems followings only explicit incentives depersonalize human beings by demotivating them from performing a personal action. Such systems do not develop the technical and moral values of a person.

Technical values can stimulate the actualization of a person's potential i.e. useful for firms' distinctive competence. When a person does not feel self-actualized in technical values, he will demand more explicit incentives for satisfaction. In reaction, if the organization increases the extrinsic benefits, the organization will become successful in depersonalization. However, on the other hand, when the organization wants to change the desired results according to the changing situation, the employee will become unable to adapt to the interventions. The system that links the personal goals with the ultimate purpose of satisfying other people's needs will develop the potential strength of moral values. Those moral values can help to alter the technical values in the future according to the changing situations.

Simons calls such control systems as "Diagnostic Control System". Where the management takes control through diagnoses rather than the measurements. Further, he says that "belief systems" are closely related to understand personal action.

This means that "trainees" come with "inference" connected with extrinsic incentives and cannot fully gear the learned knowledge in the form of distinctive competence for their organization. Moreover, in the case of public sector organizations where the compensation systems are fixed and not related to the performance and where there is no punishment in case of project failure, people attend the training just as a part of the job.

A Pluralistic Learning Theory for Project Knowledge Transfer

In the present study, the major problems of knowledge transfer are resolved through pluralism with SOR theory. The learning theories were considered for pluralism with the aim of manipulating the knowledge transfer problems. Figure 3 shows how pluralism help to resolve knowledge transfer problems.

Figure 3. How pluralism help to resolve Knowledge transfer problems

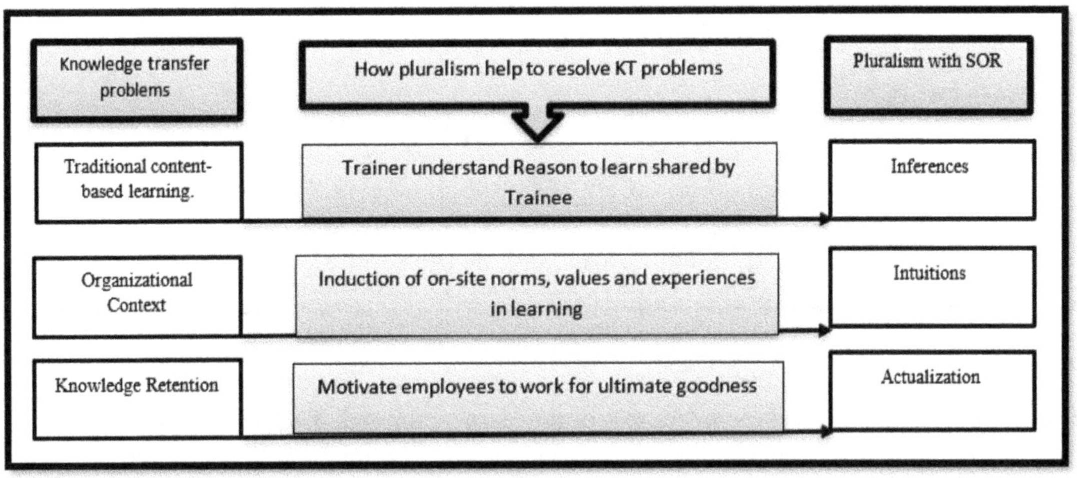

The sharing of inferences by learners from the trainers in a proper way by concerned organizations is helpful to make the trainers focus on gaps and to impart the knowledge among the trainees that are applicable. The trainers should prepare the content according to the trainees' inferences so that it could lead to resolving their problems. The purpose-based or problem-solving-based learning gets trainees' attention and response more effectively. Hence, trainers should focus on inferences to avoid the traditional content-based learning.

The intuitive training methodologies should induce a part of implicit knowledge in the training sessions through role-playing, group discussions, fictitious projects, and other physical learning activities. Such learning approaches encourage the trainees to add their implicit knowledge that needs to develop and highlight some invisible problems to the trainers that need to be resolved on site. Such activities also give the trainers to understand their on-site norms and values that limit them for knowledge exploration and exploitation. The learning through intuitive training methodologies enables the trainees to apply the learned knowledge in real-time situations.

Actualization helps in resolving the problems regarding knowledge retention by enabling the employees to recognize the value of the ultimate good instead of looking towards the organizational response and support. The employees should be motivated to resolve their problems. They should be made able to recognize the importance of their work towards the betterment of the masses with respect to their religious obligations. The aim of actualization in training sessions should be to develop the consciousness of employees to become flexible towards interventions within the organizations for the greater good of public and ultimate goodness for self.

Based on the above discussion, the suggested 'Pluralistic learning theory for Project Knowledge Transfer', through SOR learning theory modeled as shown in figure 4.

Figure 4. Pluralistic learning theory for Project Knowledge Transfer

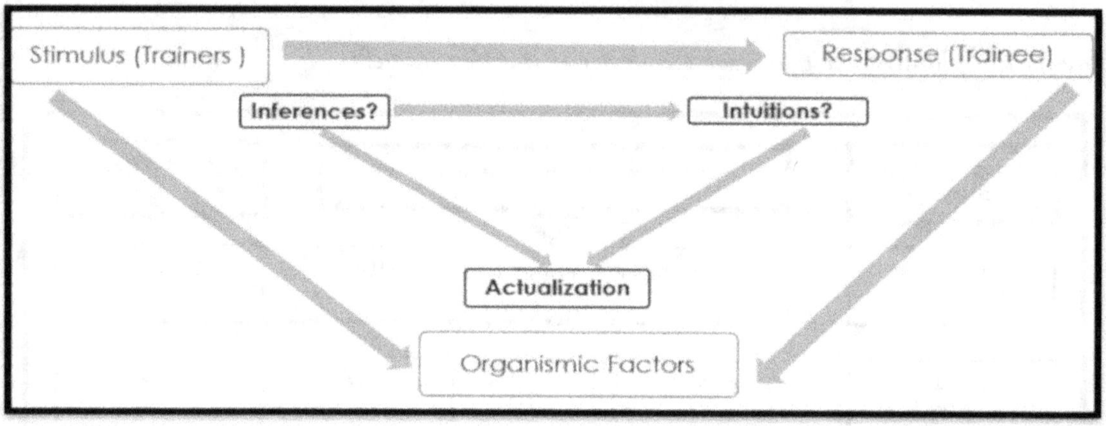

The trainers can understand the inferences of the learners through TNA (Training Needs Analysis) survey. It can improve the training activity by setting targets according to the existing gaps. TNA is an ongoing process of gathering data to determine the gaps, then to develop relevant content and finally to evaluate against the set targets. TNA is assumed to be a success for any training program as it gives a direction to develop required competencies. If the training program does not incorporate TNA first, then there is a risk of overdoing training, doing too little or missing the point completely (Brown, 2002). On the other hand, the survey can help to understand the intuitive knowledge of the project actors which can help in contextualizing the project management training.

Furthermore, the training institutes also need to train the trainers through (TOT certifications). The TNA survey highlights the needs, gaps, and learning behaviours. Hence, the trainers should train to fulfill those gaps and needs according to the situational needs of the trainees of the organizations. The modules should be developed according to the inferences and intuitions to make the training relevant and to develop applicable competencies. To utilize the intuitions in the training, the best approaches suggested and recommended in the literature are case studies, role-playing, group discussions, interactive sessions, hands-on exercises, and outcome-based training. Finally, to clear the path for applying knowledge by new skilled project managers in organizations, there is a need to conduct TNA at the executive level, to develop realized absorptive capacity of organizations. Because of the bureaucratic style of most of the organizations, and in-depth interviewing style is suitable for obtaining relevant insights on the training needs. Apart from some basic skills training, there is a need to actualize the project work value at an executive level. The actualization of the project work value at this stage can be helpful to retain the knowledge transferred through training.

The actualization of project work value can help the organizations to elevate the collective will of their workforce towards projects success. The actualization leads to develop employers/leaders who understand the value of work towards the greater good i.e. beneficial for self as well as for others. This leads to collective hard and smart work. In doing so, the employer starts learning the most efficient ways to complete the tasks in time and avoid irrelevant tasks and targets. Self-actualized employers motivate their employees to grow and learn by giving them space to apply their knowledge (Olomolaiye, & Egbu, 2004).

Conclusively, the proposed pluralistic learning theory for project knowledge transfer is helpful for trainers and training institutes to develop their modules and training methodologies according to need and behaviour of the learner. This needs training needs assessment (TNA) from training institutes, the coordination and collaboration of host organizations and training institutes, and trainers.

FUTURE RESEARCH RECOMMENDATIONS AND LIMITATIONS

This study provides new avenues for future research in three dimensions; first, the application of a proposed theory needs continuous research and survey. The results of research and surveys can emerge new phenomena to resolve and make the system more efficient. The emergence of new phenomena will help to explore the social problems connected with the organizational ineffectiveness of developing countries.

The proposed Pluralistic learning theory for Project Knowledge Transfer can be validated in different contexts facing similar learning barriers through field interviews and observations. The theory needs to be tested empirically through experiments. The experiments aid in refining the model to be used across various contexts.

Although the proposed theory has some limitations because of being context-specific. The study conducted in the context of developing countries considering Pakistan as a sample area. The theory is highly dependent on behaviours because of environment, basic education and social backgrounds. Every country has a different context; hence generalizability of this theory is not possible. However, the proposed theory can be refuting and alter according to changing context following the '*Phenomenon before theory approach*'. In different contexts, this theory can be considered as a base to study the first phenomenon then reviewing theories.

REFERENCES

Aggestam, L., Durst, S., & Persson, A. (2014). Critical success factors in capturing knowledge for retention in IT-supported repositories. *Information*, *5*(4), 558–569. doi:10.3390/info5040558

Ali, A. (2010). *Investigating project management practices in public sector organizations of a less developed country*. Academic Press.

Anatoliivna, V. (2013). *Absorptive Capacity in Organizational Theories: Learning, Innovation*. Managerial Cognition. Marketing and Management Innovation.

Anderson, J. R., Reder, L. M., & Simon, H. A. (1996). Situated learning and education. *Educational Researcher*, *25*(4), 5–11. doi:10.3102/0013189X025004005

Badilescu-Buga, E. (2013). Knowledge behaviour and social adoption of innovation. *Information Processing & Management*, *49*(4), 902–911. doi:10.1016/j.ipm.2013.02.001

Bessick, J., & Naicker, V. (2013). Barriers to tacit knowledge retention: An understanding of the perceptions of the knowledge management of people inside and outside the organisation. *South African Journal of Information Management*, *15*(2), 1–8. doi:10.4102ajim.v15i2.556

Boddez, Y., Houwer, J. D., & Beckers, T. (2017). The inferential reasoning theory of causal learning: Towards a multi-process propositional account. In *The Oxford handbook of causal reasoning* (p. 53). Oxford University Press.

Bredin, K., & Söderlund, J. (2006). Perspectives on human resource management: An explorative study of the consequences of projectification in four firms. *International Journal of Human Resources Development and Management*, *6*(1), 92–113. doi:10.1504/IJHRDM.2006.009751

Burleson, W. (2005). Developing creativity, motivation, and self-actualization with learning systems. *International Journal of Human-Computer Studies*, *63*(4-5), 436–451. doi:10.1016/j.ijhcs.2005.04.007

Buxbaum, O. (2016). *Key insights into basic mechanisms of mental activity*. Springer International Publishing. doi:10.1007/978-3-319-29467-4

Chouhan, V. S., & Srivastava, S. (2014). Understanding competencies and competency modelling—A literature survey. *IOSR Journal of Business and Management*, *16*(1), 14–22. doi:10.9790/487X-16111422

Cohen, W. M., & Levinthal, D. A. (1990). Absorptive capacity: A new perspective on learning and innovation. *Administrative Science Quarterly*, *35*(1), 128–152. doi:10.2307/2393553

Coombs, C. R. (2009). Improving retention strategies for IT professionals working in the public sector. *Information & Management*, *46*(4), 233–240. doi:10.1016/j.im.2009.02.004

Crossan, M. M., Lane, H. W., & White, R. E. (1999). An organizational learning framework: From intuition to institution. *Academy of Management Review*, *24*(3), 522–537. doi:10.5465/amr.1999.2202135

Dalkir, K. (2013). *Knowledge management in theory and practice*. Routledge. doi:10.4324/9780080547367

Donate, M. J., & Guadamillas, F. (2011). Organizational factors to support knowledge management and innovation. *Journal of Knowledge Management*, *15*(6), 890–914. doi:10.1108/13673271111179271

Dretske, F. (2004). Psychological vs. biological explanations of behaviour. *Behavior and Philosophy*, *§§§*, 167–177.

Egginton, W. E. (2010). *Project management education, training, working & learning: A longitudinal study into the experiences of British army officers in UK defence projects* (Doctoral dissertation). University of Bristol.

Epignosis, L. L. C. (2014). *E-learning concepts, trends, applications*. Author.

Epstein, R. M., & Hundert, E. M. (2002). Defining and assessing professional competence. *Journal of the American Medical Association*, *287*(2), 226–235. doi:10.1001/jama.287.2.226 PMID:11779266

Fekete, A. C., Fekete, M. C., & Greculescu, A. (2014). Factors influencing psychosocial communication in the reason/emotion society. *Procedia: Social and Behavioral Sciences*, *128*, 123–127. doi:10.1016/j.sbspro.2014.03.129

Filipatali, T. (2013). *Learning Theories. Their Influence on Teaching Methods.* Munich: GRIN Verlag. https://www.grin.com/document/293498

Gonzalez, R. V. D., & de Melo, T. M. (2018). The effects of organization context on knowledge exploration and exploitation. *Journal of Business Research, 90*, 215–225. doi:10.1016/j.jbusres.2018.05.025

Grant, R. M. (1996). Prospering in dynamically-competitive environments: Organizational capability as knowledge integration. *Organization Science, 7*(4), 375–387. doi:10.1287/orsc.7.4.375

Hollan, P. C. (2008). Cognitive versus stimulus-response theories of learning. *Learning & Behavior, 36*(3), 227–241. doi:10.3758/LB.36.3.227 PMID:18683467

Ilya, S. (2011). *Eight-window sampling model for the research interview.* Available at: http://fieldsociology.blogspot.ru/2011/12/blog-post.html

Keating, E., Oliva, R., Repenning, N., Rockart, S., & Sterman, J. (1999). Overcoming the improvement paradox. *European Management Journal, 17*(2), 120–134. doi:10.1016/S0263-2373(98)00072-3

Khan, K., Turner, J. R., & Maqsood, T. (2013, June). Factors that influence the success of public sector projects in Pakistan. In *Proceedings of IRNOP 2013 Conference* (pp. 17-19). Academic Press.

Khan, K. A., & Sheikh, R. A. (2005). Project management in Pakistan government. *Proceedings of the PMI Asia-Pacific Global Congress.*

Klima, R. (1972). Theoretical pluralism, methodological dissension and the role of the sociologist: The West German case. *Social Sciences Information. Information Sur les Sciences Sociales, 11*(3-4), 69–108. doi:10.1177/053901847201100303

Ko, D. G., Kirsch, L. J., & King, W. R. (2005). Antecedents of knowledge transfer from consultants to clients in enterprise system implementations. *Management Information Systems Quarterly, 29*(1), 59–85. doi:10.2307/25148668

Kumar, A. A. (2017). Knowledge Retention: A Key attribute in Organizational growth. Advances in Applied Science Research, 1-9.

Kuura, A. (2011). *Policies for projectification: support, avoid or let it be?* Academic Press.

Mainga, W. (2017). Examining project learning, project management competencies, and project efficiency in project-based firms (PBFs). *International Journal of Managing Projects in Business, 10*(3), 454–504. doi:10.1108/IJMPB-04-2016-0035

Maskell, P., & Malmberg, A. (1999). Localised learning and industrial competitiveness. *Cambridge Journal of Economics, 23*(2), 167–185. doi:10.1093/cje/23.2.167

McLeod, S. A. (2017). *Kolb - learning styles.* Retrieved from www.simplypsychology.org/learning-kolb.html

Mia, M.B., & Ramage, M. (2011). *IT project management in developing countries: Approaches and factors affecting success in the microfinance sector of Bangladesh.* Academic Press.

Milhem, W., Abushamsieh, K., & Pérez Aróstegui, M.N. (2014). Training Strategies, Theories and Types. *Journal of Accounting, Business & Management, 21*(1).

Minbaeva, D. B., Makela, K., & Rabbiosi, L. (2007). *Absorptive capacity and knowledge transfer: Individual level.* unpublished paper, Copenhagen Business School.

Mir, F. A., & Pinnington, A. H. (2014). Exploring the value of project management: Linking project management performance and project success. *International Journal of Project Management, 32*(2), 202–217. doi:10.1016/j.ijproman.2013.05.012

Nan, N. (2008). A principal-agent model for incentive design in knowledge sharing. *Journal of Knowledge Management, 12*(3), 101–113. doi:10.1108/13673270810875895

National Research Council. (2000). *How people learn: Brain, mind, experience, and school: Expanded edition.* National Academies Press.

O'Sullivan, K. (Ed.). (2007). *Strategic knowledge management in multinational organizations.* IGI Global.

Pak, Y. S., Ra, W., & Lee, J. M. (2015). An integrated multi-stage model of knowledge management in international joint ventures: Identifying a trigger for knowledge exploration and knowledge harvest. *Journal of World Business, 50*(1), 180–191. doi:10.1016/j.jwb.2014.02.006

Pan, W., & Hawryszkiewycz, I. T. (2004). A method of defining learning processes. In *Annual Conference of the Australasian Society for Computers in Learning in Tertiary Education.* ASCILITE.

Rauterberg, M. (2019). *Expert Opinion.* Retrieved April 24, 2019, from http://www.idemployee.id.tue.nl/g.w.m.rauterberg/lectures.htm

Reilly, A. J. 1998. *Three approaches to organizational learning.* The Pfeiffer Library.

Rosanas, J. M., & Velilla, M. (2005). The ethics of management control systems: Developing technical and moral values. *Journal of Business Ethics, 57*(1), 83–96. doi:10.100710551-004-3826-1

Salsali, M., Cheraghi, M. A., & Ahmadi, F. (2009). Organizational factors influencing knowledge transfer into practice in Iranian nursing context: A grounded theory approach. *International Journal of Nursing Practice, 15*(5), 426–436. doi:10.1111/j.1440-172X.2009.01775.x

Schilling, J., & Kluge, A. (2009). Barriers to organizational learning: An integration of theory and research. *International Journal of Management Reviews, 11*(3), 337–360. doi:10.1111/j.1468-2370.2008.00242.x

Simon, H. A. (2000). Bounded rationality in social science: Today and tomorrow. *Mind & Society, 1*(1), 25–39. doi:10.1007/BF02512227

Slaughter, S. A., & Kirsch, L. J. (2006). The effectiveness of knowledge transfer portfolios in software process improvement: A field study. *Information Systems Research, 17*(3), 301–320. doi:10.1287/isre.1060.0098

Smith, A. (1999). International briefing 4 Training and development in Australia. *International Journal of Training and Development, 3*(4), 301–313. doi:10.1111/1468-2419.00088

Söderlund, J. (2008). Competence dynamics and learning processes in project-based firms: Shifting, adapting and leveraging. *International Journal of Innovation Management*, *12*(01), 41–67. doi:10.1142/S1363919608001911

Söderlund, J. (2011). Theoretical foundations of project management. In The Oxford handbook of project management. doi:10.1093/oxfordhb/9780199563142.003.0003

Spreitzer, G. M. (1995). Psychological empowerment in the workplace: Dimensions, measurement, and validation. *Academy of Management Journal*, *38*(5), 1442–1465.

Sveiby, K. E., & Simons, R. (2002). Collaborative climate and effectiveness of knowledge work–an empirical study. *Journal of Knowledge Management*, *6*(5), 420–433. doi:10.1108/13673270210450388

Szulanski, G. (1996). Exploring internal stickiness: Impediments to the transfer of best practice within the firm. *Strategic Management Journal*, *17*(S2), 27–43. doi:10.1002mj.4250171105

Tellis, G.J. (2017). *Interesting and impactful research: on phenomena, theory, and writing*. Academic Press.

Trivellas, P., Akrivouli, Z., Tsifora, E., & Tsoutsa, P. (2015). The impact of knowledge sharing culture on job satisfaction in accounting firms. The mediating effect of general competencies. *Procedia Economics and Finance*, *19*, 238–247. doi:10.1016/S2212-5671(15)00025-8

Underwood, B. J. (1982). *Studies in Learning and Memory: Selected Paper*. Praeger Publishers.

Webster, J., & Watson, R. T. (2002). Analyzing the past to prepare for the future: Writing a literature review. MIS Quarterly, xiii-xxiii.

Williams, M. K. (2017). John Dewey in the 21st Century. *Journal of Inquiry and Action in Education*, *9*(1). Retrieved from https://digitalcommons.buffalostate.edu/jiae/vol9/iss1/7

Wilson, S. M., & Peterson, P. L. (2006). *Theories of Learning and Teaching: What Do They Mean for Educators?* Working Paper. National Education Association Research Department.

Yanwen, W. (2012). The study on complex project management in developing countries. *Physics Procedia*, *25*, 1547–1552. doi:10.1016/j.phpro.2012.03.274

KEY TERMS AND DEFINITIONS

Absorptive Capacity: The extent to which a firm can recognize the value of new external information, assimilate it, and apply it toward achieving organizational goals (Miles, 2012, p. 17).

Competence: Refers to the ability to do something. Competence demonstrates the ability to apply learnt knowledge (Epstein & Hundert, 2002).

Eight Window Model: The expert window was located as a quadrant along with two-axis 'think-knows', the four windows of quadrant were named according to the special characteristics of experts in that window, that is, the typical experts, key experts, theoretical experts, and false experts.

Expert Interviews: Obtaining additional unknown and reliable information on the topic under study from experts. Such interviews are always open ended in order to allow the experts to tell their point of view on the issues and processes of research (Ilya, 2011).

Inferences: An effortful cognitive process that is typically based on proportional premises and conclusions.

Intuitions: The intuitive thinking is developed by contemplation without the use of analytical reasoning.

Knowledge Base: The prerequisite of absorptive capacity.

Knowledge Transfer: Ko et al. (2005) defined knowledge transfer as "the communication of knowledge from a source that is learned and applied by a receipt" (p. 62).

Motivation: Refers to inferences to have a positive learning behaviours.

Phenomena: An observable fact or event.

Phenomenon Before Theory: The researcher immerses him/herself into the phenomenon, understands it deeply, and views it with fresh eyes without the bias of prior theories and the interpretations of prior models and methods (Tellis, 2017, p. 3).

Pluralism: In case of absence of presence of relevant theories to real problems, it is possible in principal to construct several different theories for the explanation of the same 'data', and by doing so may contribute to the growth of knowledge (Kalima, 1972).

SOR (Stimulus-Organism-Response): A theory by Mehrabian and Russell describes how people react to stimuli in the environment.

Training: Training includes giving of information and knowledge, through speech, the written word or other methods of demonstration in a manner that instructs the trainee.

Willingness: Confidence or readiness to learn.

ENDNOTE

1 The 'stickiness' can be considered as an 'organizational barrier', but it is separated because of 'Process model of knowledge transfer', as Sazulanski (1996) discuss barriers to knowledge transfer as stickiness.

Chapter 6
Malaysian Tourism Industry Achievement From a Knowledge Management Perspective and Emergent Trends

Naeem Hayat
https://orcid.org/0000-0003-0756-5226
University Malaysia Kelatan, Malaysia

Abdullah Al Mamun
https://orcid.org/0000-0002-9713-742X
UCSI University, Malaysia

ABSTRACT

This chapter deals with the tourism industry in general from the Malaysian context. Tourism industry (TI) across the globe is gaining momentum and is respected as an emerging business sector bringing the much needed foreign exchange and employment opportunities for the countries. Tourism as a service sector depends on the knowledge assets of the firms for building innovate services on the management of knowledge. Introduction and the evolution of knowledge management (KM) are discussed as enabling forces for the industry and making competitive advantage for Malaysia. Malaysian achievements in TI bring much needed foreign revenues and bring employment opportunities as well. KM systems stem from the need to manage service orientation of the tourism industry associated with a higher level of diversity in services requirements of the customer coming to Malaysia from across the globe. Finally, a particular set of empirical results are reported to establish the effects of KM on service innovation and performance of Malaysian tourism players. Moreover, emerging trends in the TI are highlighted as well.

DOI: 10.4018/978-1-7998-3142-6.ch006

INTRODUCTION

The tourism industry across the globe experienced rapid growth in the last two decades (Tang & Tan, 2015). More countries intended to attract more tourists as tourism is a lucrative earning opportunity for the countries to develop economically and sustainably (Zakariah, Zainal & Sharif, 2018). Tourism is associated with multiple business opportunities to travel industry, hotel industry, and retail industry (Wardi, Abror & Trinanda, 2018). The nature of the tourism industry is mainly a service orientation that prioritises customers for the success of the industry (Shafaei, 2017).

The tourism industry regarded as highly competitive in nature and customer persistent increase in demand challenging the management and employees in the tourism industry. Knowledge of business practices and having a good understanding of the business environment help the tourism firm to survive in the highly competitive business landscape (Hallin & Marnburg, 2007). Therefore, it becomes crucial to manage knowledge of the firm and keep updating the firm's offering to meet the ever-increasing customer demands better than the competitors. Moreover, meeting customer demand also requires updating the existing inventory of knowledge from outside to remain competitive and sustain progress.

In recent times, technology drives innovation and facilities the management of the business organisation in multiple ways. Knowledge management concept associated with the product and process innovation for decades (Hallin & Marnburg, 2007). However, the service sector, like tourism, still needs to realise the benefits of knowledge management to innovative and delivery tourism services delighting the customers (Shaw & Williams, 2009). Another challenge associated with tourism as a sector is the small size of the firm offering associated and comparable services and unable to take advantage of technology to perform well (Hallin & Marnburg, 2007). Therefore, inter-organizational and intra-organizational factors are essential to understand the mutual working of the industry players for the delivery of services to the customers. The inter and intra-organizational working of tourism players brings challenges to reap the benefits of knowledge management initiatives by small firms or joint efforts taken by small firms (Cooper, 2006).

Social media users reportedly increase over time for exploring tourism purposes and help the tour operators to engage with the current and prospective customers as well (Mohammad, Rashid & Tahir, 2013). Social media effectively enable the tourism services providers to understand and design services for what prospective customers are looking for and to develop a competitive advantage over their respective competitors across the globe (Muniandy, Rahim, Ahmi & Rahman, 2019). However, effective management of social media is a must to understand and satisfy customer needs (Li Sa, Choon-Yon, Chai & Joo, 2019). However, effective management of social media by small tourism players is much challenging and requires policy and resources too effectively manage and reap the economic benefits.

Malaysia situated in Southeast Asia, with 13 states and three federal territories (MAH, 2018). Malaysia is a multi-ethnic and multi-cultural country having 30 million populations (Department of Statistics Malaysia, 2018). During the last five decades, Malaysia made impressive economic growth instead of having traditional reliance on agriculture. Moreover, impressive achievements made in the science, tourism, and other service sectors (SME Corporation Malaysia, 2018). The tourism industry contribution to the Malaysian economy is about 12% of GPD and 15% of employment (Isa, Chin & Mohammad, 2018). Technology adoption and moving towards industry 4.0 are the key targets for 10th and 11th Malaysian economic five-year plans (Muniandy et al., 2019). Technology advancement and moving towards a knowledge economy are key goals set in these programs. Average small hotel in Malaysia generates revenue of about US$ 150,000 per year (SME Corporation Malaysia, 2018).

Hotel and accommodation services are a significant part of the hospitality industry. Malaysia gains the status of a low-cost traveling destination, having an innovative supply of accommodation and travel services (Tang & Tan, 2015). Increased globalisation, technological innovation, and low-cost travel increased the tourism demand on one hand and increased customer demands put pressure on the business to remain competitive by the provision of innovative services to their respective customers (Shafaei & Mohamed, 2017). The hotel industry in Malaysia operates under extremely hostile, competition-driven, and exceedingly vulnerable international hotel market (Set, 2013).

Eating out is on the rise in Malaysia, as 64.1% of Malaysians like to eat out. About RM 8.0 billons of the annual contribution made by the small food business to the Malaysian economy (Wardi et al., 2018). Moreover, it supports the tourism culture in Malaysia as quality food is available at low and reasonable prices with verities of food, Like Malaysian, Indian, Thai and western food as well (Oktadiana, Pearce & Chon, 2016). The food business is an essential part of the tourism sector.

Health tourism is the key economic area in the 12th Malaysian economic development projects (Department of Statistics Malaysia, 2018). Currently, Malaysia stands at 26th position world in as ranked by the Travel & Tourism competitiveness index 2018 (Malaysia Tourism, 2018). Thailand, Singapore, Indonesia, and Vietnam are the major countries from where tourist visits Malaysia in 2017 and 2018 (Yousaf, Rosnan & Zamzuri, 2019). However, the Middle East (ME) tourists are among the top-spending tourists in Malaysia. Holiday making, shopping, and businesses are significant purposes of the tourist entering Malaysia (Malaysia Tourism, 2017; 2018).

21st century regarded as the era of knowledge, and society became very much receptive towards the notion of knowledge in every sphere of their life. Traditional product orientation of business moved into service orientation, where innovation is a vital ingredient of the business process (Davenport & Prusak, 1998). Knowledge became a source of competitive advantage for firms, and knowledge management takes the priority business process (Yousaf et al., 2019). Mainly, the tourism industry is primarily a service-dominant industry and service design, service delivery and service updating to deliver the customer satisfying services (Shafaei, 2017). Knowledge management in the tourism industry is at infancy, and difficulty associated with the knowledge sharing, knowledge creation, and management of knowledge is difficult in the tourism industry (Muthuveloo, Shanmugam & Teoh, 2017).

MALAYSIAN TOURISM INDUSTRY HISTORY

Ministry of Tourism Malaysia (MoTOUR)

MoTOUR starts operating from 20 May 1987 and known as the Ministry of culture and Tourism (MAH, 2018). From 2004, the two Ministries separated into the Ministry of Tourism and Ministry of culture, arts, and heritage (Giap, Gopalan & Ye, 2016). The core objective MoTOUR pursing is the development of the Malaysian National culture according to the national culture policy for the strengthening of national unity and reservation of the national identity while enriching humanity and spirituality with the balanced socio-economic development of Malaysia (MAH, 2018). The Malaysian tourism industry became a significant contributor to national economic development.

Tourism Malaysia

Tourism Malaysia (TM) as a statutory body established in 1992 under the Malaysia tourism promotion board act for the achievement of the significant objective of promotion of Malaysia as the first choice of foreign tourist increasing the tourist in number and revenue earned (Malaysia Tourism, 2018). TM is the corporation working under MoTOUR.

TM is working with many divisions and having an international office as well, and more than 1000 employees are working in all states of Malaysia as well as worldwide offices promoting the visit Malaysia product (Giap et al., 2016). Malaysian tourism centre (TIC) working as a one-stop information centre facilitating tourists with multiple services (Malaysia Tourism, 2018). TIC has its knowledge management system facilitating the customers as well as management provided with relevant information for decision making (MAH, 2018). Moreover, TM has partner programs, where all the critical partners like Malaysia Tourist Guide Council (MTGC), Malaysian Hotel Association (MAH), Malaysian Association of Hotel Owners (MAHO), Malaysia Airport Bhd. (MAB), Malaysia Duty-Free Operator Association (MDFOA), Homestay Association (HMA), Malaysia Budget Hotels Association, Malaysia Complexes Management Association (PKK), Council of Bukit Bintang Shopping Centres (CBBSC), Malaysia Retailers Association (MRCA), Car Rental Chain Association (CRAM), and Malaysia Bumiputra Travel and Tour Guide Association (BUMITRA) registered and management as the industry stakeholder and get guidance and promotion from the TM (Malaysia Tourism, 2018). The achievement made by the TM is the courtesy of the KMS and the close working of industry players to provide satisfaction to the foreign tourists visiting Malaysia (Isa et al., 2018).

Moreover, it is essential to signify that the success of TM enabled with the strategic intent taken by the Government and a collaborated effort facilitates these achievements (Set, 2013). KMS interlink all the services require by the tourist coming with any purpose but require multiple services during their stay. From airport services to the hotel, all the major service providers work in co-ordination with each other, and technology enables the smooth delivery of services to the customers. The excellent co-ordination facilitates customer trips in every manner and delights them and triggers the willingness to visit again (Shafaei, 2017). Malaysia can achieve proper growth in the tourism industry by the joint efforts taken under the supervision of the state-run tourism policymaker (Department of Statistics Malaysia, 2018).

With the participation of partners, Malaysia able to deliver quality and timely tourism services (Set, 2013). In 2018, 25.8 million global tourists visited Malaysia and spent about 24 billion US$ (See Table 1). With a customised product designed for the customers, Malaysia remains among top choice among tourists from the Middle East (ME), South East Asian countries. Most tourists came from Singapore in the year 2018 (See Table 2). However, the top five most likable activities of the tourist coming to Malaysia are holiday making, visiting friends and family, shopping, healthcare, and business (See Table3). The achievement made by the TM stems from the collective efforts undertaken by all the players in the Malaysian tourism sector under the leadership of MoTOUR. Moreover, the extensive use of IT tools and interacting with the customer and providing the best available services are the hallmark of the TM. However, the role of knowledge management facilitates TM in every possible way.

KNOWLEDGE MANAGEMENT

Traditionally 'knowledge' associated with the meaning of the justified true belief and later classified as a dynamic human process enabling justified belief towards truth (Zehrer, 2001). However, Drucker (1993) view knowledge as valuable assets that trigger the actions for the success of the individuals or firms. Moreover, knowledge regarded as the farming of the experiences, values, and relevant information from an expert lens (Davenport & Prusak, 1998). All these concepts merge to one point that knowledge is something important for humans and firms and enabling action for doing something new with value.

Different approaches exist to conceptualise management of the knowledge. However, the common understanding towards Knowledge management (KM) that it is a process of capturing and sharing knowledge among the people to add value (Zehrer, 2011). KM drive started with the rise of knowledge organisation and worker creating values in products and services since the 1980s. Moreover, with the increased consciousness that knowledge is the core of the business success and source of competitive advantage for nations as well (Hallin & Marnburg, 2007). KM enabled the firms to achieve long term success based on the knowledge reside inside the organisation and transformed it into intellectual properties to assist the firm to remain competitive and successful (Zehrer, 2011).

Multiple knowledge process exists in an organisation that permits the KM process, i.e., customer-driven knowledge; innovation & knowledge creation; intellectual assets management, use of research to understand the processes and practices, and assimilation and distribution of knowledge (Grant, 2005). These all knowledge processes allow the organisations to use the knowledge to satisfy customer needs and gain competitive advantage among the competing firms (Cooper, 2006). Moreover, knowledge exists in multiple forms and follows in implicit and explicitly with an organisation. Knowledge sharing and capturing involve social processes based on the inclined self-organising feature of one (Hallin & Marnburg, 2007). However, supporting policies and programs enrich the KM activities.

Grant model (2005) delineates the six-step process of KM based on the activities of knowledge integration, knowledge sharing, knowledge replication, knowledge storage & organisation, knowledge measurement, and knowledge identification. However, success in KM projects rests with the firms' ability to capitalise on their intellectual assets employing research diffusion and commercialisation (Melton & Hartline, 2013). The reported difficulty about the KM activities related to making the implicit knowledge of the firm an explicit one, as the most knowledge exists in the implicit form in organisations (Cooper, 2006). Moreover, information technology can contribute to store, analyse, or structure the existing knowledge. However, the real challenge exists, making the implicit knowledge an explicit one.

The Knowledge Management Process in the Hotel Industry

Customer information enables the efficient working of the service industry (Mohammad et al., 2013). The collection and application of customer information from internal and external sources emerged by the dynamic spiral process. Knowledge creation goes through processes of the socialisation, externalisation, combination, and internalisation (Nonaka & Toyama, 2005). New product or service development initiated with the socialisation process that facilitates the externalisation of the customer needs into requirements that help to the conceptualisation of the services or products. The combination is a knowledge creation process that helps to screens the concepts that realised intangible efforts at a later stage of service development (Muthuveloo et al., 2017). Internalisation is the process helps to convert explicit knowledge into tacit one with the development of standard operating procedures which organisation member keep in mind before every action taken towards achieving dual goals of customer satisfaction and business success (Xin et al., 2014). Customers are essential stakeholders that help to enrich the business knowledge towards new product and service development.

Knowledge creation is the strategic source of competitive business advantage (Nonaka & Toyama, 2005). Knowledge creation capability is the response to the intense international competition in the hospitality industry (Yu, Yu-Fang & Yu-Cheh, 2013). However, it is crucial to understand and explore how the resources and processes lead to performance. Besides, it also necessary to document processes and work procedures as it helps to move beyond the already achieved performance standards (Muthuveloo et al., 2017). Furthermore, KM viewed with three diverse perspectives, namely information-based, technology-based, and culture-based. Information based perspective associated with real-time, accessibility, and actionable information (Yu et al., 2013). Information based view of KM regarded as sorting of information that holds the knowledge and how to locate the knowledge holder. In the technology-based view of KM, technology-based tools integrate the cross-functionalities of the organisation (Muniandy et al., 2019). Technology-based KM associated with the high dependence on enterprise-wide systems with enabled support for decision-making tools (Muthuveloo et al., 2017). However, culture-based KM perspective associated with the conceptions of the learning organisation, intellectual property cultivation, and innovation as a core outcome of the organisational success (Nonaka & Toyama, 2005). Generally, the technology component is only accounted for 20% of the KM, whereas cultural and informational aspect accounts for the rest of the KM (Yu et al., 2013).

The challenge of KM in the tourism industry relates to the macro-level management of knowledge (Hallin & Marnburg, 2007). Tourism industry players need to collaborate and manage the knowledge in a multidisciplinary manner to formalise the existing knowledge. Multiple small players working in the tourism sector and integrating knowledge between and among them is a compelling task (Giap et al., 2016). The tourism industry players are unable to utilise the existing technology tools to bring the existing knowledge into explicit form; this happens due to size, capacity, and inclination towards making the knowledge explicit one (Muthuveloo et al., 2017). This results in little benefits to the industry, and the use of technology for KM practices remain insignificant (Set, 2013). Moreover, tourism players need to work together and provides reliable, relevant information of destination to the customers to bring tourists to the destinations. To seize the benefits of the KM, the government need to intervene and work with small tourism players to reap the benefits of the KM bring more tourist to the destinations, and making business opportunities for the tour operators as well (Muniandy et al., 2019).

Table 1. International Tourist arrivals and receipts

Year	Visitors (Millions)	Receipts (RM in Billions)
2014	27.44	72.0
2015	25.72	69.1
2016	26.76	82.1
2017	25.95	82.1
2018	25.83	84.1
2019 (Target)	28.1	92.2

Source: Tourism Malaysia (http://mytourismdata.tourism.gov.my/)

Tourism became the world's one of the major economic sectors, and current technological changes well absorbed by the tourism industry (Mohammad et al., 2013). Recent knowledge management drives with technological innovation aids the global tourism industry to grow and facilitates its customers (Li Sa et al., 2019). The tourism industry heavily relies on social media to utilise the technology at their disposal and manage the existing product and service knowledge as well (Shafaei & Mohamed, 2017). Nevertheless, the opportunity offered by social media has its associated challenges as well, like the management of social media, working with partners and customers at the same time (Shaw & Williams, 2009). These challenges require the trained human resources and the right policy to facilitate the smooth working of the industry players to deliver the best possible services to their customers.

The tourism industry across the globe regarded as slow in adopting knowledge culture. The primary reason is that SMEs are working in the tourism sector and always short of resources to adopt technology and embrace the change (Muthuveloo et al., 2017). Moreover, tourism is associated with the marketing of destinations. The destination knowledge management is difficult for one single firm and entails collective and joint effort to integrate the knowledge at different levels to organise the knowledge for the destination (Cooper, 2006). The organisation of destination knowledge facilitated with technological tools and enables the availability of up-to-date information about the destination (Chen & Cheng, 2012). Therefore, the tourism firms must work in coordination with each other to facilitate the tourists as well as helps each other.

Knowledge Management in Malaysian Tourism Industry

Knowledge management acknowledged and recognised as key to the economic success of the organisations (Muthuveloo, Shanmugam & Teoh, 2017). Higher return of investment reaped by the organisations by the implementation of KM. KM regarded as the strategic lever of economic success by 94% of US firms. It is a highly recommended business process for every kind of industry (Nonaka & Toyama, 2005). Organisation competitiveness associated with the adoption of KM practices as failing to use KM triggers inefficiency and inability to stay competitive. Knowledge recognised as key to organisation competitiveness and economic sustainability.

Multimedia development corporation (MDeC) was the first Malaysian corporation, incepted KM, as strategic input for effective decision making in early 1999 (Muthuveloo et al., 2017). Another early example was PERODUA; Malaysian car manufacturers embarked on KM initiatives in 2001 (Xin, Ramayah, Soto-Acosta, Popa & Teoh, 2014). These early efforts represent the Malaysian corporate sector realisation to utilise KM as a strategic entity for the business progress of the respective corporations in the knowledge era (Li Sa et al., 2019).

KM is not about technology only; much is about the people enabling the organisation to thrive and be successful (Muthuveloo et al., 2017). In general, only 7% of the organisational information analysed by the corporations and a fraction of collected information converted into knowledge that reused effectively (Nonaka & Toyama, 2005). Many barriers are reported that hinders the transformation of employee knowledge into organisational knowledge, and further utilisation of organisational knowledge is minimal (Melton & Hartline, 2013). Organisational culture is one of the most significant barriers of knowledge sharing, as job insecurity and perception of the quality of people greatly hinder the sharing of knowledge among colleagues (Casidy, Wymer & O'Cass, 2018). Moreover, information quality is one of the known barriers of KM (Muthuveloo et al., 2017). Lack of information quality resulting in the absence of interest to consider content valid to send to knowledge repositories (Xin et al., 2014). The establishment of quality measurement for the quality of information facilitates the abstract data branded as comprehendible knowledge (Nonaka & Toyama, 2005).

Furthermore, technology also brings a barrier for the users as well. On the one hand, difficulty in delivering the knowledge content into organisational repositories and also finds it difficult to retrieve the appropriate knowledge for reuse (Chen & Cheng, 2012). However, enabling technology facilitates to overcome the barriers of knowledge to be categorised, organise, store, retrieve, and use for organisational purposes (Muthuveloo et al., 2017).

American Airlines developed their computer systems with the help of IBM in 1953. Travel industry of the USA using the SABRE on-line system since 1985 (Nonaka & Toyama, 2005). The travel industry realised quite early the benefits of using information technology. Today the tourism industry is driven by the customer expectation and demands as industry players facing the rising customer demands and loyalty from the customer (Evangelista, Lucchese & Meliciani, 2013). Traditional ICT in the industry consists of the computerised reservation systems, teleconferencing, video brochures, management information systems, electronic fund transfers, emails, internet, and digital telephone networks (Muthuveloo et al., 2017).

Air Asia regarded as the first Malaysian private airline offering cheaper airline tickets than the state-owned Malaysian Airline (MAS) (Ali & Buang, 2016). Air Asia welcomed in the region and easily penetrated in the market segment due to delivering price advantage to its customers. However, the value delivered to the user with the use of online selling through their websites (MAH, 2018). The information technology (IT) allows the achievement to reduce the cost reduction and price penetration. IT facilitates the services of tourism and related surrounding economic partners for the achievement of the business economic objectives and better customer services (Li Sa et al., 2019).

Information flows at both upstream (supplier) and downstream (customers) in the tourism industry. Tourism products and services are subject to the variability of customer choice and liking, such as airline, hotel, transportation, and many more (Master Card & CrescentRating, 2018). Moreover, the complex nature of the relationship and huge no. of tourism services partners actively linked through the KMS in Malaysia (MAH, 2018). On the one hand, customers have relevant information for their touring needs, and service suppler can market their services accordingly (Malaysia Tourism, 2018).

Table 2. Top five Tourist nationality and no. of tourist

Countries	2018	2017
Singapore	12,441,713	10,615,986
Thailand	1,836,522	1,914,492
Indonesia	2,796,570	3,277,689
Brunei	1,660,506	1,382,031
Philippines	370,559	396,062

Source: Tourism Malaysia (http://mytourismdata.tourism.gov.my/)

Barriers of Knowledge Management Systems (KMS) in Malaysian Hotel Industry

Knowledge sharing occurs with the environment of knowledge sharing, where individuals ask other employees to share and show a willingness to receive assistance in solving daily business problems (Yu et al., 2013). Monetary incentives regarded as the critical work-related barrier for knowledge sharing among Malaysian hotel industry employees (Zakariah et al., 2018). The hierarchical organisational structure makes it difficult for the employees to share knowledge among the fellow employees as well as new ideas from lower hierarchical employees are not welcome in the majority of the large hotels (Li Sa et al., 2019). Organisational culture is also a significant barrier as most hotels follow the autocratic styles than the participative management styles of decision making (Mohammad et al., 2013).

Moreover, knowledge hoarding reported barriers among the hotel industry employee in Malaysia (Chen & Cheng, 2012). Fellow employees never ask questions regarding business activities, and work engagement among them reported as low (Ali & Buang, 2016). Cultural and personality issues are among the highest reported factors that restrict knowledge sharing and encourage the employee to hoard value knowledge (Giap et al., 2016). Knowledge sharing capacity also reported low among the Malaysian hotel industry employees (Li Sa et al., 2019). Suitably sharing knowledge is a unique skill and requires an appropriate level of speaking and comprehension skills to deliver knowledge for facilitating the fellow employees for service innovation and delivery of service to the customers (Grant, 2005).

Moreover, the sensitive nature of the job at senior level employees also reportedly causes the knowledge of hiding behaviours among hotel industry employees (Muthuveloo et al., 2017). Lack of interest shown by the fellow employees to acquire new knowledge is the critical barriers reported by the critical position holders' employees affect the knowledge sharing in the Malaysian hospitality industry (Casidy et al., 2018). Nevertheless, the attitude to employees in the hospital industry revamped, and organisations facilitate knowledge sharing by formal and informal channels. Old employees left the organisation without transferring their knowledge to fellow employees or precious knowledge lost in the transfer process as management does not have an actual process of knowledge documentation or knowledge transfer program in place (Nonaka & Toyama, 2005). These are some of the critical areas that cause the loss of knowledge within organisations and key performance areas to improve to take full advantage of the KM practices.

Table 3. Purpose to Visit Malaysia (In Percent)

Purposes	2018	2017
Holiday making	55.3	55.5
Visiting friend & Family	19.8	22.5
Shopping	13.4	10.5
Health treatment	3.7	4.0
Business	2.4	1.6

Source: Tourism Malaysia (http://mytourismdata.tourism.gov.my/)

Enhancing tourism employees' capacities in terms of assessing knowledge and sharing knowledge facilitates the KM activities (Hallin & Marnburg, 2007). Moreover, employees need to provide soft and hard skills training to deals with people and to rightly use the technology to maximise the KM (Yu et al., 2013). Knowledge storage and retrieval of knowledge in offering customer services enhance customer satisfaction and firm performance as well (Set, 2013). However, the use of technology that empowers the KM initiatives provided to every employee that not only facilitates the employee working but also empowers the knowledge accessibility for the employees. Moreover, intra-organizational sharing and use of knowledge is the emerging field of study in the tourism sector that enrich the customer trip experience and facilitates the working of industry player to work as partners (Muthuveloo et al., 2017).

Achievement of the Malaysian Tourism Industry

KM is the key that fosters growth and achievements in the Malaysian Tourism industry. KM enables innovation in the hotel industry in Malaysia. Innovation management with the KM systems helps the employee performance as well. Service innovation ($\beta = 0.297$, $p = 0.00$) and management innovation ($\beta = 0.186$, $p = 0.02$) significantly affect the hotel employee performance (Zakraiah et al., 2018). Up taking innovation and managing it well facilities employee performance in the hotel industry. Innovation is a joint effort from management and employees. The management provides support for innovation and employees' participation in the generation of innovative ideas and improvements in the product and services (Xin et al., 2014). A clear understanding of the customer need and availability of resources enables the innovation that helps the customer and brings necessary job performance for the firm employees (Zakriah et al., 2018).

In another study, Mohammad et al., (2013) recognised that Malaysian hotel industry Knowledge management ($\beta = 0.17$, $p = 0.050$) and technology-based CRM ($\beta = 0.17$, $p = 0.000$) positively and significantly impact the financial performance of the hotels. Moreover, knowledge management practices of hotel significantly impact the hotel learning and growth performance in Malaysia (Mohammad et al., 2013). Malaysian hotels are working to improve customer satisfaction and enhance their profitability as well (Set, 2013). Though implementing a comprehensive CRM technology requires investment and skills to manage it well, and many Malaysian hotels find it difficult to reap the real benefits of the KM technologies (Munuandy et al., 2019). However, the Malaysian hospitality industry needs to focus on the employees as well. The role of employees in bringing the performance and financial advantage is a must and provides a competitive advantage as well.

Among the Malaysian hotel industry, employee knowledge sharing behaviour positively impacts service innovation performance (Ali, 2015). Knowledge creation and transforming knowledge into economic opportunities by the use of information technologies requires the aptitude, skills, and culture to gain advantages from knowledge management processes in the Malaysian hotel industry (Giap et al., 2016). Moreover, the hospitality industry needs to build the employees' capacities to work with new technological tools that enable the firm to compete and gain a competitive advantage in the highly competitive tourism industry (Muthuveloo et al., 2017). Furthermore, close coordination between employee and business managers enable to reap the benefits of KM earlier than the competitors to perform well (Li Sa et al., 2019).

Moreover, the Knowledge creation process captured and organised in a manner that realised the benefits of the KM. Li Sa et al. (2019) establishes that the knowledge creation process in small Malaysian hotels positively and significantly ($\beta = 0.215$, $p = 0.000$) affects the hotel business performance. Malaysian tourism industry significantly reaps the advantages of the KM in the parental role played by the TM (MAH, 2018). Knowledge creation process documented and created knowledge transferred to the right personnel to utilise the created knowledge into valuable knowledge assets of the firm and firmly turns these into innovative products/services (Muthuveloo et al., 2017). However, employees' contribution to the process of knowledge creation and the creation of value from the knowledge need to emphasise. Moreover, reliable interconnectivity helps to bring performance to the firm. All the key players are organised and get connected to work as a team, and massive growth Malaysian tourism witnessed in terms of tourists visiting Malaysia as well as revenue targets achieved by the tourism industry Malaysia.

Studies exploring the KM in the tourism sector of Malaysia are scant. However, tourism players in Malaysia taking a keen interest in KM practices that enable them to perform well in increasing global tourism competitive landscape among countries. Therefore, it is necessary to explore the KM practices in Malaysia and its impact on customer satisfaction and firm economic and non-economic performance.

The Emerging Trends in Malaysian Tourism

Last decade associated with the few emerging trends in the Malaysian tourism industry. Halal or Islamic tourism is the emerging trend in the Malaysian hospitality industry. 'Halal' as word has its roots in Arabic associated with the Islamic faith and engulfs the Muslim population's daily activities and more than just about the food (Kim, Im & King, 2015). Halal defined that is lawful or permissible under Islamic Shari'ah or Islamic laws (Henderson, 2016). Halal products are not merely associated with the product quality but it the philosophical assumption associated with the products or services that developed according to Islamic teaching in all aspects encompassing different disciplines of culture, anthropology, sociology, and management (Shafaei, 2017). After the September 11 incident, and with much-restricted travel rules from West for Muslim tourists (Battour & Ismail, 2016). Malaysia experienced an influx of 5.44 million Muslim tourists in 2012, and Malaysia remains among the top five destinations by Muslim travelers since 2012 (MasterCard & CresentRating, 2018).

Table 4. Tourist Expenditure in %

Activities	2018	2017
Shopping	33.4	32.7
Accommodation	25.7	25.6
Food & Beverages	13.4	13.3
Local transport	6.1	5.8
Medical	3.4	3.7
Entertainment	3.6	3.9

Source: Tourism Malaysia (http://mytourismdata.tourism.gov.my/)

Growing affluent class in the Muslim communities across the globe shows interest in tourism; the Muslim travel market stands at about US$ 200 billion, with 131 million Muslim tourists travelled in 2017 (Master Card & CrescentRating, 2018). The provision of Islamic attributes like halal food, the Islamic way of life and respecting the Islamic teaching while travelling conceptualised as Islamic or Halal tourism (Hendserson, 2016). Malaysia also stands among the top four most favoured destinations among Muslim tourists as well. Malaysian tourism industry played a leadership role in promoting halal tourism services and setting standards for halal tourism that brings satisfaction to Muslim travellers (Kim et al., 2015; Henderson, 2016). Muslim travel market estimated to reach revenue of US$300, with 156 million Muslim travellers by 2026 (MasterCard & CrescentRating, 2018).

The use of social media in general on the rise around the globe and Malaysia is not an exception at all (Xin et al., 2014). Hotel industry players and hotel customers are using social media. Industry players market their products & services on social media, and customers of the hotel industry shared positive as well as negative comments openly on social media (Mohammad et al., 2013). One way social media benefits the hotel industry players to attract and provides the offering customers are looking for. On the other hand, the customer demands and helps the hotel industry players to develop an understanding of customer needs and requirements. This emergent trend based on the high-level engagement between customer and service providers and builds on the knowledge creation and management of the knowledge process (Giap et al., 2016). However, this level of engagement between customers and service providers is not without problems. As customers, positive comments and suggestions are easy to adjust with and handle as well (MAH, 2018). However, the adverse and demanding nature of comments needs to handle with care and in a professional manner. High-level communication and project knowledge require dealing with negative comments, and only skilled professionals need to handle the social media account (Li Sa et al., 2019). Moreover, knowledge created on social media is available to all; it needs to tackle with care and caution.

Tourist friendly policies of the Malaysian Government since 2001 attracted lavishly spending tourists from the Middle East (ME) (MAH, 2018). An average tourist from ME spending about RM 7000 per trip in Malaysia. Provision of ME culture, food, and Arab speaking hotel staff attracts a large tourist from ME (Malaysia Tourism, 2018). Malaysia achieves top position itself among the famous Islamic tourism country that offers precious Islamic values, such as culture & heritage, with a friendly Muslim population and having Islamic travel facilities (MasterCard & CresentRating, 2018). Provision of halal food ensured with halal certification to satisfy Muslim tourists looking for halal amenities as well. Shopping malls in Malaysia as a requirement have to provide prayers room, ablution facilities, and halal restaurants (Isa et al., 2018; Malaysia Tourism, 2018).

Halal tourism is an emergent trend, and an increasing number of Muslims around the world like to travel for leisure and business purposes (Kim et al., 2015). The role of KM extended to offer products and services according to the requirement of the customers. Halal certification in Malaysia helps to streamline Islamic travel service offerings and enables the hotel industry players to achieve customer satisfaction (Henderson, 2016). Moreover, the knowledge created for Muslim travellers also helps the non-Muslim countries willing to cater to the ever-increasing Muslim travellers market with halal services offering from food to travelling arrangements following the teaching of Islam.

Homestay program imitative taken by the Malaysian Government in 2012 and 3 fold increase reported after the start of the program (Ghapar, Otham, and Jamal, 2014). The government offers policy guidelines and training with the subsidies that charged the Malaysian homestay program. The success of the homestay program tells the story of increasing resilience achieved through the public-private partnership (Ghapar et al., 2014). Government support is the bloodline for the homestay program providing the social protection, eradication of poverty, and reducing the vulnerability of the poor communities.

Another emergent trend in Malaysian tourism is the eco-friendly hoteling (Muniandy et al., 2019). Tourists like to stay in green hotels to justify environmental concern and responsible consumption to reduce their impact on the environment. Generally, the hotel industry regarded as the most resource wasting industry (Mohammad et al., 2013). Malaysian national green innovation policy guides the Malaysian tourism industry and attracted good numbers of eco-friendly tourists to visit Malaysia (Department of Statistics Malaysia, 2018). The green hotel helps to mitigate the environmental issues while making tourists more, acknowledging the fact environment is vital to take good care of it. Moreover, creating knowledge from the customer requirement towards the eco-friendly hoteling services empowers the hoteling services providers to learn and develop an innovative way to delight the customers and mitigate the ecology issues as well. In this way, customers attracted to Malaysia's place of eco-friendly hotel services and knowledge created enables the countries as well.

Malaysia is also pursuing to create its mark in health tourism. Malaysian health tourism council (MTHC) registered the hospitals willing to offer health services to the health services to the tourist coming to Malaysia for health purposes (Yousaf et al., 2019). About 4% of the total tourist coming to Malaysia came for health purposes (Malaysia Tourism, 2017; 2018). Currently, health tourists mainly came from Indonesia and neighbouring countries of Malaysia (Yousaf et al., 2019). However, the Malaysian government willing to increase the tourist coming to Malaysia for health purposes and earn foreign exchange as well. Nevertheless, health tourism in Malaysia is at its early stages, and the role of KM is the key to project well Malaysia as among top health tourist places.

These emerging trends in the Malaysian tourism landscape demand more exploration as well as enabling strategies to get benefits with the use of the KM initiatives at the firm level as well as at the industry level.

CONCLUSION

Management of knowledge became an integral part of organisational management in today's knowledge-based economy. The tourism industry is no expectation as well and regarded as a knowledge-intensive industry. Taking stock of existing literature on the tourism industry shows that the tourism industry is working to incorporate knowledge management practices and reaping benefits of knowledge management practices. However, the adoption of knowledge management is still slow in the tourism industry.

Achievements of the Malaysian tourism industry pave the way for others to follow and reap the benefits of technology as a knowledge management tool. Malaysian Government strategically working on tourism for three decades and gain the top position among south-east Asia and Islamic countries. As revenue from tourists and no. of tourists multiplied over time. The critical stakeholder at the supply-side managed well by the Malaysian government with the right use of policies and technology connecting all tourism players by the use of KMS. The close integration of the player helps the Malaysian tourism industry to play a dominant role in the region as a leader in service offerings and concentrating niche market segments.

The role of KM in tourism research needs to further explored with increasing the scope of inquires at the interaction between firms enabling the KM and taking advantage of KM initiatives. Moreover, the role of KM linking the industry players that enable the customers to have a good awareness of market offering and its impact on tourism industry player need further investigation. It helps to investigate in full the emerging trends in the tourism industry to technology-driven KMS to bring benefits for themselves as well as for the customers. However, tourism Malaysia facing challenges of knowledge sharing and knowledge utilisation as knowledge management practices in tourism is multidimensional and multilateral as well. Moreover, the right human capital and technological assets also require fully utilising knowledge management practices.

Moreover, effective use of technology likes IT, social media, and IOT enable the tourism suppliers to effectively deliver the services required by the prospective customer by offering customised services to diversified customer from the globe. The close association and guidance provided by the TM and willingness among the vital player delighted the tourist that they like to come again to visit Malaysia. Furthermore, cohesive policy guidelines from government agencies and high interactivity among industry players help the industry to achieve economic development for the country with improving efficiency with the right use of KMS. There are many questions remain to explore in future research endeavours, the role of organisation internal knowledge management practices creating the competitive advantage for the tourism firm, as well as externally integrated knowledge management practices between the tourism firms, working to delight the tourism customers, as it is essential to manage knowledge inside the frim as well as managing at the industry level. However, still, many challenges are emerging that need collective actions, and a delicate mix of technology use can deliver the most effective results for the development of economies like Malaysia.

REFERENCES

Ali, F. (2015). Service quality as a determinant of customer satisfaction and resulting behavioural intentions: A SEM approach towards Malaysian resort hotels. *Tourism – International Journal (Toronto, Ont.), 63*(1), 37–51.

Ali, K. A. M., & Buang, M. (2016). Study on Factors that Influence Innovation in Malaysian Public Sector. *Journal of Advanced Research in Business and Management Studies*, *4*(1), 60–73.

Battour, M., & Ismail, M. N. (2016). Halal tourism: Concepts, practises, challenges and future. *Tourism Management Perspectives, 19*(1), 150–154. doi:10.1016/j.tmp.2015.12.008

Casidy, R., Wymer, W., & O'Cass, A. (2018). Enhancing hotel brand performance through fostering brand relationship orientation in the minds of consumers. *Tourism Management*, *66*(June), 72–84. doi:10.1016/j.tourman.2017.11.008

Chen, W. J., & Cheng, H. Y. (2012). Factors affecting the knowledge sharing attitude of hotel service personnel. *International Journal of Hospitality Management*, *31*(2), 468–476. doi:10.1016/j.ijhm.2011.07.005

Cooper, C. (2006). Knowledge management and tourism. *Annals of Tourism Research*, *33*(1), 47–64. doi:10.1016/j.annals.2005.04.005

Davenport, T., & Prusak, L. (1998). *Working Knowledge – How Organizations Manage What They Know*. Boston, MA: Harvard Business School Press.

Department of Statistics. Malaysia. (2018, January 29). *Economic census 2016: Tourism statistics*. Retrieved 4 November 2018 from https://www.dosm.gov.my/v1/ index.php?r=column/ctheme&menu_id=b0pIV1E3RW40VWRTUkZoc hyZ1pLUT09&bul_id=VmZsbTU4NDlFcFZRdVF6ZDF3OW4zZz09#

Drucker. (1993). *Post-capitalist Society*. Butterworth-Heinemann.

Evangelista, R., Lucchese, M., & Meliciani, V. (2013). Business services, innovation, and sectoral growth. *Structural Change and Economic Dynamics*, *25*(June), 119–132. doi:10.1016/j.strueco.2012.02.005

Ghapar, A., Othman, N., & Jamal, S. (2014). Community resilience and knowledge on vulnerability in tourism development. Tourism. *Leisure and Global Change.*, *1*(1), 140–148.

Giap, T. K., Gopalan, S., & Ye, Y. (2016). Drivers of growth in the travel and tourism industry in Malaysia a geweke causality analysis. *Economies*, *4*(1), 1–15. doi:10.3390/economies4010003

Grant, R. M. (2005). *Contemporary Strategy Analysis*. Oxford: Blackwell.

Hallin, C. A., & Marnburg, E. (2007). Knowledge management in the hospitality industry: A review of empirical research. *Tourism Management.*

Henderson, J. C. (2016). Halal food, certification, and halal tourism: Insights from Malaysia and Singapore. *Tourism Management Perspectives*, *19*(1), 160–164. doi:10.1016/j.tmp.2015.12.006

Isa, S. M., Chin, P. N., & Mohammad, N. U. (2018). Muslim tourist perceived value: A study on Malaysia Halal tourism. *Journal of Islamic Marketing*, *9*(2), 402–420. doi:10.1108/JIMA-11-2016-0083

Kim, S., Im, H. H., & King, B. (2015). Muslim travellers in Asia: The destination preferences and brand perceptions of Malaysian tourists. *Journal of Vacation Marketing*, *21*(1), 3–21. doi:10.1177/1356766714549648

MAH. (2018). *Hotel Statistics, Malaysian Association of Hotels*. Available at www.hotels.org.my/ home.asp?hdnMRef1/457

Mastercard and CrescentRating. (2018). *Global Muslim travel index 2018*. Available at www. crescentrating.com/halal-muslim-travel-market-reports.html

Melton, H. L., & Hartline, M. D. (2013). Employee collaboration, learning orientation, and new service development performance. *Journal of Service Research*, *16*(1), 67–81. doi:10.1177/1094670512462139

Mohammad, A. A., Rashid, B. B., & Tahir, S. B. (2013). Assessing the influence of customer relationship management (CRM) dimensions on organisation performance: An empirical study in the hotel industry. *Journal of Hospitality and Tourism*, *4*(3), 228–247. doi:10.1108/JHTT-01-2013-0002

Muniandy, K., Rahim, S. A., Ahmi, A., & Rahman, N. A. A. (2019). Factors that influence customers' intention to visit green hotels in Malaysia. *International Journal of Supply Chain Management.*, *8*(3), 994–1003.

Muthuveloo, R., Shanmugam, N., & Teoh, A. P. (2017). The impact of tacit knowledge management on organisational performance: Evidence from Malaysia. *Asia Pacific Management Review*, *22*(4), 192–201. doi:10.1016/j.apmrv.2017.07.010

Nonaka, I., & Toyama, R. (2005). The theory of the knowledge-creating firm: Subjectivity, objectivity, and synthesis. *Industrial and Corporate Change*, *14*(3), 419–436. doi:10.1093/icc/dth058

Oktadiana, H., Pearce, P. L., & Chon, K. (2016). Muslim travellers' needs: What don't we know? *Tourism Management Perspectives*, *20*(1), 124–130. doi:10.1016/j.tmp.2016.08.004

Sa, L. M. L., Choon-Yon, S., Chai, Y. K., & Joo, J., H, A.(2019). Knowledge creation process, customer orientation and firm performance: Evidence from small hotels in Malaysia, Asia Pacific. *Management Review*. doi:10.1016/j.apmrv.2019.07.002

Samori, Salleh, & Khalid. (2015). Current trends on Halal tourism; Cases on selected Asian countries. Tourism Management perspective. doi:10.1016/j.tmp.2015.12.011

Set, K. (2013). Tourism small and medium enterprises (TSMEs) in Malaysia. *International Journal of Business and Social Science*, *4*(16), 58–66.

Shafaei, F. (2017). The relationship between involvement with travelling to Islamic destinations and Islamic brand equity: A case of Muslim tourists in Malaysia. *Asia Pacific Journal of Tourism Research*, *22*(3), 255–271. doi:10.1080/10941665.2016.1232741

Shafaei, F., & Mohamed, B. (2017). Malaysia's branding as an Islamic tourism hub: An assessment. *Geografia: Malaysian Journal of Society and Space*, *11*(1), 97–106.

Shaw, G., & Williams, A. M. (2009). Knowledge Transfer and Management in Tourism Organisations: An emerging research agenda. *Tourism Management*, *30*(3), 325–335. doi:10.1016/j.tourman.2008.02.023

SME Corporation Malaysia. (2018). *Small and medium enterprises*. http://www. smecorp.gov.my/index.php/en/small-and-medium-enterprises

Tang, C. F., & Tan, E. C. (2015). Does tourism effectively stimulate Malaysia's economic growth? *Tourism Management*, *46*, 148–163. doi:10.1016/j.tourman.2014.06.020

Tourism, M. (2017). *Tourism Malaysia-Statstics*. available at: http://corporate.tourism. gov.my/research.asp?page1/4facts_figures

Tourism, M. (2018). *Tourism Malaysia-Statstics*. Available at: http://corporate.tourism. gov.my/research.asp?page1/4facts_figures

Wardi, Y., Abror, A., & Trinanda, O. (2018). Halal tourism: Antecedent of tourist's satisfaction and word of mouth (WOM). *Asia Pacific Journal of Tourism Research*, *23*(5), 463–472. doi:10.1080/10941665.2018.1466816

Xin, J. Y., Ramayah, T., Soto-Acosta, P., Popa, S., & Teoh, A. P. (2014). Analysing the use of Web 2.0 for brand awareness and competitive advantage: An empirical study in the Malaysian hospitality industry. *Information Systems Management*, *31*(2), 96–103. doi:10.1080/10580530.2014.890425

Yousaf, N., Rosnan, H., & Zamzuri, N. H. (2019). Internalisation process of the medical tourism industry in Malaysia, A sequential approach. *Revista Publicando*, *6*(19), 448–462.

Yu, C., Yu-Fang, T., & Yu-Cheh, C. (2013). Knowledge sharing, organisational climate, and innovative behaviour: A cross-level analysis of effects. *Social Behavior and Personality*, *41*(1), 143–156. doi:10.2224bp.2013.41.1.143

Zakariah, S. H., Zainal, A., & Shariff, F. M. (2018). Enhancing the Role of Innovation towards Employee Job Performance at Malaysian hotels. *International Journal of Academic Research in Business and Social Sciences*, *8*(15), 146–159.

KEY TERMS AND DEFINITIONS

Eco-Friendly Tourism: Tourism services offered in such a manner that the eco-waste reduced from the travelling to lodging activities. Moreover, direct or indirect effects of tourism mitigated intending to reduce the impact on the environment.

Halal Tourism: Tourism services offered in such a manner to follow the teaching of Islam in all the tourist services offered from travelling to lodging activities. Permissibility of Islam is paramount of the tourism service offerings.

Homestay: A offering to let the part of one home for rent for a few days. The offering is associated with food and housing facilities, and some time offers the transport and guide services as well.

Knowledge Identification: Knowledge identification is the adopted practices to facilitate the identification of knowledge assets of the firm.

Knowledge Integration: The tourism firm makes a person-in-charge to identify, store, share, and integrate firm knowledge with individual employee knowledge. Knowledge integration is the outcome of the firm's knowledge strategy and goals.

Knowledge Management: Firm's business strategy to formalised firm's existing information at the disposal of the firm by the use of technology to store, retrieve, and utilise the organisational knowledge with the participation of the firm's employees.

Knowledge Measurement: Development of knowledge measurement metrics for assessment knowledge assets within the tourism firm could be useful for the management of the firm's knowledge.

Knowledge Sharing and Replication: Knowledge sharing between tourism firm employees formally or informally. Replication of knowledge achieved with the help of the knowledge stored in knowledge portals or databases.

Knowledge Storage: A repository decided by the tourism firm to store knowledge, mostly an electronic database for storing internal and external communication. This also helps to retrieve for later usages as well.

Chapter 7
Fostering Knowledge-Sharing Behavior:
A Proposed Model for the Hospitality Industry of Pakistan

Afsheen Fatima
University Institute of Management Sciences (UIMS), PMAS-Arid Agriculture University, Rawalpindi, Pakistan

Sahrish Saba
University Institute of Management Sciences (UIMS), PMAS-Arid Agriculture University, Rawalpindi, Pakistan

ABSTRACT

The present research is aimed at exploring the link between corporate ethical values (CEV) and knowledge sharing behavior (KSB) by building and empirically examining a CEV-KSB model. The goals of the current research were multi-fold. The study was aimed at understanding the impact of CEV on organizational commitment and KSB. The effect of organizational commitment on KSB was also examined. Using a sample of 250 employees of the hotel industry in the Pakistan, a structural model was tested. It was hypothesized that CEV would be positively related to organizational commitment and KSB. Moreover, it was also hypothesized that commitment partially mediates the link between CEV and KSB. All the hypotheses were accepted. Limitations and implications of the study are also discussed.

INTRODUCTION

Knowledge is related to the practical or theoretical understanding of the subject which can either be implicit or explicit. It can also be less or more systematic or formal. A well-developed knowledge sharing system enables service organizations to obtain new knowledge and improve customer service. The successful implementation of such a system offers an opportunity for mutual learning and sharing between employees. Rowley (2000) termed this process of acquiring, distributing and retaining as knowledge

DOI: 10.4018/978-1-7998-3142-6.ch007

management. Through motivational programs aimed at encouraging knowledge sharing, the behavior may bring overall service quality.

Knowledge management (KM) is a process to create, share, use and manage the information and knowledge of an organization. It refers to a multidisciplinary approach to achieving organizational objectives by making the best use of knowledge. Knowledge sharing (KSB) was also studied as a tool for knowledge management (Kang et al., 2017) and organizational learning (Almeida & Soares, 2014). It also helps in the expansion of knowledge networks in the organization (Saifi et al., 2016; Reinholt & Foss, 2011). It turns out to be an individual behavior not just a decision of the organization. Knowledge management (KM) in hospitality is especially relevant for attaining competitive advantage (Musulin, Gamulin & Crnojevac, 2011).

The present research seeks to validate the concept of knowledge sharing which is needed to be included in the hospitality management and human resource fields in the present knowledge era. Such behavior is conceptualized as the extent to which an employee shares his acquired job-related knowledge, with others in the organization. This knowledge can either be explicit or tacit. Knowledge management guide hotel managers and marketers in their marketing analyses by elevating internal customers' competences in dealing with their guests and clients. This results in increasing customer satisfaction and loyalty, leading to organization effectiveness (Chang & Huang, 2010).

The hospitality industry is now becoming knowledge-intensive due to the great influence and use of ICT and the nature of the service product (Baldwin-Evans, 2006). This industry is services based which is immaterial so its quality is based on an interaction between customers and employees who have to be knowledgeable of customers' needs to achieve customers' satisfaction. Therefore, companies could obtain significant benefits from KM by boosting and facilitating knowledge sharing across the organization which would increase the quality of output service. KM is particularly important for hotel chains that have to deliver a consistent service in all of their hotels distributed around the world.

KSB has received focus from the industry like Knowledge-Intensive Business Services (KIBS) and academia both. This type of industry mainly depends on professional knowledge. Less knowledge has been created in the hospitality sector, thus, it can be explored further. KBS has been considered as an important aspect which supports the hospitality service level (Bounvken, 2002). The industrial prospects also depend on personal communication and interaction because of being labor-intensive mechanism. However, tacit knowledge is found to be mainly prevailed here (Rao, Yang and Yang, 2018). If there is less motivation, the level of personal knowledge sharing would have been decreased or it would lead to invalid knowledge creation. Hence, a knowledge management system may facilitate the tourism and hotel industry to be a third party factor to motivate the knowledge sharing process.

Current research has also focused on the knowledge systems role and KSB, which also comprises of management role in personal knowledge and information (Hwang, Lin & Shin, 2018). Knowledge sharing has been defined as "the process of creating knowledge available to the others in the organization" (Ipe, 2003).

KSB has also been explored as a moral challenge in organizations and as an aspect of high performance also (Wang, 2004). Lin (2007) has also studied the denial of employees to share their knowledge with other staff members although this sharing behavior is very important for the industry survival and its denial would be referred to as a violation of moral norms. As this aspect is important, it would be necessary to explore the aspects which affect the motivation of employees and their willingness to get involved in knowledge sharing with other employees (Bock, Zmud, Kim & Lee, 2005).

Above all, due to the complexity of the knowledge sharing mechanism and a lack of appropriate research framework to integrate the current research in the hospitality sector, it is important to do synthesized research in order to better solve the actual problem during practice. This study proposes and examines a framework and hopes to highlight the constructive role that organizational commitment and corporate ethical values play to lead to the knowledge sharing in the hospitality business

The management may energize its workforce in different ways for example by supporting ethics (Cameron, Bright & Caza, 2004) which may, in turn, enhance a range of positive employee and organizational outcomes. During the last few years, workplace ethics has received sufficient attention mainly because of the recent scandals at organizations like Enron and Tyco. These few examples show the adverse outcomes of an "informal" unethical culture developed by higher-level hierarchical positions in these organizations. Corporate ethical value (CEV) model is referred to as a virtue as it overcomes the unethical behavior in the organization.

The current study has practical and theoretical significance. It adds knowledge to the CEV and management of knowledge literature. KSB has been referred to as the moral and ethical aspect of the organization (Brief & Motowildo, 1986; Wang, 2004). There is a lack of empirical study on KSB using the ethical values aspect. Therefore, this research is among the few to study CEV as a significant predictor of KSB. Secondly, this study identified organizational commitment as a pathway through which CEV is linked with employee KSB. Thus, this research progresses theory by explaining the role of commitment in the process CEV affects KSB. Furthermore, prior research on KSB was mostly conducted in the western context ignoring the non-western and collectivist societies like Pakistan. The present study is of enormous importance as it emphasizes the hospitality industry in Pakistan where this industry is still in its initial growth stages.

Practically, this research stresses the role which CEV plays to foster KSB in organizations. CEV is likely to have a positive impact in terms of employees' KSB. The studies on knowledge sharing also have shown that if there is a healthy environment for knowledge sharing, it generates innovative ideas and enhance commitment. The management of the organizations should motivate the communication between employees on a regular basis so that they can share ideas, information and knowledge they have.

LITERATURE REVIEW

Corporate Ethical Values

These are defined as "a subset of organizational culture, representing a multidimensional interplay among various 'formal' and 'informal' systems of behavioral control" (Trevino, 1998, p. 451). Fraedrich (1993) takes a constructionist view and expresses ethical behavior as prescribed by the workplace. In this view, the workplace is considered as the source for shaping the right and wrong. And the deviance from workplace norms is treated as unethical (Fraedrich, 1993). However, corporate ethical values (CEVs) are a combination of an organization's formal and informal ethical policies and collective ethical values of its workforce (Trevino, 2006). Corporate ethical values, no matter properly documented or not, do not necessarily associate with organizational mission statements and lists of core values. These values may be considered external to the domain of workplace standard operating procedures (Organ, 1988). Baker, Hunt, and Andrews (2006) observed the predictors and different effects of ethical behavior in the companies. The research suggested that organization citizenship is an important outcome of ethical behavior in the workplace (Turnipseed, 2002).

Leonard and Haines (2007) examined the effect of computer-mediated group on the workers' ethical behavior. They supposed that employee attitude may impact behavioral intention (toward ethical behavior). It can also be affected by the personal normative beliefs, the strength of ego, locus of control and gender. Leonard and Cronan (2001) suggested that scenario and gender are important predictors of intention to act ethically or otherwise in any organization.

Andreoli and Lefkowitz (2009) found that ethical values help to reduce unethical behavior or misconduct of students serving in various firms. Also, Mulki et al. (2009) found an ethical climate of an organization indirectly influenced turnover intention. Fu and Deshpande (2012) examined those aspects which are influential to the ethical behavior of the employees in the Chinese state-owned steel company. Deshpande (2010) concluded the aspects which impact the practices regarding the ethics of those managers who are successful in China. Martin and Cullen (2006) in a meta-analysis concluded that the corporate ethical culture of the employees was considered as one of the important aspects which impacts ethical behavior.

Barraquier (2011) studied the ethical behavior in practice through examining decision outcomes and strategic implications. Many researchers have tried to understand ethical decision-making in any organization by identifying its components. These components assist in determining why organizations are responsible and have corporate ethical culture (Campbell, 2007). Philmore (2016) has studied the influence of organizational commitment and CEV in the employees who work in the corporate organizations and who also work in organizations with strong perceived CEV, are more likely to have high commitment level.

Organizational Commitment

In the literature review, organizational commitment is conceptualized in various ways. For instance, "the relative strength of an individual's identification with, and involvement in a particular organization" (Mowday, Steers, & Porter, 1979, p. 226). The current study operationalized this construct into three dimensions of organizational commitment i.e. affective, normative and continuance commitment. Affective commitment is an employee attachment to his or her workplace (Allen & Meyer, 1990). Oliver (1990) suggested that affective commitment refers to an attitude based on the quality and nature of the association between an employee and his workplace. An emotional attachment is a built-in affective commitment, where an employee works hard to achieve the organizational objectives as he considers these objectives as his own.

Becker (1960) concluded that workers invest in their workplaces in the form of time, work amity, job-related expertise, and job effort and all these factors discourage them to go for turnover. This conceptualization of Allen and Meyer (1990) grew and another dimension came into existence i.e. continuance commitment, which can be described as an individual's desire to keep on working with his workplace in response to prior investment at the workplace. Several authors (Becker, 1960; Oliver, 1990) suggested that individuals develop an attachment to their own actions rather than to the organization. Continuous commitment refers that individuals who wish to continue with current work due to costs related to leaving the job (Meyer & Allen, 1997). Mayer and Allen (1984) developed a continuous commitment scale and discovered that it is unrelated to AC. Mayer and Allen (1984), a continuous commitment scale was further divided into two distinct dimensions namely high sacrifice commitment and low alternatives commitment (McGee & Ford, 1987).

Normative commitment emphasizes on employees' perceptions of their obligation towards their organization. Normative commitment refers to a professed obligation to stay with the workplace (Meyer & Allen, 2001) while upholding membership in the workplace. For instance, when an individual feels that his organization is concerned with his interests, he tries to oblige his organization in return i.e. normative commitment. Individuals having a high level of NC continue working with the workplace as they think they are bound to do so (Allen & Meyer 1990).

Yang and Chang (2008) examined that individuals with greater organizational commitment were highly satisfied and more expected to indulge in their firm's competitive advantage as compared to temporary individuals. Absence of organizational commitment may lead to several adverse effects, for instance, greater absenteeism and unpunctuality; increased turnover rate and turnover intention; an inferior performance that has a negative impact on the effectiveness and organizational efficiency (Cohen & Hudacek, 1998).

According to Meyer and Allen (2001), effective, normative and continuance must be taken as components of organizational commitment as an employee association with his organization might have varying levels of these three components. At a moment an individual can experience a sentiment of compulsion to stay with the workplace and he may have a firm attachment to an organization together. Another individual may have a significant amount of will, need and obligation at a time to linger with the current organization. Yet another individual might like his job but also susceptible to the adverse consequences of the parting current workplace.

Organizational commitment has been extensively studied in the United States (Vallejo & Langa, 2010). However, this model has been widely suggested to test in contexts other than the US and Europe. For instance, Doktor, Tung, and VonGlinow (1991) recommended that "It is time to move beyond the exclusive emphasis we have had in our research and writing on North American perspective to include other parts of the global arena" (p. 260). It is also suggested to do more studies on organizational commitment outside North America (Meyer, Stanley, Herscovitch & Topolnytsky, 2002).

KNOWLEDGE SHARING BEHAVIOR

This behavior is known for the sharing of knowledge or information within the workplace (Barao, et al., 2017). Researchers have also found that this behavior in recent years has been explored as an interpersonal construct (Yong et al., 2013). Aspects of KSB have been studied at different levels in different organizations which include environmental, team and individual characteristics (Wang et al., 2011).

Knowledge can be in two forms, tacit or explicit. The idea of tacit knowledge that 'we can know more than we can tell' was introduced by Polanyi (1966, 4), a concept further expanded by Nonaka. It is hard to communicate and formalize. Tacit knowledge is enormously rooted in action, involvement and commitment (Kirsch .et. al, 2014; Polanyi, 1962). The aim of knowledge sharing is to bring the needed and desired knowledge to the right individual (Shabrinaa & Silvianitab, 2014; O'Dell, Hubbert, 2011). Sharing knowledge is one of the techniques in a learning organization, which is utilized to facilitate an organization to share information, technique, and experiences (Setiarso, 2009).

Concurring to the hypothesis of the Knowledge-Based View, an individual is the fundamental storehouse of learning (Grant, 1996), and the information level of the group is influenced by the conduct of people in connection to KSB (Shabrinaa & Silvianitab, 2014; Bock, Zmud, Kim & Lee, 2005; Wang & Noe, 2010).

"Information is achievement and achievement is knowledge". The reason, the more you have applicable taking in the more you will be fruitful. We ought to move to knowledge sharing. In today's extreme business environment knowledge sharing is one of the middle components for accomplishment which constructs participation within individuals furthermore within divisions through legitimate arrangement on the reason of information sharing.

Hypotheses Development

The importance of life quality has given development to ethical values (Kareiva, 2005). It can also be said that the ethical values of a corporation (sub-dimension of corporate culture) includes a mutual association between formal and informal systems in the controlling behaviors (Baker, Hunt, & Andrews, 2006). Ethical behavior is a basic aspect to reduce unethical norms from an organization and is used to examine the difference between right and wrong (Baker et al., 2006). Employees in an organization, state their behaviors right learning from the behaviors of leaders (Mayer et al., 2009). It can also be said that ethical behavior is an extension of an ethical value system. In this way, managers and academicians try to make ways to form ethical organizations (Wells et al., 2001). This study has been supported by the studying ethical activities in hotels also. This can be achieved from ethical training. There are three ways behind ethical learning, i.e., firstly it leads to improving the ethical. Considerations of employees towards their organization, secondly it helps the employees to get, engage in ethical activities and finally it also improves the ethical atmosphere of an organization, which leads to the success level (Wells et al., 2001). Organizations have started the process of knowledge management for some reason like a merger or acquisition could share their expertise to encourage knowledge-sharing behavior and intentions. The recruitment drive shows knowledge management wisdom in training the new employees also. Mason and Pauleen (2003) have studied the knowledge management execution factor in qualitative research. This was further supported by Cabrera, Collins, and Salgado (2006) by acknowledging personality and commitment to the organization also contributes to KSB. A study on knowledge sharing behavior has also shown that it is an important aspect of the Chinese hotel industry also (Akram, Lei, Haider & Hussain, 2019). They have studied the recent paradigm shift of knowledge sharing intentions in hotel employees. The findings have shown the positive effects of these types of intentions in the employees which increase the effectiveness of the organization also.

Furthermore, it was studied that knowledge running is a social procedure that shows that knowledge sharing is explored as companies' strategy to assimilate knowledge, expertise, and skills in a dynamic environment or culture (Clarke and Rollo, 2002). Dynamics like technologies (Malhotra & Majchrzak, 2004), cooperative environment (Yang, 2007), inspiring behavior (Ardichvili, page & Wentling, 2003) and communication in the organization (Moffet, McAdam & Parkinson, 2003) enhance the KSB. Hu (2009) has studied the environment (composite) of hotels and communication benefits, which generates the recognition of individuals and organizations both.

A very recent study has come across about the knowledge sharing behavior at the workplace (Zhang, Song & Song, 2019). They have studied the aspects of knowledge sharing effectiveness to maintain its influences on the employees. From the positive characteristic, KSB occasionally reliefs innovation prospects and decision making of organization, however, any single negative aspect can be count up as knowledge-reserve and minus organization commitment from the perspective of employee (Cabrera et al.,2006). "Knowledge is power"- employees in the hospitality industry may stand-in their knowledge for their peculiar personal development due to the tough race in the hospitality industry itself, and an intention on the road to knowledge sharing behavior within themselves are not frequently noticed. Partial knowledge program is found to be more mutual (Goh, 2002; Hu, 2009) as an organization faces trials while removing the knowledge-reserve behavior from the employees. A mainstream of organizations tries to realize team behavior in sharing the knowledge between them, which has been establishing to be one of the vital aspects to understand the team culture within them. Additionally, prior research showed a want for knowledge sharing and knowledge management in demand to generate new strategies and fresh innovation possibilities in the hospitality and tourism industry (Sungsoo, 2005; Al-Refaie, 2015; Grissemann, Plank & Brunner-Sperdin, 2013; Jiménez-Zarco, González-González, Martínez-Ruíz & Izquierdo-Yusta, 2014).

Hotels may rally their service quality by improving their employees 'knowledge through an understanding of customers' fondness and therefore improve their service. Empirical studies advise that sharing of knowledge within and by the team returns a well-developed "team process" behavior and grades for team performance (Finnegan & Willcocks, 2006; Paton & McLaughlin, 2008; Yu et al., 2013).

Organizational commitment (OC) specifies the level of commitment in the employees towards their organization (Angle & Perry, 1981). Meyer and Allen (1991) have studied the components of OC i.e. continuance, normative and affective commitment. Robbins and judge (2007) have explored the dimensions and called as tails of OC. Yang et al., (2012) have suggested that this level of commitment is necessary to predict either they would be loyal to their organization or not. Cho et al., (2009) have claimed that OC has a negative association with turnover in employees, while Gunlu et al., (2010) stated that OC has a significant relation with KSB in the employees.

OC of employees and its association with many work domains and other outcomes like performance (Atmojo, 2015), engagement level in employees (Albdour & Altarawneh, 2014), quality of service (Dhar, 2015), work-life balance (Wayne et al., 2013) and turnover in employees (Yin-Fah et al., 2010; Zopiatis et al., 2014) have been explored. The OC affect in the hotel industry has rarely been studied, hence the aim of the study has also examined the effects of OC in employees' attitudes and behavior in the hotel.

The individuals, who want to share their knowledge, are affected by many factors such as incentives, efficacy, group identity level and personal responsibilities (Cabrera, 2002). This study focused on incentive roles, which support knowledge sharing among firms (Taylor, 2006). It can also lead to enhance the commitment level among the hotel employees (Cabrera, 2002). It also gets affected by ethical considerations. It establishes good work relationships among employees.

The propensity of employees is to publicize their knowledge with the other colleagues depends not only on CEV but there are also some psychological components including organizational commitment which plays a pivotal role in prophesying an employee extra-role behaviors i.e. knowledge sharing (Joo, Yoon, and Jeung 2012). Employees who are more committed always stay loyal not only in role behavior but also in extra-role behavioral intentions like knowledge sharing for the organizations' betterment (Naeem, Mirza, Ayyub, & Lodhi, 2019). Camelo-Ordaz et al. (2011) claimed that OC is considered as a link of employees with the organization and their effort to complete the tasks and it leads to knowledge sharing behavior also. In the same way, some researchers have examined that commitment is also considered as a predictor of KSB in the employees (Biswanth Dutta et al., 2015).

Fu, Deshpande, and Zhao (2011) have studied that the ethical behavior of both employees and successful managers have a significant impact on OC. Prior studies have shown the significant relationship between OC and the ethical climate of the organization (Koh & Boo, 2004). Malik (2010) suggested that there are greater chances of ethical employees to stay in the organization.

On the basis of the above discussion, the following hypotheses are developed.

H_1: Corporate ethical values are significantly and positively associated with knowledge sharing behavior.
H_2: Corporate ethical values are significantly and positively associated with organizational commitment.
H_3: Organizational commitment is significantly and positively associated with knowledge sharing behavior.
H_4: Organizational commitment mediates the relationship between corporate ethical values and knowledge sharing behavior.

Figure 1. Theoretical framework

METHODOLOGY

Sample and Procedures

Employees at several four and five-star hotels in Pakistan were invited to participate in the study. The author made contact with the human resources manager of the hotels to introduce the research project. Those hotels which were agreed to participate visited for data collection. All the respondents were assured of the conðdentiality of their responses.

Measures

- **Knowledge Sharing Behavior:** It was measured with a self-rated nine-item scale. Based on Bock et al. (2005). This study measured both explicit knowledge sharing and tacit knowledge sharing.
- **Corporate Ethical Values***:* This construct was measured through a scale developed by Hunt, Wood, and Chonko (1989). This scale is involved in three items.
- **Organizational Commitment***:* It was measured using a scale developed by Allen and Meyer (1990). The construct consisted of three dimensions, i.e, affective, normative and continuance commitment and 11 items.

Analytic Strategies

Considering the nested nature of data, all the hypotheses were tested in current investigation using the structural equation modeling as it allows simultaneous estimation and provides the model ðt indices. For this purpose, the software was used is AMOS. Following Anderson and Gerbing's (1988) recommendation, this research is ðrst tested hypothesized measurement model, followed by the path assessments to test each hypothesis in the theoretical model.

Table 1. Participants' profile

Demographic variables	*f*	%
	N = 250	
Gender		
Male	188	78.3
Female	52	21.7
Age (Years)		
20 – 25	110	45.8
26 – 35	105	43.8
36 – 45	22	9.2
46 and above	3	1.2
Education		
Graduation	138	57.5
Masters	94	39.2
M. Phil/MS/Equivalent	8	3.5
Experience		
5 years or less	157	65.4
6-10 years	59	24.6
11-20 years	19	7.9
21 or above	5	2.1

Results

Table 1 shows the demographic characteristics of the respondents of research. It also shows that the majority of respondents were males, i.e. 52%. The majority were young people among the age group of 20-25 and 26-35 (46% and 44% respectively).

Since items were reduced and little variations were made in the comprised ones, inter-item consistency was ensured concluded construct reliability. The reliability values of all instruments are explored as in acceptable range i.e greater than 0.60 (Tang, 2008). The researcher has found them convergent. The values of standardized loadings were significant as less than 0.05 (Fraj, 2006). This table also shows that all three variables are positively and significantly correlated. For details see Table 2.

While assessing the model fitness of both measurement and structural values, absolute and incremental fit indices were also explored (Ho, 2006). Table 3 has shown the appropriateness of hypothesized associations.

Table 4 shows the results of hypothesized associations.

CEV has caused significant positive variation in organizational commitment and KSB, i.e., 48% and 51% respectively (H1 and H2 are supported). Whereas, organizational commitment explains significant positive variation in KSB that is 24.3% (H3 is supported).

After testing that all variables are significantly related, the mediation hypothesis was tested. For this purpose, two models were tested, one without a mediator (Model 1) and second with a mediator (Model 2). Mediation analysis was performed following the procedure of Barron and Kenny (1986). In the first step, it was found that CEV has a positive and significant impact on KSB (B=0.68). On the second step mediator (KSB) was added. Doing this, the beta weight reduced from 0.679 to 0.517 and was still significant as shown in the below table. Thus, it was a case of partial mediation (H4 is supported).

Table 2. Correlations, Standardized Loadings, and Reliabilities

Measures	**Standardized Loadings**	**Cronbach's Alpha**	**Correlations**		
			1	**2**	**3**
1. *Corporate Ethical Values*		0.658	-		
CEV1	0.560*				
CEV2	0.644*				
CEV3	0.576*				
2. *Organizational Commitment*		0.813	0.385**		
OC1	0.472*				
OC2	0.502*				
OC3	0.659*				
OC4	0.416*				
OC5	0.517*				
OC6	0.400*				
OC7	0.597*				
OC8	0.555*				
OC9	0.567*				
OC10	0.618*				
OC11	0.526*				
3. KSB		0.691	0.455**	0.382 **	-
KS1	0.506*				
KS2	0.576*				
KS3	0.657*				
KS4	0.635*				
KS5	0.538*				

$*p < 0.05$, $**p < 0.001$

Table 3. Model Fitness Indices

Fit indices	Measurement Model		Structural Model	
Absolute Fit Indices				
χ^2 / df	32.523 / 162 = 1.985*	(good fit)	321.523 / 162 = 1.985*	(good fit)
GFI	0.890	(acceptable fit)	0.890	(acceptable fit)
RMSEA	0.064	(good fit)	0.063	(good fit)
Incremental Fit Indices				
IFI	0.856	(acceptable fit)	0.856	(acceptable fit)
TLI	0.827	(acceptable fit)	0.827	(acceptable fit)
CFI	0.852	(acceptable fit)	0.852	(acceptable fit)

* $p < 0.001$

CONCLUSION

The current study has focused on the importance of corporate ethical values in enhancing the knowledge sharing behavior among employees of the hospitality sector. It also included an important part of the commitment level to their organization also.

Table 4. Hypotheses

Models			
Model 1 – Relationships	*B*		Results
Corporate Ethical Values ® KSB	0.679	***	Supported
Model 2 – Relationships			
Corporate Ethical Values ® Organizational commitment	0.482	***	Supported
Organizational commitment ®KSB	0.243	***	Supported
Corporate Ethical Values®Organizational commitment ®KSB	0.517	***	Supported

*$p < 0.10$, **$p < 0.05$, ***$p < 0.01$, ns = not significant

Knowledge is a necessary organization resource and has a critical impact on business operations. Knowledge sharing inside of an organization has frequently has declared at this very moment rehearse for achievement and manageability of its implementation. Knowledge sharing means exchange, spreading, and trade of information, experience, abilities, and important data from one person to different individuals inside of an organization. It might be formal and casual. It may happen by means of composed correspondence, about interchanges or by utilizing electronic learning frameworks. It may happen when a partner separately imparts his or her slight to different companions. It might likewise happen among colleagues or inside and between distinctive reliable units. Knowledge sharing may vary as per who offers his or her insight and with whom (e.g., managers, partners, subordinates). These can impact the goals for sharing (or not sharing) learning. In all cases, knowledge sharing incorporates dynamic correspondence, conversing with associates, gaining from them, teaming up, helping other people, analysis and tutoring.

It is generally recognized that knowledge sharing has a beneficial outcome on authoritative achievement and aggressiveness. Nonetheless, supportive information sharing is troublesome.

The results showed that the proposed hypotheses are said to be accepted as the findings of analysis seemed to be significant. There is a positive effect of CEV on KSB, the same as commitment shows a positive association with KSB. Organizational commitment is showing the partial mediation between the association of CEV and KSB. The results are also supported by the previous literature, discussed in the literature review section. The results also showed that an increase in corporate ethical values leads to an increase in knowledge sharing behavior among employees, hence organizational commitment played a vital role in their relationship between CEV and KSB. It enhances the efficiency of employees as well as the productivity of the organization also. The goal of that study was to examine and encourage knowledge sharing in the employees. Knowledge management could enhance the performance of employees in organizations. It also explores the awareness and learning of best practices and improvement in organizations also.

The current study has some implications for theory and practice. The results show that CEV act as a significant factor in the development of KSB. For practical implications, findings also show that the management of hotels should know the practice of CEV to increase their commitment level. This is significantly associated with the hospitality sector where the KSB is said to be less in the employees. Hence, management should implement those characteristics in their own selves and also provide such an environment in the hotels which support CEV and hence enhances KSB and also commitment among employees.

FUTURE RESEARCH QUESTIONS

Every study has its limitations. So, there are always chances of value addition in it. Future studies can explore the other factors involved in enhancing knowledge sharing behavior. The current research has used the sample of the hospitality sector; it can further be generalized by studying these components and their association in some other sector also, even in other cultures also. Knowledge management needs more time to form and it must be organized in such a way that knowledge should be easy to find and it must be up to date.

REFERENCES

Abu, S. (2015). Fostering Knowledge Sharing Behaviour among Public Sector Managers: A Proposed Model for the Malaysian Public Service. *Journal of Knowledge Management, 19*(1), 121–140. doi:10.1108/JKM-11-2014-0449

Akram, T., Lei, S. Haider, M. J. & Hussain, S. T. (2019). The impact of organizational justice on employee innovative work behavior: Mediating role of knowledge sharing. *Journal of Innovation & Knowledge*, 1-13.

Al-Refaie, A. (2015). Effects of human resource management on hotel performance using structural equation modeling. *Computers in Human Behavior, 43*, 293–303. doi:10.1016/j.chb.2014.11.016

Albdour, A. A., & Altarawneh, I. I. (2014). Employee engagement and organizational commitment: Evidence from Jordan. *Int. J. Bus., 19*(2), 192–212.

Almeida, M. V., & Soares, A. L. (2014). Knowledge sharing in project-based organizations: Overcoming the informational limbo. *International Journal of Information Management, 34*(6), 770–779. doi:10.1016/j.ijinfomgt.2014.07.003

Angle, H. L., & Perry, J. L. (1981). An empirical assessment of organizational commitment and organizational effectiveness. *Administrative Science Quarterly, 26*(1), 1–14. doi:10.2307/2392596

Atmojo, M. (2015). The influence of transformational leadership on job satisfaction, organizational commitment, and employee performance. *Int. Res. J. Bus. Stud., 5*(2), 113–128. doi:10.21632/irjbs.5.2.113-128

Baker, T. L., Hunt, T. G., & Andrews, M. C. (2006). Promoting ethical behavior and organizational citizenship behaviors: The influence of corporate ethical values. *Journal of Business Research, 59*(7), 849–857. doi:10.1016/j.jbusres.2006.02.004

Baldwin-Evans, K. (2006). Hilton highlights link between staff loyalty and e-learning. *Human Resource Management International Digest, 14*(1), 36–38. doi:10.1108/09670730610643990

Barão, A., Vasconcelos, J. B. D., Rocha, Á., & Pereira, R. (2017). A knowledge management approach to capture organizational learning networks. *International Journal of Information Management Processing, 37*(5), 735–740.

Bock, G. W., Zmud, R. W., Kim, Y. G., & Lee, J. N. (2005). Behavioral intention formation inknowledge sharing: Examining the roles of extrinsic motivators, social-psychological forces, and organizational climate. *Management Information Systems Quarterly, 29*(1), 87–111. doi:10.2307/25148669

Bouncken, R. (2002). Knowledge Management for Quality Improvement in Hotels. In R. Bouncken & P. Sungsoo (Eds.), *Knowledge Management in Hospitality and Tourism* (pp. 25–59). New York: The Haworth Hospitality Press. doi:10.1300/J162v03n03_03

Brief, A. P., & Motowidlo, S. J. (1986). Prosocial organizational behaviors. *Academy of Management Review, 11*(4), 710–725. doi:10.5465/amr.1986.4283909

Cabrera, A., Collins, W. C., & Salgado, J. F. (2006). Determinants of individual engagement in knowledge sharing. *International Journal of Human Resource Management, 17*(2), 245–264. doi:10.1080/09585190500404614

Cabrera, E. F., & Cabrera, E. F. (2002). Knowledge-sharing dilemmas. *Organization Studies, 23*(5), 687–710. doi:10.1177/0170840602235001

Camelo-Ordaz, C., García-Cruz, J., Sousa-Ginel, E., & Valle-Cabrera, R. (2011). The Influence of Human Resource Management on Knowledge Sharing and Innovation in Spain: The Mediating Role of Affective Commitment. *International Journal of Human Resource Management, 22*(07), 1442–1463. doi:10.1080/09585192.2011.561960

Dhar, R. L. (2015). Service quality and the training of employees: The mediating role of organizational commitment. *Tourism Management, 46*, 419–430. doi:10.1016/j.tourman.2014.08.001

Finnegan, D., & Willcocks, L. (2006). Knowledge sharing issues in the introduction of a new technology. *Journal of Enterprise Information Management*, *19*(6), 200–221. doi:10.1108/17410390610708472

Foss, N. J., & Pedersen, T. (2002). Transferring knowledge in MNCs: The role of sources of subsidiary knowledge and organizational context. *Journal of International Management*, *8*(1), 49–67. doi:10.1016/S1075-4253(01)00054-0

Goh, S. C. (2002). Managing effective knowledge transfer: An integrative framework and some practice implications. *Journal of Knowledge Management*, *6*(1), 23–30. doi:10.1108/13673270210417664

Grissemann, U., Plank, A., & Brunner-Sperdin, A. (2013). Enhancing business performance of hotels: The role of innovation and customer orientation. *International Journal of Hospitality Management*, *33*(1), 347–356. doi:10.1016/j.ijhm.2012.10.005

Holste, J. S., & Fields, D. (2010). Trust and tacit knowledge sharing and use. *Journal of Knowledge Management*, *14*(1), 128–140. doi:10.1108/13673271011015615

Hu, H.-H., Kandampully, J., & Juwaheer, T. D. (2009). Relationships and impacts of service quality, perceived value, customer satisfaction, and image: An empirical study. *Service Industries Journal*, *29*(2), 111–125. doi:10.1080/02642060802292932

Jiménez-Zarco, A. I., González-González, I., Martínez-Ruíz, M. P., & Izquierdo-Yusta, A. (2014). New service innovation success: Analyzing the influence of performance indicator nature. *Computers in Human Behavior*, *51*, 1024–1031. doi:10.1016/j.chb.2014.09.046

Joo, B.-K., & Yoon, H. J. (2012). The Effects of Core Self Evaluations and Transformational Leadership on Organizational Commitment. *Leadership and Organization Development Journal*, *33*(6), 564–582. doi:10.1108/01437731211253028

Kang, Y. J., Jin, Y. L., & Kim, H. W. (2017). A psychological empowerment approach to online knowledge sharing. *Computers in Human Behavior*, *74*, 175–18. doi:10.1016/j.chb.2017.04.039

Kareiva, P. (2005). Is the key to conservation changing ethical values or policing unethical behavior? *Current Biology*, *15*(2), 40–42. doi:10.1016/j.cub.2004.12.048 PMID:15668150

Mason, D., & Pauleen, D. J. (2003). Perceptions of knowledge management: A qualitative analysis. *Journal of Knowledge Management*, *7*(4), 38–48. doi:10.1108/13673270310492930

Mayer, D. M., Kuenzi, M., Greenbaum, R., Bardes, M., & Salvador, R. B. (2009). How low does ethical leadership flow? Test of a trickle-down model. *Organizational Behavior and Human Decision Processes*, *108*(1), 1–13. doi:10.1016/j.obhdp.2008.04.002

Meyer, J. P., & Allen, N. J. (1991). A three-component conceptualization of organizational commitment. *Human Resource Management Review*, *1*(1), 61–89. doi:10.1016/1053-4822(91)90011-Z

Musulin, J., Gamulin, J., & Crnojevac, I. H. (2011). Knowledge management in tourism: The importance of tacit knowledge and the problem of its elicitation and sharing. In *2011 Proceedings of the 34th International Convention MIPRO*, (pp. 981- 987). IEEE.

Naeem, A., Mirza, N. H., Ayyub, R. M., & Lodhi, R. N. (2019). HRM practices and faculty's knowledge sharing behavior: Mediation of affective commitment and affect based trust. *Studies in Higher Education, 44*(3), 499–512. doi:10.1080/03075079.2017.1378635

Paton, R. A., & McLaughlin, S. (2008). Services innovation: Knowledge transfer and the supply chain. *European Management Journal, 26*(April), 77–83. doi:10.1016/j.emj.2008.01.004

Philmore, A. (2016). The influence of organisational commitment and corporate ethical values on non-public accountants' whistle-blowing intentions in Barbados. *Journal of Applied Accounting Research, Emerald Group Publishing, 17*(2), 190–210. doi:10.1108/JAAR-12-2013-0118

Rao, Y., Yang, M., & Yang, Y. (2018). Knowledge Sharing, Organizational Learning and Service Innovation in Tourism. *Journal of Service Science and Management, 11*(05), 510–526. doi:10.4236/jssm.2018.115035

Reinholt, M., & Foss, N. J. (2011). Why a central network position isn't enough: The role of motivation and ability for knowledge sharing in employee networks. *Academy of Management Journal, 54*(6), 1277–1297. doi:10.5465/amj.2009.0007

Robbins, S., & Judge, T. A. (2007). *Organizational Behavior* (12th ed.). Upper Saddle River, NJ: Prentice-Hall.

Saifi, S. A. A., Dillon, S., & Mcqueen, R. (2016). The relationship between face to face social networks and knowledge sharing: An exploratory study of manufacturing firms. *Journal of Knowledge Management, 20*(2), 308–326. doi:10.1108/JKM-07-2015-0251

Sungsoo, P. (2005). Knowledge map for tourist destinations needs and implications. *Tourism Management, 26*(4), 583–594. doi:10.1016/j.tourman.2004.03.001

Taylor, E. Z. (2006). The effect of incentives on knowledge sharing in computer mediated communication: An experimental investigation. *Journal of Information Systems, 20*(1), 103–116. doi:10.2308/jis.2006.20.1.103

Vera-Muñoz, S. C., Ho, J. L., & Chow, C. W. (2006). Enhancing knowledge sharing in public accounting firms. *Accounting Horizons, 20*(2), 133–155. doi:10.2308/acch.2006.20.2.133

Wang, C. (2004). The infiuence of ethical and self-interest concerns on knowledge sharing intentions among managers: An empirical study. *International Journal of Management, 21*(3), 370–381.

Wang, S., Noe, R. A., & Wang, Z. M. (2011). Motivating knowledge sharing in knowledge management systems: A quasi-field experiment. *Journal of Management, 40*(4), 978–1009. doi:10.1177/0149206311412192

Wayne, J. H., Casper, W. J., Matthews, R. A., & Allen, T. D. (2013). Family-supportive organization perceptions and organizational commitment: The mediating role of work family conflict and enrichment and partner attitudes. *The Journal of Applied Psychology, 98*(4), 606–622. doi:10.1037/a0032491 PMID:23565896

Wells, D., & Schminke, M. (2001). Ethical development and human resources training an integrative framework. *Human Resource Management Review, 11*(1-2), 135–158. doi:10.1016/S1053-4822(00)00044-9

Yin-Fah, B. C., Foon, Y. S., Chee-Leong, L., & Osman, S. (2010). An exploratory study on turnover intention among private sector employees. *International Journal of Business and Management*, *5*(8), 57–64.

Yong, S. H., Kim, B., Lee, H., & Kim, Y. G. (2013). The effects of individual motivations and social capital on employees' tacit and explicit knowledge sharing intentions. *International Journal of Information Management*, *33*(2), 356–366. doi:10.1016/j.ijinfomgt.2012.10.009

Yu, C., Yu-Fang, T., & Yu-Cheh, C. (2013). Knowledge sharing, organizational climate, and innovative behavior: A cross-level analysis of effects. *Social Behavior and Personality*, *41*(1), 143–156. doi:10.2224bp.2013.41.1.143

Zhang, Z., Song, F., & Song, Z. (2019). *Promoting knowledge sharing in the workplace: Punishment v. reward. Chaos, Solitons & Fractals.* Academic Press.

Zopiatis, A., Constanti, P., & Theocharous, A. L. (2014). Job involvement commitment, satisfaction and turnover: Evidence from hotel employees in Cyprus. *Tourism Management*, *41*, 129–140. doi:10.1016/j.tourman.2013.09.013

Chapter 8
Religious Tourism as a Source of Knowledge Sharing and Trading:
A Case From Russia

Emilia Alaverdov
Georgian Technical University, Georgia

ABSTRACT

Religious tourism in the 21st century is presented in one form or another in many segments of the tourism business, and even in those that are not directly related to it. The goals of religious tourism at present are the knowledge of traditional cultures, the history of religions, national religious traditions, rituals, ceremonies, and ceremonies, participating in them as spectators, acquaintance with unique architectural monuments, including the traditional methods of building religious buildings, with masterpieces of icon painting, sculptures, mosaics, and decoration. In the modern scientific community, there is a stable definition of tourist trips associated with visiting religious sites for various purposes. This chapter explores religious tourism as a source of knowledge sharing and trading through a Russian case.

INTRODUCTION

A person traveling for religious purposes any place of his/her permanent residence for not a long period to visit sacred places and religious centers is considered a religious tourist. Speaking about the religious tourism we have to bear in our minds that these activities are associated with the provision of services and the needs of tourists traveling to the Holy places and religious centers outside the usual Wednesday for them. Religious tourism is divided into two main varieties: Pilgrim tourism and religious tourism excursion-cognitive orientation. Religious tourism itself, excursion-cognitive orientation for the contingent of the participants can be probably divided into two groups: the first group gives chance to people whose aim is the study of religious artifacts, ceremonial objects with scientific purposes to participate in scientific expeditions. As for the second group, which is considered much more affluent in the number

DOI: 10.4018/978-1-7998-3142-6.ch008

of exhibitors includes mass tourists different e.g. attending religious-cultic objects composed of tour groups for purely informative purposes.

This work is dedicated to the expansion of geographical view, statistics and development trends of Islamtourism in Russia. However, it is not possible to do research in any religious field without study and understand the religion itself and its impact on the country's political and social system. So, the work gives a certain picture of religious Islam in Russia.

The work aims to study the current state situation of Islam in Russia's socio-political circles and tourism industry and the prospects for its further development.

The object of the research is the tourism industry, modern Islam and their cooperation with each other.

LITERATURE REVIEW

The tourism industry is undergoing not only quantitative but also qualitative changes. The engine of these changes is the steady desire of tour operators to introduce a new tourism product to the market that meets the changing needs of consumers. Due to the heterogeneity of the interests of consumers of the tourism product, new types of tourism, religious tourism has emerged. Today, scientists in many areas pay great attention to the study of these types of tourism, but in modern science, there are still a considerable number of unresolved problems. And one of the most serious is the correlation of religious tourism and pilgrimage.

There is still no consensus on whether the pilgrimage is subordinate to religious tourism (it is a variety of it), or whether the pilgrimage should be considered as a separate phenomenon of human life. The opinions of scholars vary widely: travel industry practitioners' opinion has split. The first group includes researchers Yu. Alexandrova, A.Babkin, A. Skorokhodov, E. Dracheva, etc., the second - mainly ministers of the Churches and theological aspects - T. Khristov, A. BabkinV. Yakunin, S. Jitenevand others. According to Russian tourism researcher T. Khristov, more than 200 million people make pilgrimages a year. However, this point of view seems too optimistic, since the global number of tourists in 2003 (the year of writing of T. Khristov's work) amounted to 730 million (data from the World Bank). Thus, if we follow the opinion of T. Khristov, it turns out that 27% of tourists around the world traveled precisely for religious purposes, which is quite difficult to believe.

As for Islam and its past and present situation and its impact over Russia's political and social system the has analyzed the works of following authors: I.Tekusheva and K. Shevchenko, R. Silantiev, L. Syukiyainen, M. Yakubov, and other prominent Islam researchers, where they describe the history of Islam in Russia, Islam in North Caucasus its historical and modern situations, about the influence of Sharia in North Caucasus, Islamic renaissance in the XIX-XX cc, etc.

THE ESSENCE OF RELIGIOUS TOURISM

The history of religious tourism is undoubtedly very rich. It is a pilgrimage that can rightfully be considered the most ancient form of tourism. The history of religion has preserved for posterity a huge number of religious cults and monuments. In the modern world, religious tourism is in great demand, and therefore, the theme of pilgrimage as a cultural phenomenon is becoming increasingly important.

Naturally, the reason for the appearance of the pilgrimage was religious motivation. Decades of religious destruction have passed, and now we have a huge demand for religious sacred values. This trend is now observed in almost all areas of human life. All this suggests that the phenomenon of religiosity is brewing.

It was religious consciousness that prompted people to do things like a pilgrimage, for example, on foot to Mecca. We can say that religious tourism has its roots in ancient times. Information about travel with religious purposes has been available since the antiquity (Dallen & Olsen, 2006). An indispensable custom among the ancient Greeks and Romans, there was a visit to the temples and tombs of the kings. In which they could ask for help, about something (Alexandrova, 2004).

Religious tourism served only as a cover for the realization of their interests, sometimes very far from religion. Someone went to join the knights, someone for political and military espionage, etc.

The reason for the widespread of religious tourism may also be that people are gradually beginning to understand that a multipolar and safe world can only be created through a productive dialogue of civilizations, as a common space of multifaceted spirituality. Therefore, one of the main means of improving society continues to be a cognition process, which assists a person in a developed perspective to the universal values of culture and based on them to develop their worldview. The traditional values of different cultures are a blessing that at the current juncture of the socio-economic development of society allows people to realize the priority of spiritual and moral goals over material motivation. Modern religious tourism repeats all the basic models of mass tourism, which is subject to commercialization and economic criteria for evaluating its effectiveness. It contains all the uniformly defined signs of tourism: planning, security, transportation, accommodation, food, cultural services, additional services, trade, escort, and others. Often, when choosing the direction of religious trips by modern tourists, the motive is the information received about the historical routes of ancient and current pilgrims. In modern Russia, tourism is successfully reviving, and religious tourism is booming. Many tourists come from trips to holy places by truly believing people and later return to them as pilgrims. Of course, a literal repetition of the past is impossible, but the revival of the pilgrimage tradition at a new stage in the development of modern society is real and feasible. Pilgrimage develops and becomes the 7th mass religious movement, which, relying on the ancient traditions of pilgrimage, uses for its purposes the modern achievements of science and technology, especially in the transport sector(Jitenev, 2016).

Religious Tourism in Islam

Islamic tourism is considered a type of religious tourism since Muslims are inspired to visit religious ceremonies, gatherings, conferences and functions at local, regional and international levels at religious centers. Even though Islamic tourism is a component of the above-mentioned tourism, it is also linked to the other types of tourism such as holidaymaking and social ones. Nevertheless, the conjunction of these types of tourism should follow Islamic laws. Islam does not create any obstacles for a perfect and amazing journey, so we can claim without any shadows that if one follows the Islamic rules he/she can have a pleasant journey.

Islamic tourism is flexible because it does not have a certain purpose only. Usually, Muslims have various purposes of traveling, such as shopping, medical, sports, visit religions and educational. Islam encourages Muslims to visit not only its religious places but the holy places of other religions as well. For instance, families usually enjoy seeing the world's historical places, children's attractions, visiting relatives, etc. It is known, Islam loves peace and persuades to seek harmonization by looking at the beauty of creations and also visit historic sites. Islam is a simple religion because it does not oblige its followers to do certain things; usually, they do it by their wills and with great pleasure. For example, the Muslim responsibility to pray five times per day is not a burden at all, but it is actually may help humans to feel calm physically, emotionally and mentally without spending much. What's more if one is not possible to pray at a certain time, he/she may miss the praying and combine it with the next praying. Also, the most important point is that by prying Muslims are expressing their thankfulness to God. So we can acclaim that, praying is human's natural treatment(Babkin, 2008).

Furthermore, the mandatory of Muslim duties during traveling is the same as usual, but there is some lightness (Rukhsah) for tourists to perform duties such as shortening and combining prayers (Jama' and Qasr), tayammum, and they even can break their Ramadan fasting. We can say that religious (Islamic) tourism is a cross between the hedonic conceptualization of tourism and the Koranic pilgrimage itself. Despite the "calls" in Islam for travel, World Tourism Organization data show some passivity of Muslim citizens in terms of outbound tourism. Ultimately, we see that tourism in Islam is balanced by a dual life, which is in the world now and hereafter. Also, traveling strengthens the relationship among humans (Habl Min An-Nas) and also the relationship towards Allah (Babl Min Allah). Aforesaid makes it obvious that Islamic tourism, like its rules, bring happiness to the world and the hereafter simultaneously (Laderlah &Rahman, 2011).

Religious Objects of Islam

Islamic tourism bares certain components of pilgrimage. Since, for every Muslim, it is necessary at least once in his/her life to make a pilgrimage (hajj) to the holy cities - Mecca and Medina. Another characteristic feature is a sharp surge in the number of Muslim pilgrims during the Hajj, which happens once a year, and with dates changing depending on the lunar calendar. While talking about Mecca and Medina, we have to say that this Holley can have many more visitors if no restrictions: only Muslims are allowed to visit them. So this limits the excursion, the Zion component, sightseers - representatives of other religions cannot visit the main shrines of Islam.

A pilgrimage in Islam is a visit to two holy cities - Mecca and Medina. These pilgrimages: 1) Hajj, 2) death. Visits to other places where there may be graves of Muslim saints or famous mosques are not considered pilgrimages. Sunni Islam emphasizes that a person does not come to worship this place, but Allah in this place. In Shiite Islam, veneration of the relics of imams is great. Other Shiite pilgrimage centers are Najaf (Iraq), where the tomb of Caliph Ali, Karbala (Iraq) is located, the place of death and burial of Imam Hussein, the youngest son of Caliph Ali, Mashhad (Iran), where the tomb of Imam Ali ar-Reza is worshiped. The residences of the most respected Mujahideen (religious authorities with the right to make independent judgments on religious and legal issues) are located in the cities of Qum (Iran) and Samarra (Iraq).

Religious Centers of Islam

There are a large number of mosques in completely different countries in the world. There is a growing influence of Islam in countries where Muslims constitute a minority of the population. As we know the pilgrimage centers in Islam are Mecca and Medina, located in Saudi Arabia. Mosques and other religious sites in other cities and countries of the world are also visited by tourists.

For example, Mecca is annually visited by about 2 million pilgrims. There is built one of the largest mosques in the world - Haram beit Ullah. Inside the mosque is the Kaaba sanctuary. The Kaaba is made of stone slabs and has the shape of a cube 15 meters long. According to the news agency, in 2018 Saudi Arabia issued visas for more than 1.7 million foreign pilgrims wishing to make hajj this year. Of these, 22.5 thousand visas were issued to residents of Russia (Hajj to Saudi Arabia, 2018).

As for the year 2019, the Ministry of Hajj Affairs of Saudi Arabia has published statistics according to which in the current year, the Saudi authorities issued over 7.5 million visas for a pilgrimage to Mecca. Saudi news agency Asharq Al-Awsat notes that the largest number of pilgrims arrived from Pakistan, their number exceeded one and a half million people. It is reported that most Muslims making a small pilgrimage arrived in Saudi Arabia by air, their number reached six million. About a million more foreign citizens arrived in the Kingdom by land and sea. Pakistan takes the lead in the number of pilgrims (1,530,855), followed by Indonesia (934,827), India (635,675), Egypt (524,728), Algeria (353,425), Yemen (323,083), Turkey (300,517), Malaysia (26 9554), Iraq (268 179) and Jordan (194 908), Russian (25 000)(Hajj to Saudi Arabia, 2019).

The pilgrimage to Medina is usually combined with the pilgrimage to Mecca. Between Mecca and Medina is 300 km, pilgrims overcome this distance by car or plane. The main mosque in Medina is smaller than the mosque in Mecca. The prophet Muhammad is buried in it. Nearby is the al-Baqi cemetery, where many family members and associates of the prophet are buried (his daughter Fatima, grandson Hassan and Caliph Osman). In Medina, pilgrims climb Mount Jabal al-Nour to visit the cave, where the prophet was sent the first vision, to Mount Jabal al-Thawr, where he was hiding from the troops of the inhabitants of Mecca. Pilgrims visit famous battlefields and mosques. For example, Takva Camii, was built during the first entry of Muhammad into Medina. Its interesting feature is that it has two Mihrabs, one of them facing Jerusalem.

As we already know, the hajj is one of the five pillars of Islam. Every Muslim should at least once in his life perform Hajj. However, he should not commit it with the money of others and in damage to the family budget. A pilgrimage to Mecca will be considered a hajj if it ended on the Eid al-Adha holiday and a person performed all the necessary rituals within ten days (Khristov, 2005).

Religious Tourism in Russia

Educational activity in the tourism sector is an important factor in maintaining the spiritual and cultural processes at various levels of the functioning of society. This strengthens direct and reverse intergenerational ties in cultural systems and determines the original combination of traditions and innovations in the Russian national culture. An ethnocultural community, finding itself in a different cultural landscape, can noticeably change, adapting to new circumstances. In Russia, the cultural landscape over the past hundred years has changed dramatically under the influence of ideological factors. In particular, confessional architecture has undergone significant changes - from almost destruction - to some even excessive restoration. So far, the processes of returning religious buildings to the cultural landscape are perceived ambiguously. Religious tourism should help restore the traditional image of the region in regional culture. The sphere of tourist interest in such trips can be architectural ensembles, monasteries, churches, icons, relics of saints, holy sources. The mainstream of religious tourism in as well as throughout Russia falls on domestic directions. The determining factor here, of course, is not so much the financial component of travel. Regular visits to such holy places have already become a part of daily life. At this level, a person satisfies his physical needs, the need for communication, etc. The Russians (Orthodox) have developed specific cultural elements that allow them to build cultural algorithms for "nourishing" vital energy and strengthening fortitude. It is the traditional culture of the Russian regions, but the social culture from the very beginning of life rises ethnic and ideological (confessional) stereotypes that make a person a representative of a particular ethnic group and faith. But the principal point is that, regardless of ethnic and religious affiliation, a follower will hold in his/her heart the image of a small homeland that, as you know, begins, including, from the natural and cultural landscape. Religious tourism helps people to touch ancient traditions, restore the connection of times and to some extent block the negative forms of relieving emotional stress. Now a unique situation is being created for many Russian regions in using its rich cultural resources in a new round of civilization. It is impossible to restore and revive the former traditional culture (Suvorova, P & Ovsyannikov, V, 2014). However, reality shows that the best way to retain the achievements of the past is to accumulate and turn them into social values to create a fundamentally new culture that preserves through the institutions of culture its external similarity (factuality) to its prototypes. Speaking of traditional values, one cannot say about the many years of work on spiritual and moral education carried out by various public organizations in Russia, among which is highlighted the Russian Cultural Center, due to which an entire generation of young people has grown up who sees Russia as a Power, strong in all respects, based on traditional Orthodox values. The Russian cultural center will be able to gather all the best experience in reviving Russian culture in Russia. The main cultural center for Russian religious tourism can be considered Samara and Tolyatti where are residing quite a long number of Tatars, Muslims. The plans of the Russian Cultural Center in include a project of a unified information portal that will accumulate all information about the activities of public organizations and national cultural centers, thus providing residents and all interested people with the most complete information about spiritual and moral initiatives implemented in the city, as well as support for projects promoting pilgrimage and religious tourism. The goal of the Center in the field of spiritual and moral education is the development of high spirituality in society, the formation of active citizens with positive values and qualities that can manifest them in a creative process in the interests of the city and the Fatherland. The functioning of the system of spiritual and moral education of the Center should result in a spiritual and cultural upsurge, a decrease in the severity of the spiritual and moral crisis in the city, activation of the spiritual and cultural consciousness of the population, and an increase in the level of moral education of children and adolescents. The activities of the Russian Cultural Center are designed to help restore the spiritual, moral and intellectual potential of the bearer of the core of Russian civilization - the Russian people, as well as the indigenous peoples of the Volga region and representatives of all nationalities, all citizens, both living in the territory of Tolyatti and the entire region.

Islam as a Religious System

Islam is the third largest religion in the world. It is spread among many nations and nationalities. In the world's many eastern countries, domestic and foreign policy, human existence, moral and legal standards are governed by Islamic law, Sharia. Even nowadays the influence of Islam on the socio-political life of particular countries is enormous.

Like any other religion, Islam has its flexibility as well. There are different social levels in this religion. There are also certain parties and regimes with different, sometimes even conflicting oppositional political goals and aims. Many international Islamic organizations have been established in eastern countries. Today Islam is very active in the process of its revival, and this can be explained not by the specificity of Islam, but by the social and political changes in Asia, Africa and the Middle East and the development of national consciousness of these peoples.

Islam, as a religious system, has its ways and means of influencing people. It includes elements of worldview, cult-ritual and religious presence. Its work is being conducted in active mosques, by local Mollas, active believers, and religious families.

One of the specific features of Islam is that it is a kind of domestic religion. The religion of Islam is rooted in the human spirit, and as for its branches, they take us to world view and politics. In Islam, unlike other religions, religious and secular aspects are not separated. We can say that all aspects of human life are governed by the Qur'an and Sharia.

From this, it becomes clear that, if the present part of public life is Islamized, the possibilities of its influence over youngsters are great. Moreover, much of the Islamic traditions of life is centered around youth, which has a certain influence on their outlook and lifestyle. It is well known that every new generation automatically adopts ready-made traditions from the old customs and of course religion as well. The same we can say about religious traditions and Islamic cult-rituals. Certain youngsters consider them as their national traditions and national cultural elements. Such performances are often based on false identities of religion and nationality; it means that nowadays young generation in some cases even cannot draw a line between national and religious values.

Any religion always has had more or less influence on the relations between peoples and of course Islam is not an exception. Here we can name like the examples religious-nationalist and communal animosity in some of India states in the 1990s, religious-political groups and their interconnectedness in the Middle East, etc.

While investigating Islamic traditions we came across some political significance and aspects. Several political organizations in the eastern countries are trying to use Islam for their political purposes. Unfortunately, some terrorist organizations also operate under the cover of Islam. Terrorist acts are carried out on both Islamic and non-Islamic territories, the example, are terrorist attacks in the US, September 11, 2001; the Moscow theater hostage crisis or Nord-Ost siege 2002; terrorist attack at Beslan School on September 1, 2004; the Terrorist Act of March 29, 2010, in Moscow, etc.

Islam as a Political Religion

Islam as a political ideology has pre-twentieth-century roots. For example, in the first millennium, the Muslim Empire stretched from Spain to India. Eventually, the Muslim Empire expanded greatly its authority and owing to a variety of factors fragmented first into smaller empires and then into small, amorphous states that were easily penetrated by Western colonial powers. The period during and following the West domination of the Middle East and the Islamic world saw Islam reborn as a political ideology. However, this is not the same Islam that legitimated the conquests of the early Islamic period, the reign of the Umayyad or Abbasid dynasties, or even the rule of the Ottoman Empire. At this time, power relations have shifted; mechanization and increased contact with Western thought have modified Islam to a great degree. Thus, it is worth to know the nature of Islam's political ideology in the second half of the twentieth century and its current features.

Islamic scholar Bassam Tibi writes, "The wholeness of a given society is the creation of the intimate intertwining of the economics, culture, and politics; and any attempt to divide it schematically into a socioeconomic base and a political or ideological-superstructure, as is the desire of most Marxian writers, is a violation of its natural integrity". This concept raises perhaps the most fundamental issue when attempting to understand Islamic political ideology. To understand the success of political Islam in the last half-century one must not only examine socioeconomic conditions but also look at the cultural situation in which the Islamic world currently finds itself (Tim Meyer).

Understanding the essence of the relationship between Islam and politics requires the testimony of the origin of this religion. Because, first of all, Islam emerged as a political religion of Arab tribes, with the ideology of unification and formation of small states and cities into a single Arab state. Secondly, the method of historicism, the historical approach to the issue and the explanation of its genesis in the cognition of social events, is one of the basic methodological principles in social science.

As we have already mentioned above in the preceding sub-chapter, Islam is a policymaker world religion (or influenced by it) across many countries in North Africa and Asia. And it has followers in many countries around the world and Russia is not an exception. While writing my Ph.D. dissertation I found out that, "There are approximately 1 billion 126 million believers of Islam in the world, which is approximately 19% of the world's population. In Russia, there are more than 20 million Muslims. In modern Russia there are about 3,500 mosques; about 30 spiritual administrations operate, including the Central Spiritual Administration of Muslims (TsDUM) of Russia. Many Islamic universities and madras's are functioning and not less important issue is that numerous Islamic parties are operating in Russia" (Alaverdov, 2013).

History of Islam in Russia

As Akhmetova writes, "The history of Islam in Russia provides a unique story of co-existence of two great religions, Islam and Christianity, for centuries long, their journey of learning to live side by side in peace and harmony. This story is not merely one of conflict, conquest, and resistance. Rather, relations between the Christian Russians and Muslims, both within Russia and between Russia and other Muslim countries, have also marked by periods of co-existence, tolerance, accommodation, and even cooperation. The period of the Russian Empress Catherine the Great's reign (r.1762-1796) provides an early and good example of such mutual accommodation and cooperation"(Akhmetova, 2013).

Roman Silantiev, Russian scholar of Islamic religion in his book, "The Newest History of Islam in Russia" states that it is very difficult to name the exact date of the spread of Islam in Russia, as various sources give different information: for example, some scholars confirm that the history of Islam began in Russia in 642, when the first Muslim invasions started in Dagestan; and others claim that the spread of Islam in Russia began in Volga Bulgaria basin in 922 when the population start to convert to Islam. He also notices that he is not sure about both of the dates since he thinks that relations with the civilization of Islam date back to the 15th century. And divides the history of Islamic society in Russia into three phases: 1 - Before the Soviet Union; 2 - Soviet period; and 3 - the post-Soviet period.

The first the period as it is called as the period of Muslims "golden age" in Russia, it starts from the period of Catherine II reign when she issued several more decrees that certainly made life easier for Muslims and subsequently, the Muslim clergy began to enjoy even greater support from the authorities. The policy of Catherine II concerning Islam was continued by her son Emperor Paul the first. In 1797 he ordered to print 3,600 copies of Alcoran (the Koran) in the treasury of Asian Printing House and ordered to send them for sale to Muslim populated provinces. It is noteworthy that the Korans published in Russia were famous for their quality far beyond its borders and received high reviews from European orientalists. Since 1800, the publication of Muslim religious literature was not restricted at all. By the late XIX century, the Muslim clergy was finally integrated into the system of government. Its representatives had a very good salary, were free for the Hajj (in the first half of the XIX century, the number of pilgrims annually amounted to 8-10 thousand people), built mosques, found madrassas and publishing houses. But the Spiritual Directorates were required strictly monitor the moral character of the clergy, as well as guarantee their loyalty to the Emperor. The end of XIX and the early 20th century is often called the "golden age" of Russian Islam, which at that time became a highly respected and influential community throughout the Islamic world. It should be noted that the first mosque in Moscow, now called Historic, was built to commemorate feats of the Tatar and Bashkir regiments, valiantly fighting in the Patriotic War of 1812.

Also Akhmetova, in her publication, "Islam in Russia: Historical Facts and Modern Developments" mentions that the restrictions placed on Muslim trade and entrepreneurship were partly lifted, the cultural life of Muslims received a boost from the proceeds of international trade and manufacturing. New mosques were built and religious schools were opened in big cities such as Kazan, Ufa, and Orenburg. The emergence of such relative religious, intellectual and economic freedom due to the imperial reforms of the late eighteenth and also in the nineteenth centuries initiated an Islamic renewal and reformist movement among Russia's Muslims. The Muslim intellectuals of Tatar origin such as Abd alNasr al-Qursawi (1776-1812), Shihab al-Din Marjani (1818-1889), Abd al-Qayyum Nasiri (1824-1902), Musa JarullahBigiyev (1875-1949) and many others advocated the creative and flexible potential of Islam and its compatibility with modern progress. At the beginning of the twentieth century, this religious-educational movement evolved into a broader socio-political and cultural phenomenon which was also referred to as Judaism (Akhmetova, 2013). The policy of the Russian Empire towards Islam was formulated from the decree of the State Council in 1887. "Muslims are free in worship, provided that it will not be to harm the Orthodox Church ", it was said. We have to highlight that during the reign of Emperor Catherine II, mosques were built in other cities as well: 1768 a two-storied mosque was built in Tatarstan, then in the city of Yatsyk (present-day Uralsk in Kazakhstan) and the new Khanski Mosque in Kasimov. During the reign of Ekaterina II, the number of Muslims rose sharply, resulting from the accession of the Crimean Khanate. In 1783, Catherine II took the responsibility to protect temples, customs, and faith itself (Silantyev, 2007). The attitude of Catherine II resulted from the establishment of a Muslim national community in Russia as an organized national-political force that strengthened Islamic integrative functions in the country's political life, but despite this approach, Muslims had to deal with many difficulties.

However, this religious freedom came to the end soon after the Bolshevik Revolution in 1917. Tekusheva and Shevchenko in their publication "Islam in the North Caucasus: History and Present" wrote that "Antagonistic policies of the Communist regime disrupted the traditional religious, social, economic and political foundations of the Muslim society". But we can say that the Russian Muslims activities became more intensified. In May of that year, Moscow has held All-Russian Muslim Congress, which was attended by 900 delegates from all over Russia. At this congress, the first time in the history of the Russian Muslim community was elected Mufti - one of the leaders of Russian Judaism (education) - GalindjanBaharuddin. Also, at this congress was established a clerical council of Russian Muslims, which was later transformed into a Muslim clerical council of the European part of the USSR and Siberia during the Soviet period.

Independent states on Russian territory existed for over 1,000 years from the Volgograd-Bulgaria to the end of the Caucasus War. The era of Islam and Islamic state relations flourished during the period of the existence of three powerful Muslim states - Bulgaria, the Volga, the Golden Horde, and the founding of the Kazan Khanate (X-XVI centuries).

As for the North Caucasus, Islam originated here during the Arab conquest. In this region, Islam was first adopted by the people of Dagestan. Although the Arabs in Dagestan could not stand their ground and take roots, Islam spread here quite quickly and then in any other region (Tekusheva & Shevchenko, 2011). The spread of Islam in the North Caucasus is linked to the military actions of the Ottoman Empire. It is noteworthy that the Russian authorities prevented the establishment of an independent North Caucasian Muslim clerical body in this region (Alov & Vladimirov, 2001). In the mid-1920s, the government suppressed Islamic religious life and culture. By 1923, the Muslim higher education system was no longer exists, but we can say that the anti-religious attitude had subsided during World War II (1941-1945) and several mosques were reopened.

Islamic Renaissance in Russia

The number of Muslims in Russia in the 21st century is about 15-35 million In 2001, scientists from the Russian Academy of Civil Service published information indicating that the number of people following the Muslim rules and traditions - 15 million(Makarenko, 2000). Experts on Russia's demographic development M. Sharpe predicts, "Muslims will be in majority in Russia in the second half of the twenty-first century. They observe the decline in the ethnic Russian population and the rapid growth among the country's Muslim population. A significant gap exists between the birthrates of ethnic Russians and Russia's predominantly Muslim ethnic groups: 1.7 births per 100 women for ethnic Russians annually, which is below the replacement rate, and 4.5 births per 100 women for Muslim ethnic groups. Besides, Muslim immigration from Central Asia and the South Caucasus is increasing inexorably due to a higher demand for the labor force in Russia" (Sharpe, 2004). This is just the known number of people living in Russia, however, the number of Muslims in the Russian Federation is higher than it is mentioned, which is due to a large number of legal and illegal immigrants.

The Islamic uprising, which began in the early 1990s, has opened the eyes of Muslims living in Russia that they belong to the Muslim world, helping them overcome the "younger brother" complex. This complex was first established in "Tsarist" Russia and later in the Soviet period. The main reasons for the Islamic Renaissance can be considered:

- Religious awareness and pride in the fact that they are Tatars, Bashkirs, car accidents, Chechens, etc. That they belong not to the ethnic minority group but the world civilization;
- The revival of Islamic customs and the increase in the number of mosque constructions (by 2000 their number had reached 4,658 (Mukhametshin & Dubkov, 2001). For Muslims, the mosque means much more than a religious structure; it is also a place of political gathering for them;
- Establishment of a religious legal system that consolidated the knowledge and traditions of Islam. (For example, in 1999 more than 100 Muslim secondary and tertiary institutions were registered in Russia);
- Restoration of the Islamic clerical elite;
- The disappearance of the "Iron Curtain" that allowed the Muslims of the Soviet Union to connect with their fellow believers abroad;
- Finally, the politicization of Islam (Gainutdin, 1997).

These last points and mainly the events in Chechnya have made Islam a popular subject in Russian media.

Islam's behavior is predominantly a part of Russian politics ahead of several factors:

First: the establishment of political organizations based on the values of Islam;
Second: involvement of Muslims in politics;
Third: the transition of secular political leaders to Islam;
Fourth: membership of Muslims in political opposition parties
And finally, fifth: the behavior of Islam's foreign policy factor.

But we have to admit that the Islamic factor did not play a major role and was not a key point in Russian-Chechen both 1994 and 1999 wars. Yet, the dreadful and appalling consequences of these wars, together with the 9/11 attacks in the United States, played a crucial and fundamental role that described not only Russian but all Muslims in a very negative way and instantly created an evil image for them and the whole religion. The Russian mass-media and cinema industry also adopted a hostile attitude towards Islam and Muslims which fueled hostility.

The Impact of Religious Tourism on Easing Social Tensions in Society (on the Example of the Russian Federation)

The country's way of life is affected by the spiritual and moral state of society no less than the economy, the material sphere and the level of welfare of the population. The realization of freedoms and human rights, the affirmation of individuality and collective principles, the uniqueness of historical traditions, the stability of moral principles, ethics, personal and social behavior are inextricably linked with it. Today, Russian society is experiencing a spiritual and moral crisis. Economic turmoil, political upheaval, national contradictions, the uncertainty of life principles, the loss of positive ideals, complex environmental problems - all this deprives people of emotional balance, creates lack of spirituality, contributes to the destruction of life guidelines. The contradictions are aggravating, leading to acute social and labor conflicts, to clashes on ethnic and religious grounds, people have a feeling of fear and hopelessness, and confidence in the future is being lost.

The state of the social sphere, which is directly related to the distribution of material and spiritual wealth, the satisfaction of a person's specific needs, quality and level of life, working, living and leisure conditions, in some regions, leads to an increase in social tension. The formation of an alternative society concerning negative social experience is the goal of the development of religious tourism. While discussing religious tourism, we have to highlight that the interests of religious organizations and the state have to be combined by reducing social tension in the country and that the main point is the social factors of the nation's health and their spiritual feeling and situation. Such an approach to religious tourism will require project development to obtain in the future a streamlined mechanism for the peaceful interaction of groups of Russian society of different interests and worldviews, first of all, this should involve the young generation.

From spiritual and patriotic education, religious tourism doesn't apply only to youth; religious tourism is relatively generally accessible and has a great emotional impact on a person of any age. People have the desire to visit holy places for various reasons and motives: to find a system of values, to establish themselves on the spiritual path, to be aware of the cultural heritage of a certain state. Many after such travels return transfigured, resolving some life's contradictions, rested and in good condition. Often it is during the trip that they reassess every day, life situations, behavior, and thoughts. All these have positive effects and bring the relaxation of social tension in society. That is why the ideological component of religious tourism is the mitigation of general social tension, and not the division of society into hostile parties; establishing dialogue and peace between faiths: religious tourism cannot cause clashes or resentment on religious and social grounds.

Regulations of Religious Tourism in Russia

Religious tourism is a type of educational tourism; therefore, when organizing such trips, one should be guided by the legislative acts which are in force in Russia and determining the organization of such trips. The main legal acts related to religious tourism include the following documents such as, "Constitution of the Russian Federation", the "Civil Code of the Russian Federation", and federal laws "On Freedom of Conscience and Religious Organizations" and"Based on Tourist Activity".

For example, "The Constitution of the Russian Federation establishes the separation of all religious associations from the state and their equality before the law. Citizens have equal rights and freedoms, beliefs, any restrictions on the rights of citizens on the grounds of religious affiliation are prohibited. Everyone is guaranteed freedom of conscience, freedom of religion, including the right to profess individually or jointly with others any religion or not to profess any, freely choose, have and spread religious and other beliefs, act following them".

For instance, "The Federal Law "On Freedom of Conscience and Religious Associations" (September 26, 1997, N 125-ФЗ) determines that following the constitutional principle, a religious association is created and carries out its activities following its own hierarchical and institutional structure, selects appoints and replaces its personnel under its regulations. Religious organizations act under their internal regulations, if they do not contradict the legislation of the Russian Federation, and have the legal capacity provided for in their charters" (The Federal Law "On Freedom of Conscience and Religious Associations, 1997).

Religious organizations may own buildings, land, industrial, social, charitable, cultural, educational and other purposes, religious objects, money and other property necessary to ensure their activities, including those related to historical and cultural monuments. This point should be taken into account by representatives of the tourism business who organize trips for religious purposes. Cooperation with such organizations should be carried out on a contractual basis, taking into account for example "The provisions of the "Civil Code of the Russian Federation" and the federal law "On Non-Profit Organizations".

Besides, representatives of the tourism business using cultural heritage sites as tourism resources should not allow conflicts with members of the religious communities under whose jurisdiction they are located. They should inform tourists about the need, when visiting local shrines, temples, monasteries, to observe the norms of behavior established by religious communities, to be properly dressed, to show respect and not to interfere with the performance of religious rites and services. As you can see, at the federal level, the issues of relations between travel agencies and the peculiarities of using religious objects in tourism are reflected. Regulatory acts regulating activities in the field of religious tourism are also adopted in the constituent entities of the Russian Federation.

From the point of development of religious tourism, the Concept of solving the complex task of "Creating the conditions for the development of spirituality, high culture and moral health of the population of the Novosibirsk Region", approved by order of the Governor of the Novosibirsk Region (dated 10.06.2009 No. 150-r), is of interest. The strategic goal of the concept is to ensure wide public participation in the formation and implementation of state cultural policy. The activities of state authorities in implementing the principles of this policy are based on the priority of value-based approaches in the development of social development programs. It is connected, first of all, with the formation of stable spiritual and moral guidelines among the population, the most important of which are the processes of aesthetic, moral and spiritual education of the younger generation, as well as the creation of conditions for fruitful dialogues between secular and religious culture.

Vinokurova and Dombaeva in their paper, "Sacred Sites of the Sakha People: Custom and Law" state that "The legal and social problem is the legal maintenance of the safety of sacred places since the very concept of "sacred places" is not adopted in Russian law. The sacred places of indigenous peoples are not directly mentioned in the list of monuments of history and culture, but within the meaning of some of them fall under the monuments of archeology and the construction of religious architecture". Also, they stress that the UN Declaration on the Rights of Indigenous Peoples of the World in 2007 is the starting point for the development of legal support for the protection of sacred places and cultural heritage sites. The law related to sacred places, aimed at strengthening the respect and affection of peoples for the cultural, cultural-natural and historical-cultural heritage of their ancestors, requires a legislative settlement(Vinokurova& Dombaeva, 2014). This will provide an opportunity for the further development of religious tourism.

The Impact of Tourism on the Country's Economy

As we have already mentioned, tourism as a sphere of economic activity, it is a very important issue and has many different characteristic features. This industry draws the attention of humans and is considered as a source of certain income, both at the micro and macroeconomic levels. One of the main factors in creating additional jobs in the tourism industry, it accelerates the construction of the road and development of hotels, also it stimulates the production of all types of vehicles, preserve the national culture of regions and countries and their folk crafts. Taigibova in her presentation about "Actual Issues of Economic Science" says that, "According to the forecasts of the World Tourism Organization, the number of tourist arrivals by 2020 will amount to 1.6 billion people, world income from tourism in 2020 will increase to 2 trillion dollars".

In her presentation, Taigibova stresses that "Tourism is the fundamental basis of the economies of many developed and developing countries of the world. The basis of the modern tourism market, both qualitatively and quantitatively, is paid vacation for employees"(Taigibova, 2011). Recently, the role of business travel, as well as travel of people of retirement age, has been increasing in tourism. The size and degree of influence of international tourism in the world can be estimated by the following indicators. According to many analytical analysts, the following factors-possibilities lie at the heart of the development of international tourism: economic growth and social progress, the improvement of all types of transport made travel cheaper and made possible for many segments of the population, an increase in the number of wage workers and employees in economically developed countries and an increase in their material and cultural level have increased their desire for spiritual and cognitive values.

Economic opportunities for the development of tourism on an international scale created favorable conditions for the development of organizational and economic processes in individual countries. Tourism has an impact on the economy in almost all aspects of the fundamental definition of this area of society (Taigibova, 2011). In economic terms, tourism is considered: as a specific set of social relations in the field of production, exchange, and distribution of products; a part of the national economic complex of a given country, including certain sectoral types of production and economic activity; an economic science that studies tourism as a branch of the economy of a country or region (tourism economy).

Even though the tourist potential grows in the modern world, the Russian Federation currently leaves much to be desired and takes not a significant place in the global tourism market: its share is about 1.4% of the global tourist flow (Skorokhodov,2007.) Some experts estimate that the potential of Russia allows with an appropriate level of development of tourism infrastructure, to receive up to 40 million foreign tourists a year(Dracheva, 2010). Although nowadays, the number of foreigners arriving in Russia for holiday, business and private purposes is 7, 4 million people, which does not fit its tourism potential. However, the inbound tourism in Russia has enormous potential, despite the exogenous factors (environmental, political, environmental, etc.) that are not too close to the ideal country of the tourism business and tourism infrastructure and even more it hinders the growth of inbound holidaymakers flows.

Russia took 45th place of 130 in the global halal tourism rating According to the results of 2017, Russia dropped from 44th to 45th place in the Global Muslim Travel Index (GMTI), but experts predict that by the year 2020 the Russian Federation grow into the most popular non-Islamic country in the world for Muslim tourists. In the TOP-20 non-Islamic areas of the Russian Federation, it took 18th place, maintaining the position of 2017. In the ranking of 2018, Muslim tourists gave low assessments of the availability of places for prayers and general communications in the Russian Federation, and they rated the transport and visa availability of Russia on average (Tourism Associations, 2018). However, it is important to say that the World halal day Islamic business forum will be held for the first time in Russia.

Attracting inbound tourist flows from Muslim countries, in turn, is one of the effective ways to solve the problem of increasing the share of non-oil exports in the total export of the Russian Federation, creating conditions for increasing revenues of 53 sectors of the economy that are involved in the formation of tourism products, employment growth and self-employment improving local quality of life. Head of the Federal Tourism Agency, Oleg Safonov, in IV Kazan Tourism Forum 2017 said that "Muslim countries are one of the promising markets in terms of the development of inbound tourism in our country". He also stresses that the modern generation of Muslims is actively traveling around the world. According to expert forecasts, by 2020, the number of representatives of this religion going abroad for tourism will increase to 156 million people, and the Muslim tourism market will be $ 200 billion (Safonov, 2017).

According to the results of 2017, Russia dropped from 44th to 45th place in the Global Muslim Travel Index (GMTI), but experts predict that by 2020 the Russian Federation will become the most popular non-Islamic country in the world for Muslim tourists. Global Muslim Travel Index 2017survey results show that Muslim travelers are important to the global tourism industry. According to experts, the total costs of only young Muslim tourists by 2025 will exceed 100 billion US dollars, while the entire segment of Halal tourism in the world will reach 300 billion US dollars by 2026. The MMTR2017 study is the first comprehensive report reflecting an increase in the number of Muslim tourists who travel around the world. Its main goal is to assess the potential of the category of traveling young Muslims in the context of the development of tourist destinations, tour operators, airlines and the hospitality industry as a whole (Global Muslim Travel Index (GMTI), 2018).

CONCLUSION

The analysis of historical development, the state of the international and Russian market and the prospects for the development of religious tourism allows us to draw the following conclusions. Today, many trips can be attributed to trips with religious purposes - religious tourism. Currently, about 90% of the objects of tourist interest in the world are related to religion. The annual number of tourists and pilgrims in the world is 300-330 million; they make about 3–3.5 billion trips with pilgrimage or religious purposes. The ideological component of religious tourism is the mitigation of general social tension, and not the division of society into hostile parties; establishing dialogue and peace between faiths: religious tourism should not cause clashes or resentment on religious and social grounds. Religious tourism is one of the areas of work on the spiritual and patriotic education of the population, primarily young people, familiarizing people with moral values, spiritual and cultural-historical heritage of the Russian people. Religious tourism brings not only economic benefits to the local population, state structures and private enterprises (not only tourism), religious organizations - the main result is the creation of a climate of understanding between the participants of the trips and the local population to jointly overcome existing problems and tensions in society. The development of religious tourism necessitates the regulation of tourist flows taking into account the throughput potential of the territory since random and uncontrolled visits to places of interest and monuments do not allow for their preservation and lead to the discontent of believers and the emergence of socio-economic problems among the local population. Religious tourism will help to reduce social tension, a positive role in the dialogue between cultures, religions, and civilizations, if it is based on the close cooperation of tourist, state, religious organizations with the local population of the territories visited by tourists.

As for Islam itself, based on the above-mentioned facts we can claim that domestic affairs and relations of Russia and Muslims have not been fully sorted out yet. The post-Soviet society and elite are still learning how to settle internal relations. At present, there is only one thing to say: due to the country's geopolitical location, and largely due to its historical and cultural development, the Islamic factor plays a large role in Russian domestic politics, even to some extent in the foreign policy and of course in the sphere of tourism as well.

REFERENCES

Akhmetova, E. (2013). *Islam in Russia: Historical Facts and Modern Developments*. SIS Publications.

Alaverdov, E. (2013). *Islamic Factor in History and Modern Russia*. Georgian Technical University.

Alexandrova, Y. (2004). *International Tourism*. Aspect Press.

Alov, A., & Vladimirov, N. (2001). *Islam in Russia*. Moscow: Russian Research Institute of Cupid Heritage.

Arab News. (n.d.). Retrieved from https://www.arabnews.com/node/1538466/saudi-arabia

Babkin, A. (2008). *Special Types of Tourism*. Rostov-don.

Constitution of the Russian Federation. (1993).

Dallen, T. J., & Olsen, D. H. (2006). *Religion and Spiritual Journeys*. Routledge.

Dracheva, E. (2010). Economics and Tourism Organization. Moscow: Knorus.

FoundationsI. (n.d.). Retrieved from https://indigenousfoundations.arts.ubc.ca/un_declaration_on_the_rights_of_indigenous_peoples/

Gainutdin, R. (1997). *Islam Belief in Charity*. Moscow: Russian Muftis Council.

Global Muslim Travel Index (GMTI). (2018). Retrieved from https://www.crescentrating.com/reports/global-muslim-travel-index-2019.html

Jitenev, S. (2016). Religious Tourism and Pilgrimage in Russian federation: Current State and Prospects. *Heritage Institute Journal, 1*(4).

Jitenev, S. (2019). *Educational Tourism in Russia*. Moscow: Yurayt.

Jitenev, S., & Abakumov, S. (2010). *Current State of Religious Tourism and its Social Role in Russian and Abroad*. Moscow: International Tourism Academy.

Khristov, T. (2005). *Religious Tourism*. Moscow: Publishing Center Avademy.

Kurmanaliyeva, Rysbekova, Dussenbayeva, & Izmailov. (2014). Religious Tourism as a Sociocultural Phenomenon of the Present The Unique Sense Today is a Universal Value Tomorrow. This is the Way Religions are Created and Values are Made. *Pricedia -Social and Behavioral Sciences*.

Makarenko, V. (2000). Who are the allies of Russia? Moscow: The Mental and Geopolitics: Paradoxes of Russia's Security Policy.

Malashenko, A. (1998). *Islamic Rebirth in Present-Day Russia*. Moscow: Carnegie Moscow Center.

Mukhametshin, F., & Dubkov, A. (2001). The State, Religion, and Church in Russia and Abroad. Information-analytical bulletin, no. 2. Muslim Organizations in the Russian Federation.

Refworld. (n.d.). Retrieved from https://www.refworld.org/docid/4ae8611155.html

Russion Tourism Market Report. (n.d.). Retrieved from https://mtu.gov.hu/documents/prod/RMAA-Group--ndash--Russian-Tourism-Market-Report--Trends---Analysis-and-Statistics.pdf

Sadikov, M., & Khanbabayev, K. (2009). *Religious and Political Extremism (a textbook for high schools). Makhachkala* Institute of Theology.

Sharpe, M. (2004). *Islam in Russia: The Politics of Identity and Security. Moscow*. Moscow: Cambridge University Press. Cambridge.

Silantyev, R. (2007). *The Latest History of Islam in Russia*. Moscow: Algorithm.

Skorokhov, A. (2007). *International Tourism*. Moscow: Infa.

Standford. (n.d.). Retrieved from https://web.stanford.edu/group/sjir/3.2.07_meyer.html

Standford. (n.d.). Retrieved from https://web.stanford.edu/group/sjir/3.2.07_meyer.html

Statista. (n.d.). Retrieved from https://www.statista.com/statistics/617696/saudi-arabia-total-hajj-pilgrims/

Suvorova, P., & Ovsyannikov, V. (2014). Cultural Code in Poetic Text of the XX century. *World Applied Sciences Journal, 29*(4), 489-192.

Syukiyanen, L. (2015). *Islam and Human Rights (coursebook)*. Moscow: St. Petersburg State University.

Taigibova, T. (2011). Actual Issues of Economic Science. In *Materials of the International Scientific Conference* (pp. 125-128). Ufa.

Tekusheva, I., & Shevchenko, K. (2011). Islam in the North Caucasus: History and Present Medium Orient... Prague: Academic Press.

"The Federal Law "On Freedom of Conscience and Religious Associations". (1997, September 26 N 125-F3).

The National Center for Biotechnology Information. (n.d.). Retrieved from https://www.ncbi.nlm.nih.gov/pmc/articles/PMC3747831/

TourismR. (n.d.). Retrieved from https://www.russiatourism.ru/news/14078/

Ulyana, A., Vinokurova, & Dambaeva, L. (2014). Sacred Sites of the Sakha People: Custom and Law. *Anthropology & Archeology of Eurasia.*

World Tourism Organization Annual Report. (2016). World Tourism Organization.

Yakubov, M. (2011). *Constructive Potentials of Islam on the Example of the Muslim Revival in Russia at the Kazan*. Academic Press.

Yakunin, V. (2015). Specific features and classification of tourism in the domestic historiography. *Science and Tourism: Strategies and Interactions, 4*(2).

KEY TERMS AND DEFINITIONS

Business: Besides being one of the main occupations in a market economy it is also a source of economic and social development of society. There are different types of business also it can have non-profit activities, such as charitable, religious, scientific, educational, etc.

Islam: The youngest and second-largest religion in the world, the followers of Islam are called Muslims and believe in Allah. Muslims live according to the Islamic law called Sharia and a holy book called the Quran.

Knowledge: The result of formal and non-formal studies. Itis information and understanding about a subject or object. It applies to certain facts, information, and acquired through experience or education skills, and either the theoretical or practical understanding of a subject.

Pilgrimage: The journey of believers to worship geographic areas and relics which are sacred in this religion. A pilgrim who makes such a journey is called a pilgrim.

Politics: The activities of state authorities and their officials, as well as issues and events of public life related to the functioning of the state. Politics can be seen everywhere, even in everyday life.

Russian Federation (Russia): The legal successor of the Soviet Union was established in 1991, shortly after the disintegration of the Soviet Union. Modern Russiasituates in Eastern Europe and northern Asia have borders with the Arctic and Pacific oceans and on the Baltic and Black seas. Moscow is one of the largest cities in the country and its capital.

Tourism: Departures of visitors to another country or place other than the place of permanent residence for various purposes.

Chapter 9
Contribution of Social Capital to Innovation:
The Mediating Role of Knowledge Embedded in Social Networks

Mustapha Bengrich
University of Ibn Zohr, Agadir, Morocco

Amina Omrane
FSEG, Sfax, Tunisia & ECTSRA, IHEC, Carthage, Tunisia

Adil Azzahidi
University of Ibn Zohr, Agadir, Morocco

ABSTRACT

In today's uncertain environment, companies should develop solutions to meet the global competition trends and the rapidly changing customer demands. In fact, to gain a competitive advantage, organizations should be equipped with knowledge, which is considered as an essential resource, rather than with traditional resources, such as durable assets or material/technical resources. In other words, investing in intangible assets that are difficult to imitate, such as social capital, promotes dissemination and exchange of accessible knowledge used by organizations. Such knowledge management processes enable firms to enhance their capacity of innovation as well as their creativity skills. The objective is therefore to develop a literature review that helps enrich and in-depth understanding the dynamics of social capital within an industrial cluster (i.e., a common geographical area).

INTRODUCTION

A particular attention has been paid to the better understanding of mechanisms implemented in order to improve the survival and development of SMEs' networks, like technopoles, clusters, service-based spaces, innovative areas, industrial districts, and learning regions (Carluer, 2006). More specifically,

DOI: 10.4018/978-1-7998-3142-6.ch009

many researches are currently focused on industrial clusters and their respective role in the local development of countries (Karlsson et al, 2005). A cluster, recognized as a group of interconnected actors/organizations within a particular domain, enables companies to gain competitive advantages. It allows them to be more innovative and creative than any other enterprises established elsewhere (Baptista & Swann, 1998; Molina-Morales & Martinez-Fernandez, 2010). Moreover, the cluster is designed to be a place that enhances innovation among the stakeholders of a company, while taking advantage of the existing proximity between them (Boschma, 2005). However, it seems that such a proximity is not enough to explain their competitive advantage (Mitchell, Burgess & Waterhouse, 2010). It appears that social connectivity and interaction between actors are the most appropriate triggering factors of their innovation capacity (Doloreux & Parto, 2005). On the other hand, social networks within a cluster promote cooperation, as well as exchange of resources and knowledge; and more precisely transfer of tacit knowledge, which better contributes to innovation. Consequently, a group of skilled entrepreneurs enhances a collective capacity of innovation, greater than the sum of the individual capacities developed by each entrepreneur working lonely within the group.

In this perspective, a strong stream of studies emerged to develop a knowledge-based cluster theory (Maskell and Malmberg, 2007). In such researches, scholars have mainly focused on the functioning of social networks within clusters and their evolution in order to understand and explain how knowledge is created and organized. These social networks have been associated with knowledge transfer and dissemination structures within clusters (Boschma, 2005).

Such research annotation turned to networking and knowledge development in clusters is relatively recent, but increasingly inspirational. Indeed, new concerns consider a cluster as a dynamic system (Cooke, Uranga, and Etxebarria, 1997), which can be explained by two factors. First of all, the rise and convergence between theories is based on knowledge, acquisition, creation and dissemination of knowledge (Maskell et Malmberg, 2007). Then, the theory of social networks focuses mainly on the interactions between different actors of a cluster (Eklinder-Frick, Eriksson and Hallen, 2012). Moreover, these two theoretical fields merged and combined have the effect of associating the capacity for innovation within clusters with the social networks that compose these areas and the social capital disseminated by actors evolving there (Suire and Vicente, 2014).

Nevertheless, it is worth mentioning that researches on the dynamics of social networks and knowledge within clusters are still scarce and could be better developed.

Our objective is therefore to understand how social capital affects positively the companies' innovative capacity within a cluster, by fostering the exchange of knowledge embedded in networks. Such innovation capacity-based on knowledge transfer, is pursued as a source of success and sustainability. In this perspective, we will try to answer the following central question: "To what extent does social capital enhance enterprises' innovative capability within a cluster?"

In order to provide some elements of responses to this question, we propose a literature review, which enables, in the first part, to explore and enrich the notion of social capital (with its three dimensions) applied to the clustering process. Then, in the second part, the focus will be on the process of knowledge, its exchange and dissemination, as promoted by social networks. Such development framework will help us to examine and understand, in the third part, the major role of knowledge embedded in social networks in boosting the capacity for innovation within companies belonging to the cluster.

On the basis of the above theoretical statements, we propose the following theoretical framework.

Figure 1. Theoretical framework

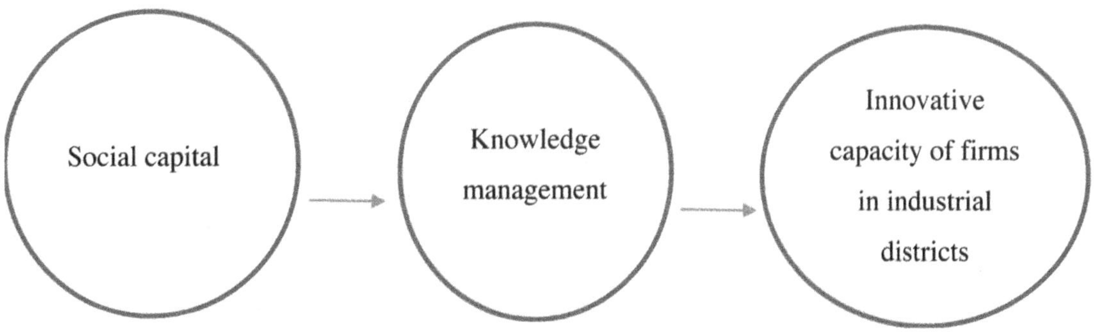

THE SOCIAL CAPITAL - CONCEPTUALIZATION AND DIMENSIONS

It is currently increasingly recognized that co-located companies sharing a common geographical space could better develop their capacity of innovation, as well as their wealth creation. In fact, according to Schmitz and Nadvi (1999), clustering can help to overcome the constraints related to scarcity of resources, and to gain competitive advantages in distant markets. Moreover, the emergence of clusters occur mostly in emerging industries known to have a large amount of tacit knowledge that requires proximity and interaction for its dissemination. For this reason, studies have largely focused on the importance of social networks and social capital, which are regarded as promising conductors to innovation, permanent evolution, adaptation to the environment and sustainability characterizing a cluster involvement.

The Concept of Social Capital Applied to the Clustering Process

Specialization of economic activities and their agglomeration in a common space were mentioned by Marshal (1919) under the name of 'industrial districts', which emerged after the disappearance of the medieval guilds. Moreover, the changes in innovative activities have prompted researchers to dig deeper into the cluster concept. For instance, Porter (1998) referred to the work of Schumpeter (1939) in order to explore this concept by introducing the notion of "industrial cluster" in the field of regional studies.

In fact, the concept of clusters emerged in 1980, firstly in the work of Porter who emphasized the importance of taking into account the environmental influence on companies' industrial activities. In this orientation, Porter (1998) defined an industrial district as *"a geographically close group of companies connected together and associated institutions in a particular field" (Porter, 1998, p. 199).* Besides, he attempted to enumerate the advantages of the spatial concentration of an industry and the knowledge diffusion it produces, for a nation/region. Since the early 1990s, a company's performance has been studied not only in terms of efficiency, but also regarding the environmental factors it can imply, such as the intensity of interactions with other companies, the geographical proximity, or the specialization of activities within a sector (Porter, 1998 ; Torre, 2006).

The authors cited earlier underlined also the role of non-market, geographically close, and organized inter-personal relationships in the coordination between actors/organisms within a cluster. The success of such a cluster is finally dependent of that of companies' involved within it.

Moreover, the geographical proximity is not sufficient in itself for explaining the development of relationships and collaboration between companies. For this reason, "social capital", as explored by Bourdieu (1980), held by actors and entities belonging to these productive configurations would be a major resource for the constitution and institutionalization of a cluster (Feldman, 2002).

Regarded as a fundamental concept in social sciences, social capital has been examined in sociology, as well as in economics and management. However, many researchers in socio-economics insist on the importance of integrating economic activities in social relations (Granovetter, 1973). These activities are at least partly linked to the committed individuals and social structures that support them. Moreover, social capital is used to identify the relationship between an individual's social assets and his integration into the economic world (Kostova & Roth, 2003). Recently, Sozbilir (2018) proved also the existence of a positive effet of social capital on creativity and organizational efficiency.

In the same perspective, Bevort and Lallement (2006) argued that two orientations of research could be distinguished in this context: the social capital as an individual resource, and social networks and their mode of constitution and governance.

Among advocates of the first approach, Bourdieu underlined a lack of economic capital in the creation of wealth, by calling upon an "extra-economic" capital. He added that social capital is associated with *"all the current or potential resources that are linked to the possession of a sustainable network of more or less institutionalized relationships of inter- and inter-recognition"* (Bourdieu, 1980, p.2). It is so a collection of assets and links (Lee and Tsang, 2001). In this sense, social capital, constituted by an individual's network of social relations, depends on the number of persons and the volume of different kinds of capital they hold, such as the economic and the cultural capital that these individuals can reach and mobilize for their own interests (Bourdieu, 1995).

Then, Bourdieu's concept, in turn, triggered Coleman's (1988) conception of other complementary shapes of capital: physical and social. Indeed, this author argued that social capital is considered as a resource that should be reached first. He assimilates it to an inalienable and unchanging "public good" that could not be the property of any particular actor; but resides in the structure of established social relationships that enable individuals to achieve common objectives. However, it is viewed as a complex resource, not easily controllable, as it requires individual strategies with a shared set of knowledge, standards and rules.

Regarding the second approach proponents, the attention should be paid to the framework related to the benefits of relationships between individuals, pursued as a collective entity emerging from building a network for its members. Therefore, such social ties enable these members to develop information exchange and knowledge creation, leading to gain a comparative advantage in the corresponding market. More specifically, network integration promotes trust in relationships, reduces transaction costs, and increases information asymmetry for these members (Burt, 1992, 2000).

On the other hand, Putnam (1995) argued that social capital refers to more or less institutionalized relationships and networks, as well as resources that flow from them. Besides, interpersonal relationships are at the origin of cooperative behaviors by forming "relational assets" for actors who are concerned (Storper, 1997).

With respect to recent research on local productive systems, some economists supplemented the various concepts explored above by drawing on the resources' theory. In consonance with Gumuchian and Pecqueur (2007), there is a distinction between "virtual" resources (composed of social capital, networks of relationships, etc.) and their "proven" productive configuration, such as financial resources or patent portfolio. For Coleman, the first category stimulates the access to the second.

In the same context, Lee (2018) studied the positive influence of enterprises' settlement and integration within a cluster on their growth and prosperity. More precisely, he pointed out that attractiveness of resources, as well as knowledge development, become enlarged whenever their networking within the cluster is highly expanded.

A consideration of all these research approaches pointed out commonly that social capital, networks of social relations, and communities of practice are all conductive to the development of economic activities.

In recent times, entrepreneurial acts, such as the creation of industrial clusters, have emerged as one of the practical developments of the theoretical approaches listed above (Aldrich & Zimmer, 1985). In this context, many researchers focused on industrial clusters and their respective role in the local development of countries (Karlsson et al, 2005); while some others have mainly put the emphasis on the networks of individual relationships developed into the cluster, and stressed the correlation between the intensity of links and "clustering" (Walker, Kogut and Shan, 1997). Other scholars have mainly studied the networks' overall effects (Westlund & Nilsson, 2005). Therefore, social capital is pursued as a particular resource, which enables access to other resources or development of existing ones (Porter, 1998). It is then an explanatory variable of performance.

The Multi-Dimensionality of Social Capital

In recent years, social capital has been largely reflected in management science due to its involvement in different management research situations. At the organizational level, most researchers consider that this concept evolved in terms of the nature of ties it encompasses, their number, as well as their quality. For this reason, multitudes of writings have dealt with it according to various aspects while taking into account several different perspectives. In addition, the literature review related to this concept emphasized its three most commonly tackled dimensions in order to measure its richness (Burt, 1998, 2000). However, despite its development over time and its importance for companies, as well as for economic and social actors, this concept remains unstable and mixed according to many scholars (Baret & Soto Maciel, 2004).

For their part, Alder and Kwon (2002, p. 17) considered social capital as *"the goodwill generated by the creation of social relationships and that can be mobilized to facilitate action"*, which can also be carried out both individually and collectively. For Nahapiet and Ghoshal (1998, p.243), it is *"the sum of the actual and potential resources included, available and derived from a network of relationships owned by an individual or social entity"*. Therefore, social capital includes both the network and the assets that can be mobilized by it. These authors operationalize social capital by referring to three essential and complementary dimensions, i.e. the structural, the cognitive and the relational dimensions, in order to analyze the role of social capital in relation to the intellectual capital. Moreover, the conceptualization of social capital remains widely discussed in the literature dealing with the problem of knowledge management.

In different circumstances, the structural social capital refers to the social interactions between the members of a community, encompassing the pattern and the strength of ties (Pearson, Car et Shaw, 2008). This dimension focuses on all the ways by which links within a social network are developed (who do you join and how do you join them?). It takes into account the manner by which the structure of a network affects the exchange of resources between its members, their place within the network and the type of relationships between them. Indicators related to this dimension consider the social network schema, its configuration, density, connectivity and hierarchy (Tichy, Tushman and Fombrun, 1979).

As reported by Granovetter (1985) and Uzzi (1996), the structure of a network can be studied through its density and the strength of the ties that compose it. In addition, other studies focus on the formal structure of relationships, which found the relational network, such as structural holes (Burt, 1992). This latter concept refers to "gaps" or places where the network is so limited, or where it is cohesive (Coleman, 1988). In keeping with Burt (2001)'s approach related to network cohesion and sparcity, Ahuja (2000) stated that direct and indirect ties as well as structural holes play various roles in the acquisition of new knowledge, additional resources, and different kinds of information, in order to achieve innovation outcomes. Another research conducted by Gubbins and Dooley (2014) argued that an optimal cohesive network structure promotes intensive non-redundant knowledge sharing.

The cognitive social capital represents the set of resources that improve mutual understanding between the different actors through shared representations, interpretations and systems of meaning between them (common languages and codes). Moreover, this dimension integrates the group's shared vision, its purpose, unique language, its deeply rooted stories and its culture (Pearson, Car and Shaw, 2008). According to Wasko and Faraj (2005), knowledge sharing within a group requires a common mutual understanding between the different members, such as a same culture and common objectives. In line with this same assumption, Nambisan and Sawhney (2011) pointed out that shared languages, narratives, vision, and common mutual understanding are predictors of innovation through knowledge creation, transfer and sharing in inter-organizational networks.

Finally, the relational social capital integrates the resources created and mobilized through social and interpersonal relationships, including trust, norms, obligations and identity. This relational dimension stresses the type of relationships that each person develops with others, such as respect, empathy, and friendship. It also describes the level of confidence developed between individuals within a group. In return, these norms, trust, obligations and identity improve the actors' awareness of their collective objectives. In this perspective, Levin et al (2016) stated that trust deriving from strong ties (family, fools, and close friends) is recognized as a necessary condition for effective knowledge exchange and sharing.

On the other hand, in the Granovetter's (1973) approach, the interpretation of relationships relied on the strength of ties existing between actors. On the one hand, long-term relationships characterized by trust and reciprocity are untitled strong ties. On the other hand, irregular and short-term relationships, which do not incorporate an emotional dimension based on a high frequency of interaction, are called weak ties. For this reason, organizations represent nodes within a network, with their economic and social links. By the way, such nodes facilitate cooperation and dissemination of resources between companies, thanks to norms and rules of reciprocity established by interactions over time (Uzzi, 1996). In this case, the performance of a company and its management will be dependent on its social networks, the characteristics of its existing links, as well as their pattern (Zaher, Gozubuyuk & Milanov, 2010).

Moreover, in the early 2000s, and since inter-organizational relationships took the form of clusters, social networks have attracted the attention of researchers in this perspective (Zaher et al., 2010). On the other hand, scholars in management sciences are becoming increasingly interested in economic sociology by exploring the social network approach. Indeed, social network can be defined as *"a completed set of actors and relationships defined between them"* (Wasserman and Faust, 1994, p.20). The relationships developed between these actors can be of a varied nature and would imply the exchange of information or friendship. Nevertheless, they remain social. In the same context, the literature on social capital emphasizes the privilege of creating intellectual capital within an organization, insofar as its three dimensions (structural, cognitive, and relational ones) enhance its ability to boost knowledge combination and exchange between individuals. It often underlines the importance of social capital in explaining the business performance (Nahapiet & Ghoshal, 1998).

In this perspective, it would be judicious to mention that social capital and network were considered, by several studies, as two closely related concepts, showing that both of them are complementary and contribute jointly to the same objectives (Montes & Pronovost, 2007). Moreover, researches also pointed out the crucial role of knowledge and its management as an organizational resource.

Indeed, knowledge is the intellectual capital of an organization, which widely contributes to gaining a competitive advantage. However, sharing knowledge among the members of teams/groups is essential for the success of a company in a highly competitive current environment. It often leads to high levels of performance, and subsequently improves productivity and innovation capacity within these groups/teams.

KNOWLEDGE EMBEDDED IN NETWORKS

Most of studies devoted to clusters have focused on the accumulation of knowledge in a limited geographical area. This is mainly explained by the evolution of this embedded knowledge and its maintenance in front of changes, so that it does not quickly become obsolete. In addition, according to Coleman (1988), social capital, assimilated to the product of the structure of social relations, is a tool serving for collecting and dispersing knowledge. However, it was argued that the evolution of a cluster is largely dependent on the knowledge embedded in its networks.

Definitions and Dimensions

Managing knowledge within an organization is often not considered as an obvious approach. Yet, while some companies have succeeded in developing a culture that is conducive to the creation and sharing of knowledge, other ones have experienced a resounding failure to meet this challenge. In their study on the failure to implement a knowledge management project within a consulting firm, Mansour and Gaha (2009) indicated that the fact of not taking into account the tacit dimension of knowledge could induce the organizational failure.

For some scholars, there is a confusion between knowledge and information. Therefore, these two concepts should be distinguished from each other. According to Dosi (1988), information has the character of a message flow while knowledge has the character of a subjective production that requires the selection, processing and interpretation of messages in order to produce other ones. In other words, knowledge is more than the sum of a set of accessible and transferable information.

More specifically, Nonaka and Takeuchi (1995) outlined the two dimensions of knowledge: explicit and tacit ones. Explicit knowledge is easily formalized and transferable to others, whereas implicit knowledge, also known as hidden and withheld, is difficult to externalize and to share with others. In his study on the conversion of existing knowledge into new knowledge in Japanese companies, Nonaka (1991) identified four different modes of interaction between tacit and explicit knowledge. He argued that tacit knowledge is at the heart of the knowledge creation process, which requires dynamic interactions between the different modes for their outsourcing and amplification. He also stressed the importance of the exchange between these two aspects of knowledge (tacit and explicit) through internalization and externalization. On the other hand, the broadening and acceleration of interactions between tacit and explicit knowledge depends on the involvement degree of organizational actors. The more they are interrelated each one to another, the more exchange of knowledge is facilitated.

Differently, according to some researchers (e.g. Tsoukas, 1996; Tsoukas & Vladimirou, 2001), tacit knowledge requires a good capacity of analysis, action and continuous training, rather than a simple, stable and finite knowledge. In fact, it is a "sixth sense" formed and transformed through conjunctures, interactions and exchanges, information gathering and daily experience. Moreover, thanks to information and communication technologies, tacit knowledge can never be left completely "frozen" (Nonaka and Konno, 1998). This hidden dimension of knowledge and expertise gives actors a distinctive potential for ingenuity and inventiveness. In line with this assumption, Correia, Mesquita and Paulos (2010) added that the implementation of new ways of thinking and behaving by creative virtual communities of practice leads to the generation of new knowledge creation processes by their members. Indeed, these communities become more innovative whenever their individuals broaden their knowledge base and share perspectives about their work practices with each other.

In the same orientation, Adama (2017) argued that a good knowledge management influences positively the capacity for innovation of enterprises, by stimulating their adaptation to techological evolutions, market fluctuations, and rapid commercialization of their products. These affirmations corroborate those advanced by Ologbo (2015), as well as Byukusenge et al (2016).

However, many researches have pointed out the importance of developing a culture that encourages the creation and dissemination of knowledge between stakeholders, leading to a best knowledge management. This fact explains how many companies fail due to their organizational culture, which does not stimulate openness and interaction. In this same perspective, Rajapathirana and Hui (2018) supported that managers should explore and develop the new ideas proposed by their employees, customers, suppliers, investors, or partners. They are also invited to exploit new sources of innovation, by an efficient top management and knowledge sharing.

Knowledge Dynamics Within the Organization

The speed of technological changes occurring in the current globalized environment encourages companies to minimize the time spent by them to launch and design new products/services in the market. For this reason, companies are always dependent on new knowledge they develop in order to invest, maintain, reproduce and adapt their technologies and expertise to the market trends. Indeed, with the globalization of exchanges, knowledge and know-how will no longer be acquired at once. However, the survival and competitiveness of companies are largely dependent on the comparative advantages they gain in acquiring, creating and sharing new knowledge. In this perspective, learning enables organizations to improve the skills and abilities that strengthen their potential for much production and innovation (Landry, Amara & Lamari, 2002).

On the other hand, the work of Lundvall (1995) showed that, in addition to the importance of knowledge as a productive resource within an enterprise, organizational learning is a key conductive element for innovation and performance. Moreover, many other studies noted that learning is an interactive and a collective process, which is often associated with its social context in its evaluation and stimulation. Admittedly, innovative companies are becoming learning organizations, which constantly create, exchange, transform and valorize new knowledge creation. According to Senge (1990), learning organizations use their learning abilities and new knowledge they grow up in order to improve their creativity and their innovativeness.

In the same context, we can cite the absorptive capacity-hereafter mentioned AC- as a concept, which is deeply rooted in the learning theory, as well as a crucial catalyst of the firms' innovative capacity and performance (Cohen and Levinthal, 1989, 1990, 1994). The AC refers to *"...the ability to recognize the value of new information, assimilate it, and apply it to commercial ends..."* (Cohen and Levinthal, 1990, p.128). It enables an organization to combine the acquired external information with the internal existing one, to convert them into knowledge, leading then to a better management and leadership.

More recent, many researchers have emphasized the AC dynamics, components, and its overall positive effects on the prosperity of the company. For instance, Lane, Koka & Pathak (2006) have underlined, in line with many previous studies, that the AC is assimilated to a three-step process. It includes: (1) exploration, which requires broad and generalized research for knowledge deployment, (2) transformation, which consists in developing organizational routines and processes, (3) analysis, interpretation, and understanding of external knowledge, and finally (4) exploitation, which is reflected into extensive research and use of combinatorial mechanisms between the existing and the newly acquired knowledge and their implementation to achieve the desired objectives.

Similarly, Lewin et al (2011), as well as Schweisfurth and Raasch (2018), have recently undermined the positive effects of AC on innovation development. They supported that organizations are called to upgrade the recognition, the valorisation, as well as the use and exploitation of external knowledge related to new technologies and emerging customers' needs and preferences.

Knowledge creation can also be served by the process of its sharing, its transfer and exchange. Concerning knowledge sharing it can be regarded as the learning process by which ideas are assimilated. Nevertheless, it may be an advantage or a disadvantage for the organization, as it risks to be externally imitated once it is codified and articulated (Winter, 1987). In addition, knowledge could be asymmetrically distributed within an organization as its exchange depends on the willingness of individuals to report their possession and to express their predisposition to share knowledge according to the needs of their organization (Nonaka, 1991). Hence, knowledge sharing requires the commitment of the actors who are concerned, as well as a direct contact between them.

The understanding of effects of knowledge embedded in networks is essential for its dynamics. However, it is not taken for its true value in the models existing in theory. We can therefore conclude that knowledge management is closer to innovation management through various processes, where the design and boundaries are not yet stabilized. In other words, detailing these mechanisms and dynamics can be illuminated with the establishment of a central concept in social sciences: the social capital. Moreover, the proximity between concepts related to knowledge and social capital is obvious. Nevertheless, our main objective is to demonstrate their contribution to the development of innovation capacities within organizations.

THE EXPECTED CONTRIBUTIONS OF KNOWLEDGE MANAGEMENT TO INNOVATION AS A COLLABORATIVE PROCESS

The conception of innovation adopted by companies has been widely used in the field of management, while underlining the firm's risk of innovation. However, Landry et al (2002) or Karlsson and Rodriguez (2015) researches confirmed that the success of innovation is henceforth ensured by a multitude of interactions maintained by an organization with its various stakeholders, rather than by traditional approaches based on a linear and simple model. Moreover, innovation is recently regarded as a systemic and open phenomenon, which develops the extension of the (inside) network approach, by including inter-organizational innovation networks (Chesbrough, 2003, 2006).

The Knowledge-Based Innovation: A Historical Background

In recent years, the innovation activity has often been linked to a discrete event resulting from knowledge developed in an isolated way by researchers and inventors.

Currently, innovation represents the result of a process based on various interactions, as well as on an exchange of knowledge between several independent actors (Du Plessis, 2007). This concept focuses more on social interactions in the innovation process and not only on the development of new technical solutions. Moreover, it is not only the combination of tangible forms of capital (physical or financial ones) that count in the innovation process; but rather the mixture of intangible forms of social capital that it encompasses.

On the other hand, innovation is generally illustrated by different sources of knowledge emerging from: science, market needs, links between stakeholders, technological networks and social networks. However, the innovation engineering theory, being the first to be built on knowledge, relies mainly on research results in order to improve products and manufacturing processes. Besides, basic research and industrial R&D are its main sources. In this case, innovation follows a linear sequence going from research to product commercialization and it combines tangible forms of capital (technological, physical, human and financial ones).

Besides, in the 1969s, the limits of engineering solutions pushed researchers to focus more on ideas coming from the market and considered as a new alternative vision for solutions. This phenomenon contributes to the emergence of the market innovation theory. In fact, in addition to taking into account factors related to research, this theory integrates organizational factors in order to achieve the development of innovation (Carter and Williams, 1957; Schmookler, 1966; Myers and Marquis, 1969). For this reason, it combines tangible forms of capital with market data appearing in an intangible form.

In the early 1980s, Mowery and Rosenberg (1978) suggested that more attention should be given to the links between research and markets through engineering, production, technological development, marketing and sales, in order to ensure successful innovation. Then, a few years later, Von Hippel (1980) focused on information generated by a company's customers and suppliers. This gave rise to the theory of innovation chain, including forms of tangible capital with those of intangible capital, such as data related to customers and suppliers and organized as information for innovators.

Subsequently, during the 1990s, a number of researchers (e.g. Dosi et al, 1988; Nelson, 1993; Lundvall, 1995) developed a concept under the name of "innovation systems", in which they assumed that innovation requires a set of links between various actors within a network of exchanges. However, in addition to technical criteria, market and network, innovation systems were focused mainly on the exchanges of information in terms of collaboration. They also supposed that interactions between the company and its various sources of information upgrade whenever its capacity to absorb such information is high. This finding gave rise to the technology network theory, which combines the different forms of tangible and intangible capital mentioned above with technology networks, as tools for acquiring and absorbing data transformed into information.

Later on, the role given to knowledge in promoting innovation witnessed an increasing importance. Innovation, relying on networks, retained previous ideas related to research (in the context of engineering theory) and interactions with actors (in relation to technological network theory). This strong interest in knowledge, as a factor of production and a determinant of innovation, can be explained by the continuous expansion of the amount of technical knowledge accumulated over time, as well as by the use of information and communication technologies (ICTs) that make knowledge available very quickly on a global scale.

However, innovation theories related to social networks emphasized the strategic importance of relational as well as technical tools and knowledge, rather than technological networks, as factors of production and innovation. The creation of competitive advantages is essentially found on relational tools and the way a business is done and projects conducted, since technical tools become easily accessible to others. In fact, the objective is to enhance the ability to transform information into knowledge.

Therefore, knowledge-based innovation requires not only one type of knowledge, but also several types held by different categories of actors (such as codified, tacit, or explicit knowledge). For this reason, business performance depends on the density and relevance of relationships, as well as on the cooperation between actors in the productive system or the innovation process (other companies, suppliers, financial institutions, research institutes, education and regional development agencies, etc.) through collaborative networks and clustering. Furthermore, innovation theory on social networks combines tangible and intangible forms of capital, which are characterized by interactions with a diverse group of actors.

Recently, with the advent of the knowledge economy, innovation is no longer conceived as a discrete phenomenon originating from isolated individuals, but rather as a problem-solving process, which takes place mainly within companies. In fact, this interactive process is based on formal and informal relationships that the company maintains with the various stakeholders in its environment, by getting involved in several networks (Kline & Rosenberg, 1986). In other words, innovation is associated with various types of learning, including learning through use, learning through practice, and learning through sharing (Rosenberg, 1982). However, this learning comes from two sources of knowledge: internal and external. The former refers to the AC of companies (Cohen & Levinthal, 1990; Lewin et al, 2011). Moreover, although the exchange of internal knowledge is essential for the company, it remains insufficient for innovation, which requires the exchange of internal and external knowledge.

Finally, we can conclude that innovation is an interactive process of learning and exchange, where interdependence between actors generates an innovative system (Niosi, 1993; Landry and Amara, 1998; Karlsson and Rodriguez, 2015). This latter could be a social innovation system, an innovative environment (Storper, 1997), a cluster, or a competitiveness cluster (Porter, 2000). Moreover, this new conception of innovation is characterized by the growing importance of social ingredients, while it was previously explained and interpreted merely by the forms of tangible capital.

Role of Social Capital in Knowledge Management

In recent decades, research has revealed the considerable role that social capital plays in creating and sharing knowledge within networks (Benedict et al, 2009), as well as its importance as a particular resource that helps exploit and develop other resources (Porter, 1998). In fact, it is an explanatory variable of performance and innovation within the company.

Generally, the characteristics and properties of social capital previously mentioned, such as shared history and experience, trust, group identity, norms and obligations, etc., are only relevant for a community that has been in existence for some time. Indeed, the dimensions of social capital become visible and meaningful when there are actors who know one another within a group or a network. This means that two individuals or entities must first get to know each other in terms of skills, characteristics and behaviors, before collaborating, exchanging trust, or interacting in a same group.

The literature developed in this orientation considers social capital within clusters or innovation networks in terms of two essential aspects. On the one hand, as a specific resource apparent or incorporated into a network of relationships. On the other hand, as a complementary resource, insofar as it enables access to other specific resources that are available in the network.

According to Coleman (1988); Alder and Kwon (2002), social capital has several specific properties compared to other types of capital. These properties encourage innovation within networks and clusters.

An analysis of the literature review enabled us to synthesize these different properties, as well as their contributions to innovation, in the table below.

Table 1. Properties of social capital and its contribution to innovation

Properties of social capital	Authors	Contributions to innovation
Appropriable: can be mobilized and used for various purposes.	Alder and Kwon (2002)	It is appropriable, convertible, substitutable and complementary to other resources for future reinvestment. The cyclical design of innovation makes it possible for resources generated by innovators to be reused for new actions and other innovations. This is a part of the systemic approach of innovation.
Convertible: the benefits it provides can be converted into other benefits and advantages.		
Substitutable or complementary: it can substitute or supplement other resources by compensating for other types of capital.		
Socio-degradable asset: it depreciates through abuse or non-use	Nahapiet and Ghoshal (1998)	The network must continue over time to generate benefits, and social capital must be mobilized so as not to depreciate, which implies the regular interaction of innovators and members of the network. Innovation is not done once and for all. It must always be renewed continuously. Social capital is a collective asset that requires at least the connection of two actors to produce it. This assumption joins the interactive model of innovation, which considers innovation as a collective action that evolves with increase of links' number. Innovation relies on the use of social capital as a resource for innovators. Social capital is based on the network of links. Innovation on its side is built on the creation of links.
Asset of a progressive nature: it develops according to its use. So, with many relationships, so much trust and so many resources.		
Collective good: it is never the exclusive property of an individual, it is owned by the community, its use will not reduce its availability to others; but we can deprive someone of resources if we exclude him from the network.	Nahapiet and Ghoshal (1998) Coleman (1988)	
Purely intangible asset: it is embedded in the network of social relations, unlike material capital, which resides in goods and human capital, which resides in skills.	Coleman (1988)	
Uncertain: not measurable, whose benefits are quantified. Neither its evolution, nor its foreseeable depreciation, can be measured. It is a function of contextual changes.	Burt (2000)	The context surrounding innovation is also uncertain because of uncertain and unmeasurable social capital.

Source: Authors

Scholars argued that the social capital, with its different dimensions, plays an important role in fostering the innovation capacity of firms, by reducing their transaction costs, including research and information, negotiation and decision costs, as well as control and implementation costs (Maskell, 2000). In fact, innovation dimensions of social capital provide a competitive advantage in terms of reduction of costs, production of reliable and relevant information, commitments' honoring, respect of agreements, and sharing of tacit knowledge.

Knack and Keefer (1997) advanced also that a high level of trust enables companies to reduce expenses related to economic transactions, eliminate disputes between partners and finally, minimize the risks and expenses related to protection and security. However, weak confidence can also discourage innovation if companies devote more time, investment and energy to monitor their partners and actors involved in the innovation process.

Following a series of interactions and exchanges between the actors of a network over time, trust develops and companies located in environments characterized by a high level of mutual confidence are more willing to innovate. In fact, standards of behavior, such as good reputation, reliability and loyalty are developing. Indeed, minimizing opportunistic behaviors contributes to exchange and productivity. Moreover, the sharing of the same culture and identity leads to the gathering and exchange of knowledge among cluster members (Lazega, 2008; Roth, 2008). In fact, the collaborative innovation networks improve the capacity for innovation by the mediating effect of the AC (Najafi-Tavani et al, 2018; Schweisfurth and Raasch, 2018).

Similarly, in the hospitality and tourism management, Kim Taegoo et al (2013) conducted an empirical investigation on 14 five-star hotels in Seoul Korea and confirmed the importance of social dynamics and capital derived from interpersonal and group relationships and their role in increasing employees' willingness to collect knowledge and share it with their colleagues. In line with this same assumption, other researchers undertaken by Grissemann et al (2013) underlined the role of knowldege management and sharing in generating innovation possibilities and new updated strategies. Liu (2018) for instance offered evidence that qualities such as capabilities, as well as social capital lead to effective knowledge transfer when cultural and creative firms have a higher level of organizational learning and AC (in the Taiwanese context).

Consequently, and after reviewing previous researches on the dimensions of social capital, we can conclude that structural social capital provides an opportunity for connection between network members while monitoring the intensity and structure of interactions. In fact, it promotes transversal and non-hierarchical communication that contributes to the development of clusters. On the other hand, relational social capital enhances knowledge, exchange within the network while developing a high degree of trust, which facilitates the action and coordination between the cluster members by supporting mutual commitment. Finally, cognitive social capital promotes proximity while sharing the same culture, objectives and vision within a network. It is then the catalyst that maintains relationships within a cluster and enables its development.

In this work, our objective was to find out how social capital within an innovation network, more particularly within an industrial cluster, intervenes at the level of knowledge management and what are its possible contributions to innovation. Moreover, on the basis of the empirical previous studies already mentioned, we can conclude that the three dimensions of social capital enhance the sharing and creation of knowledge, which is at the heart of network innovation. In fact, several hypothetical links should be taken into account by proposing a model that postulates the relationships between these concepts, including social capital, knowledge management and network innovation. In this context, a future empirical validation of the modelling of these hypothetical links would be relevant to provide further development prospects. This model can be presented as follows:

Figure 2. The proposed conceptual model

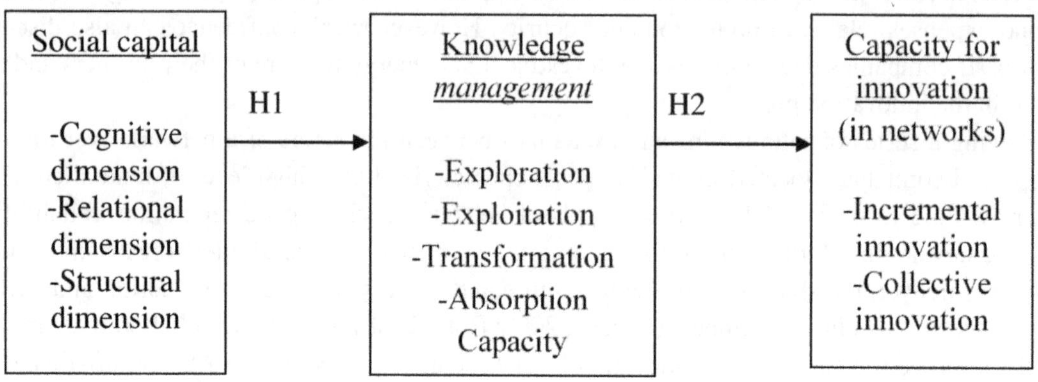

CONCLUSION

This current study is a follow-up to previous researches on the innovation capacity based on innovative based-knowledge networks. The exploration of the concept of social capital leads us to identify its role in the development of industrial clusters, as well as the intermediate role of knowledge creation and sharing.

The objective of our work was to underline the positive effects of the dimensions of social capital on the capacity for innovation within an industrial cluster, thanks to the mediating role of management knowledge. To do so, a literature review helped us to fairly reconcile the three complex concepts of social capital, knowledge management and innovation. The theoretical framework developed was then enlightening in order to shed some light on their characteristics and the interactions between them.

In a first part, we discussed the concept of social capital with its different dimensions, its application to the clustering process and its links with social networks and the improvement of innovation. Secondly, the focus was on the new trend towards a knowledge-based economy, the creation, sharing and dynamics of knowledge that enhance learning processes and absorption capacity within an organization. Finally, on the basis of several theories and prior research, the respective emphasis was put on knowledge-based innovation, the main role of the properties of social capital, and also its different dimensions within a network.

A synthetic analysis of research trends and orientations related to the main concepts of the present study and the correlation between them helped us to explore and reveal the importance of networking between individuals. Indeed, proximity and cooperation between actors promote their access to additional resources and foster creation and sharing of knowledge as well as the extension of existing relationships. All these elements are widely required in a current context of globalization and digitalization.

Accordingly, we can argue that the present study has many theoretical, managerial and institutional contributions.

Theoretical Contributions

First of all, the current research can be considered to enrich and in-depth understanding the social capital theory as well as the social exchange theory (Adler and Kwon, 2002; Donaldson & O'Toole, 2007), and the effectiveness of mutual commitments and trust between actors for enhancing the knowledge management into the organization.

Second, this study highlights the knowledge-based innovation approach and demonstrates how much the knowledge creation, updating, sharing, and exchanging increases innovation capabilities of the organization and encourage its actors to create new ideas and develop new products, services, and systems, by leveraging their absorption capacities (Mitchell et al, 2010; Najafi-Tavani et al, 2018 ; Sozbilir, 2018).

Managerial and Institutional Contributions

At this era of digitalization and global growing competitiveness, firms are conveyed to cope to the current knowldge-based economy. Therefore, it is essential that managers, as well as government responsibles, update their strategies and revise their operational systems, by integrating new applications and digital technologies in view of the international shifts. Moreover, they are called upon to cultivate an organizational culture that nurtures constructive interpersonal relationships based on networking, sharing, trust and exchange between actors (Eva and Maria, 2011 ; Tounkara and Arduin, 2014 ; Kucharska and Wildowicz-Giegiel, 2017). In this perspective, team buildings, meetings, as well as knowledge cafés, are considered as the best drivers of effective organizational cultures (Lefika and Mearns, 2015).

Limitations and Future Research Perspectives

In light of all the above observations resulting from this analysis, it is clear that the present study is basically theoretical. It is then recommended to undertake an empirical investigation and to collect appropriate data on the target field of industrial clusters (via questionnaires or interviews) in order to test the proposed conceptual model. This study will then gain in relevance and value, particularly in the Moroccan or the Tunisian context.

Another avenue of research can be suggested by exploring the linkages that may exist between innovative industrial clusters and business tourism. To do so, it would be interesting to refer to prior research conducted by Nordin (2003), Coeslikowski (2015) and Zach and Hill (2017). Such scholars stated that innovative industrial clusters contribute to the transformation of territories, regions, or countries into attractive and touristic destinations. In turn, theses shifts will improve the business tourism, and generate positive economic effects and success of these areas, while promoting jobs, service quality, cooperation processes and additional revenues for companies. Besides, many researchers, such as Dragicevic et al (2012), Hong Trinh (2016) as well as Gall and Strezo (2019), argued that the development of competitive clusters in touristic regions has a great influence on destinations' competitiveness, tourism value chain, as well as the economic growth and strategies of the regions.

REFERENCES

Adama, T. Y. M (2017). Facteurs clés de succès de la gestion des connaissances et capacités d'innovation des entreprises nigériennes. *La revue Gestion et Organisation, 9*(1), 11-24.

Adler, P. S., & Kwon, S. W. (2002). Social capital: Prospects for a new concept. *Academy of Management Review, 27*(1), 17–40. doi:10.5465/amr.2002.5922314

Ahuja, G. (2000). Collaboration Networks, Structural Holes, and Innovation: A Longitudinal Study. *Administrative Science Quarterly*, *45*(3), 425–455. doi:10.2307/2667105

Aldrich, H. E., & Zimmer, C. (1985). Entrepreneurship through social network. In D. L. Sexton & R. W. Smilor (Eds.), *The Art and Science of Entrepreneurship* (pp. 3–24). Balling Publishing Company.

Audretsch, D., & Feldman, M. (1995). R&D Spillovers and the Geography of Innovation and Production. *The American Economic Review*, *86*(3), 630–640.

Baptista, R., & Swann, P. (1998). Do firms in clusters innovate more? *Research Policy*, *27*(4), 525–540. doi:10.1016/S0048-7333(98)00065-1

Baret, C., & Maciel, S. O. (2004). *Apports et limites de mesure du capital social en recherche en gestion des ressources humaines*. Communication au XV ème Congrès AGRH, Montréal.

Batjargal, B. (2003). Social capital and entrepreneurial performance in Russia: A longitudinal study. *Organization Studies*, *24*(4), 535–556. doi:10.1177/0170840603024004002

Benedict, M., Valoggia, P., Rousseau, A., & Schmitt, C. (2009). Gestion des connaissances etcapital social: Quelles interrelations? *Management & Avenir*, *14*(27), 152–170. doi:10.3917/mav.027.0152

Bevort, A., & Lallement, M. (2006). Le capital social. Performance, équité et réciprocité. La Découverte.

Boschma, R. A. (2005). Proximity and Innovation: A critical Assessment. *Regional Studies*, *39*(1), 61–74. doi:10.1080/0034340052000320887

Bourdieu, P. (1980). Le capital social, notes provisoires. *Actes de la Recherche en Sciences Sociales*, *31*(31), 2–3.

Bourdieu, P. (1995). La misère des médias. *Télérama*, 8–12.

Burt, R. S. (1992). *Structural holes, the social structure of competition*. Cambridge: Harvard University Press.

Burt, R. S. (2000). The network structure of social capital. *Research in Organizational Behavior*, *22*, 345–423. doi:10.1016/S0191-3085(00)22009-1

Burt, R. S. (2001). Structural Holes Versus Network Closure as Social Capital. 31-56. In N. Lin, K. S. Cook, & R. S. Burt (Eds.), *Social Capital: Theory and Research*. New Brunswick, NJ: Transaction.

Byukusenge, E., Munene, J., & Orobia, L. (2016). Does knowledge management lead to innovation? An empirical study on SMEs in Rwanda. *International Journal of Management Science and Business Administration*, *2*(9), 7–19. doi:10.18775/ijmsba.1849-5664-5419.2014.29.1001

Carluer, F. (2006). Enterprises' networks and territorial dynamics: a strategic analysis » (Trad. « Réseaux d'entreprises et dynamiques territoriales: une analyse stratégique. *Geography Economy Society*, *8*(2), 193–214.

Carter, C., & Williams, B. (1957). *Industry and Technical Progress*. Oxford: Oxford Univ. Press.

Chesbrough, H. (2003). *Open Innovation: The new imperative for creating and profiting from technology*. Boston: Harvard Business School Press.

Cieslikowski, K. (2015). Business Tourism as Innovative Product For Destinations. *Conference proceedings, Innovative (Eco-) Technology, Entrepreneurship and Regional Development. Conference Proceedings.*

Cohen, W., & Levinthal, D. (1990). Absorptive capacity: A new perspective on learning and innovation. *Administrative Science Quarterly*, *35*(1), 128–152. doi:10.2307/2393553

Coleman, J. S. (1988). Social capital in the creation of human capital. In E. Lesser (Ed.), *Knowledge and social capital*. Chicago: University of Chicago Press.

Cooke, P., Uranga, M. G., & Etxebarriab, G. (1997). Regional innovation systems: Institutional and organizational dimensions. *Research Policy*, *26*(4-5), 475–491. doi:10.1016/S0048-7333(97)00025-5

Correia, A., Mesquita, A. & Paulos, A. (2010). Innovation through virtual communities of practice: motivation and constraints in the knowledge-creation process. *Projectics / Proyéctica / Projectique*, *4*(1), 39–49.

Czamanski, S., & Ablas, L. A. (1979). Identication of Idustrial clusters and Complexes: A Commarison of Methods and Findings. *Urban Studies (Edinburgh, Scotland)*, *16*(1), 61–80. doi:10.1080/713702464

Dibiaggio, L., & Ferrary, M. (2003). Communautés de pratiques et réseaux sociaux dans la dynamique de fonctionnement des clusters de hautes technologies. *Revue d'Economie Industrielle*, *103*(1), 111–130. doi:10.3406/rei.2003.3112

Donaldson, B., & O'Toole, T. (2007). *Strategic market relationships: From strategy to implementation.* John Wiley & Sons.

Dossi, G. (1988). Sources, procedures, and microeconomic effects of innovation. *Journal of Economic Literature*, *26*(3), 1120–1171.

Dragicevic, V., Jovicic, D., Belsic, I., Stankov, U., & Boskovic, D. (2012). Business tourism destination competitiveness: A case of vojvodina province (Serbia). *Economic Research-. Ekonomska Istrazivanja*, *25*(2), 311–332. doi:10.1080/1331677X.2012.11517510

Du Plessis, M. (2007). The role of knowledge management in innovation. *Journal of Knowledge Management*, *11*(4), 20–29. doi:10.1108/13673270710762684

Eklinder-Frick, J., Eriksson, L. T., & Hallen, L. (2012). Effects of social capital on processes in a regional strategic network. *Industrial Marketing Management*, *41*(5), 800–806. doi:10.1016/j.indmarman.2012.06.007

Eva, E., & Maria, E. (2011). Organizational culture As the driver of dense intra-orgnaizational networks. *Journal of Competitiveness*, *2*, 28–42.

Feldman, M., & Francis, J. (2002). Entrepreneurs and the formation of industrial clusters. 3rd day on proximity, Paris.

Gall, J., & Strezo, M. (2019). Quantitative Analysis of Environment Potential for Cluster Development in Tourist Regions of Slovak Republic. *Geographica Pannonica*, *23*(3), 195–203. doi:10.5937/gp23-21375

Granovetter, M. (1973). The strength of weak ties. *American Journal of Sociology*, *78*(6), 1360–1380. doi:10.1086/225469

Granovetter, M. (2005). The impact of social structure on economic outcomes. *The Journal of Economic Perspectives, 19*(1), 33–50. doi:10.1257/0895330053147958

Granvotter, M. (1985). Economics action and social structure: The problem of embeddedness. *American Journal of Sociology, 91*(3), 481–510. doi:10.1086/228311

Grissemann, U., Plank, A., & Brunner-Sperdin, A. (2013). Enhancing business performance of hotels: The role of innovation and customer orientation. *International Journal of Hospitality Management, 33*(1), 347–356. doi:10.1016/j.ijhm.2012.10.005

Gubbins, C., & Dooley, L. (2014). Exploring Social Network Dynamics Driving Knowledge Management for Innovation. *Journal of Management Inquiry, 23*(2), 162–185. doi:10.1177/1056492613499203

Gumuchian, H., & Pecqueur, B. (2007). *La ressource territoriale*. Paris, France: Économica Editions Anthropos.

Hong Trinh, T. (2016). The industry cluster approach for the development of central Vietnam. *International Journal of Business and Management, 11*(5), 167–178.

Karlsson, C., Johansson, B., & Stough, R. R. (2005). Industrial Clusters and Inter-firm Networks: An Introduction. In Industrial Clusters and Inter-Firm Networks. Edward Elgar.

Karlsson, C., & Rodriguez, A. (2015). *Knowledge sharing in an open innovation collaboration. A case study of the SEVS Project* (Master's Thesis). Chalmers University of Technology, Gothenburg, Sweden.

Kline, S., & Rosenberg, N. (1986). An overview of innovation. In R. Landau & N. Rosenberg (Eds.), *The Positive Sum Strategy. Harnessing Technology for Economic Growth* (pp. 275–306). Washington, DC: National Academy of Sciences.

Kostova, T., & Roth, K. (2003). Social Capital in Multinational Corporations and Micro-Macro Model of its Formation. *Academy of Management Review, 28*(2), 297–317. doi:10.5465/amr.2003.9416356

Kucharska, W., & Wildowicz-Giegiel, A. (2017). Company Culture, Knowledge Sharing and Organizational Performance: The Employee's Perspective. *Proceedings of the 18th European Conference on Knowledge Management, 1*, 524–531.

Lai, G., Lin, N., & Leung, S. Y. (1998). Network resources, contact resources, and status attainment. *Social Networks, 20*(2), 159–178. doi:10.1016/S0378-8733(97)00012-9

Landry, R., Amara, N., & Lamari, M. (2002). Does social capital determine innovation? To what extent? *Technological Forecasting and Social Change, 69*(7), 681–701. doi:10.1016/S0040-1625(01)00170-6

Lane, L. P., Koka, B., & Pathak, S. (2006). The reification of Absorptive Capacity: A critical review and rejuvenation of the Construct. *Academy of Management Review, 31*(4), 833–863. doi:10.5465/amr.2006.22527456

Lazega, E. (2008). Théorie de la coopération entre concurrents: Interdépendance, discipline sociale et processus sociaux. *Le Libellio d'Aegis, 4*(3), 1–5.

Lee, C. Y. (2018). Geographical clustering and firm growth: Differential growth performance among clustered firms. *Research Policy, 47*(6), 1173–1184. doi:10.1016/j.respol.2018.04.002

Lee, D. Y., & Tsang, E. W. (2001). The effects of entrepreneurial personality, background and network activities on venture growth. *Journal of Management Studies, 38*(4), 583–602. doi:10.1111/1467-6486.00250

Lefika, P. T., & Mearns, M. A. (2015). Adding knowledge cafés to the repertoire of knowledge sharing techniques. *International Journal of Information Management, 35*(1), 26–32. doi:10.1016/j.ijinfomgt.2014.09.005

Levin, D. Z., Walter, J., Appleyard, M. M., & Cross, R. (2016). Relational Enhancement: How the Relational Dimension of Social Capital Unlocks the Value of Network-Bridging Ties. *Group & Organization Management, 41*(4), 415–457. doi:10.1177/1059601115574429

Lewin, A. Y., Massini, C., & Peeters, S. (2011). Microfoundations of Internal and External Absorptive Capacity Routines. *Organization Science, 22*(1), 81–98. doi:10.1287/orsc.1100.0525

Liu, C. H. S. (2018). Examining social capital, organizational learning and knowledge transfer in cultural and creative industries of practice. *Tourism Management, 64*(February), 258–270. doi:10.1016/j.tourman.2017.09.001

Lundvall, G. A. (1993). Explaining interfirm cooperation-the limits of transaction cost approach. In G. Grabher (Ed.), *The Embedded Firm; On the Socioeconomics of Industrial Networks* (pp. 52–64). London: Routledge.

Lundvall, G. A. (1995). *National Systems of Innovation. Pinter Publishers.* London: Editions.

Mansour, N., & Gaha, G. (2009). L'échec d'un projet de gestion des connaissances: Cas d'une entreprise de conseil. *Management International, 13*(4), 91–101. doi:10.7202/038587ar

Marshall, A. (1919). *Industry and Trade: A Study of Industrial Technique and Business Organization and of Their Influences on the Condition of Various Classes and Nations.* London: MacMillian and Co. Limited.

Maskell, P. (2000). Social Capital, Innovation, and Competitiveness. In S. Baron, J. Field, & T. Schuller (Eds.), *Social Capital: Critical Perspectives* (pp. 111–123). Oxford: Oxford University Press.

Maskell, P., & Malmberg, A. (2007). Myopia, knowledge development and cluster evolution. *Journal of Economic Geography, 7*(5), 603–618. doi:10.1093/jeg/lbm020

Mitchell, R., Burgess, J., & Waterhouse, J. (2010). Proximity and knowledge sharing in clustered firms. *International Journal of Globalisation and Small Business, 4*(1), 5–24. doi:10.1504/IJGSB.2010.035328

Molina-Morales, F. X., & Martinez-Fernandez, M. T. (2010). Social networks: Effects of social capital on firm innovation. *Journal of Small Business Management, 48*(2), 258–279. doi:10.1111/j.1540-627X.2010.00294.x

Montes, J. C., & Pronovost, S. (2007). *La conversation sur les notions de capital social et de réseau: À la recherche d'une canne blanche. 16th AIMS Conference*, Montreal, Canada.

Mowery, D. C., & Rosenberg, N. (1979). The influence of market demand upon innovation: A critical review of some recent empirical studies. *Research Policy, 8*(2), 102–153. doi:10.1016/0048-7333(79)90019-2

Myers, S., & Marquis, D. G. (1996). *Successful Industrial Innovation*. Washington, DC: National Science Foundation.

Nahapiet, J., & Goshal, S. (1998). Social capital, intellectual capital and the organizational advantage. *Academy of Management Review*, *23*(2), 242–266. doi:10.5465/amr.1998.533225

Najafi-Tavani, S., Najafi-Tavani, Z., Naudé, P., Oghazi, P., & Zeynaloo, E. (2018). How Collaborative innovation networks affect new product performance: Product innovation capability, process innovation capability, and absorptive capacity. *Industrial Marketing Management*, *73*(25), 193–205. doi:10.1016/j.indmarman.2018.02.009

Nambisan, S., & Sawhney, M. (2011). Orchestration Processes in Network-Centric Innovation: Evidence from the Field. *The Academy of Management Perspectives*, *25*(3), 40–57.

Nelson, R. R. (Ed.). (1993). National Innovation Systems: A Comparative Analysis. Oxford University Press.

Nonaka, I. (1991). The knowledge-creating company. *Harvard Business Review*, *69*, 96–104.

Nonaka, I., & Konno, N. (1998). The concept of "Ba": Building a foundation for knowledge creation. *California Management Review*, *40*(3), 40–54. doi:10.2307/41165942

Nonaka, I., & Takeuchi, H. (1995). *The Knowledge Creating Company*. New York: Oxford University Press.

Nordin, S. (2003). *Tourism Clustering & Innovation*. European Tourism Research institute. [Ostersund, Sweden.]. *Mid-Sweden University.*, *SE-831*, 25.

Ologbo, A. C., & Noor, K. M. (2015). Knowledge management processes and firm innovation capability: A theoretical model. *Asian Social Science*, *11*(8), 10–17. doi:10.5539/ass.v11n18p10

Pearson, A. W., Carr, J. C., & Shaw, J. C. (2008). Toward a theory of familiness: A social capital perspective. *Entrepreneurship Theory and Practice*, *32*(6), 949–969. doi:10.1111/j.1540-6520.2008.00265.x

Pinch, S., Henry, N., Jenkins, M., & Tallman, S. (2003). From 'industrial districts' to 'knowledge clusters': A model of knowledge dissemination and competitive advantage in industrial agglomerations. *Journal of Economic Geography*, *3*(4), 373–388. doi:10.1093/jeg/lbg019

Porter, M. (1998). *On Competition*. Boston: Harvard Business Review.

Putnam, R. D. (1995). Bowling alone: America's declining social capital. *Journal of Democracy*, *6*(1), 65–78. doi:10.1353/jod.1995.0002

Rajapathirana, R. P. J., & Hui, Y. (2018). Relationship between innovation capability, innovation type, and firm performance. *Journal of Innovation & Knowledge, 14*(3), 44–55.

Rosenberg, N. (1982). *Inside the Black Box: Technology and Economics*. New York: Cambridge Univ. Press.

Roth, C. (2008). Réseaux épistémiques: Formaliser la cognition distribuée. *Sociologie du Travail*, *50*(3), 353–371. doi:10.4000dt.19477

Schmitz, H., & Nadvi, K. (1999). Clustering and Industrialization: Introduction. *World Development, 27*(9), 1503–1514. doi:10.1016/S0305-750X(99)00072-8

Schmookler, J. (1966). *Invention and Economic Growth*. Cambridge, MA: Harvard University Press. doi:10.4159/harvard.9780674432833

Schweisfurth, T. G., & Raasch, C. (2018). Absorptive capacity for need knowledge: Antecedents and effects for employee innovativeness. *Research Policy, 47*(4), 687–699. doi:10.1016/j.respol.2018.01.017

Senge, P. M. (1990). *The Fifth Discipline: The Art and Practice of the Learning Organization*. New York: Doubleday Currency.

Sozbilir, F. (2018). The interaction between social capital, creativity and efficiency in organizations. *Thinking Skills and Creativity, 27*(March issue), 92–100. doi:10.1016/j.tsc.2017.12.006

Storper, M. (1997). *The regional world: Territorial development in a global economy*. New York: Guilford Press.

Suire, R., & Vicente, J. (2014). Récents enseignements de la théorie des réseaux en faveur de la politique et du management des clusters. *Revue d'Economie Industrielle, 152*(4), 91–119.

Tichy, N. M., Tushman, M. L., & Fombrun, C. (1979). Social network analysis for organizations. *Academy of Management Review, 4*(4), 507–519. doi:10.5465/amr.1979.4498309

Torre, A. (2006). Clusters et systèmes locaux d'innovation: Un retour critique sur les hypothèses naturalistes de la transmission des connaissances à l'aide des catégories de l'économie de la proximité. *Région et Développement, 24*, 15–44.

Tounkara, T., & Arduin, P.-E. (2014). Aligning knowledge sharing strategy with organizational culture. *European Conference on Knowledge Management*, 983–993.

Tsai, W. P., & Ghoshal, S. (1998). Social capital and value creation: The role of intrafirm network. *Academy of Management Journal, 41*(4), 464–476.

Tsoukas, H. (1996). The firm as a distributed knowledge system: A constructionist approach. *Strategic Management, 17*(2), 11–25. doi:10.1002mj.4250171104

Tsoukas, H., & Vladimirou, E. (2001). What is Organizational Knowledge? *Journal of Management Studies, 38*(7), 973–993. doi:10.1111/1467-6486.00268

Uzzi, B. (1996). The Sources and Consequences of Embeddedness for the Economic Performance of Organizations: The Network Effect. *American Sociological Review, 61*(4), 674–698. doi:10.2307/2096399

Van Dijk, M. P., & Arni, S. (2003). Enterprise clusters in developing countries: Mechanisms of transition and stagnation. *Entrepreneurship and Regional Development, 15*(3), 183–206. doi:10.1080/08985620210159239

Von Hippel, E. (1988). *The Sources of Innovation*. Oxford: Oxford Univ. Press.

Walker, G., Kogut, B., & Shan, W. (1997). Social Capital, Structural Holes and the Formation of an Industrial Network. *Organization Science, 8*(2), 109–125. doi:10.1287/orsc.8.2.109

Wasko, M., & Faraj, S. (2005). Why Should I Share? Examining Social Capital and Knowledge Contribution in Electronic Networks of Practice. *Management Information Systems Quarterly*, *29*(1), 35–57. doi:10.2307/25148667

Wasserman, S., & Faust, K. (1994). *Social network analysis: methods and applications*. Cambridge University Press. doi:10.1017/CBO9780511815478

Westlund, H., & Nilsson, E. (2005). *Measuring enterprises' investments in social capital-a pilot study*. The 43rd Congress of the European Regional Science Association, Jyvaskyla, Finland. 10.1080/00343400500328164

Winter, S. (1987). Knowledge and Competence as Strategic Assets. In D. Teece (Ed.), *The Competitive Challenge: Strategies for Industrial Innovation and Renewal*. New York: Harper & Row, Ballinger Division.

Zach, F. J., & Hill, T. L. (2017). Network, knowledge and relationship impacts on innovation in tourism destinations. *Tourism Management*, *62*, 196–207. doi:10.1016/j.tourman.2017.04.001

Zaheer, A., & George, V. P. (2004). Reach Out or Reach Within? Performance Implications of Alliances and Location in Biotechnology. *Managerial and Decision Economics*, *25*(6-7), 437–452. doi:10.1002/mde.1200

Zaher, A., Gozubuyuk, R., & Milanov, H. (2010). It's the connections: The network perspective in interorganizational research. *The Academy of Management Perspectives*, *24*(1), 62–77.

Zeitlin, J. H. (2008). Industrial districts and regional clusters. In G. Jones, & J. Zeitlin (Eds.), Oxford handbook of business history (pp. 219–243). Oxford University. doi:10.1093/oxfordhb/9780199263684.003.0010

KEY TERMS AND DEFINITIONS

Absorption Capacity (CA): The organisation's ability to identify, assimilate, transform, and use external information, scientific and technological knowledge, as well as other practices and kinds of knowledge existing outside of the organization and enabling it to learn.

Collaborative Innovation: The innovation process involving several human individuals as creators. The collaboration is either intra- or inter-organizational (open), depending on whether the individuals belong respectively to the same organization, or to different organization(s).

Firm Capacity for Innovation (or Innovative Capacity): The organization's capacity and abilities to move beyond current ways of doing things by introducing novelty in existing/new products/services, strategies, technologies, methods, and operational systems with revolutionary, major, or minor modifications and transformations.

Industrial District: A common geographical space or area where workers, small companies, and firms, specialised in a main industry and auxiliary industries, live, work together, and can benefit from their proximity to collaborate and elaborate common ideas and outputs.

Knowledge Café: A common physical place where individuals/organizations come and meet in order to gather innovative ideas, to exchange and discuss on their different problems, searching for further interesting possibilities and solutions.

Knowledge Management: The process of critically acquiring, disseminating, conbining, selecting, exploiting, and sharing a great amount of information and existing knowledge in order to meet current needs and develop new opportunities.

Social Capital: The sum of the actual and/or potential, virtual resources (including emotional and affective support), which are linked to the possession of a durable network of institutionalized or informal relationships of mutual acquaintance or recognition.

Chapter 10
Knowledge Management Processes and Infrastructures:
A Hotel's Performance Overview

Pedro Mota Veiga
Instituto Politécnico de Viseu, Portugal & NECE Research Unit, Universidade da Beira Interior, Portugal

Filipe Ambrósio
https://orcid.org/0000-0001-6048-0542
Instituto Politécnico de Viseu, Portugal

Cristina Isabel Fernandes
https://orcid.org/0000-0001-8560-0758
NECE Research Unit, Universidade da Beira Interior, Portugal

ABSTRACT

The principal goal of this study is to identify the role of knowledge management processes and infrastructures as a determinant of financial and operational in the hotel industry, particularly how these knowledge management strategies have positive an impact on the hotels performance. The authors use a quantitative approach based on an email survey applied to 55 Portuguese hotels. The results show that the knowledge management systems and technological and non-technological infrastructure are critical for hotels competitiveness and obtaining operational and financial competitive advantages. This will help hotel managers better understand how to align knowledge management strategies linked with systems, technological and non-technological infrastructure, and improve hotel competitiveness. This chapter is one of the first studies to investigate the links among knowledge management strategy of systems and technological and non-technological infrastructure with operational and financial performance.

DOI: 10.4018/978-1-7998-3142-6.ch010

INTRODUCTION

Despite the uncertainty of the global economic outlook and the risks of terrorism, the number of international tourists has continued to expand almost continuously in recent decades (Algieri, Aquino, & Succurro, 2018). Tourism is commonly considered one of the most important industries globally due to its central role in the growth and economic development of many regions and countries (Cheraghalizadeh & Tümer, 2017; Nunes, Estevão, & Filipe, 2018). The hotel industry in particular is a core activity in the development of tourism, as it provides accommodation and rest facilities for tourists and travellers (Chen, 2013; Shamim, Cang, & Yu, 2017). This global trend is particularly felt in Europe, which leads tourism growth in absolute terms, namely in the Mediterranean countries, which have the largest share of the European market (Algieri et al., 2018). In Portugal, the tourism is also a thriving industry and the right policies could boost the country's economic growth. The Portuguese balance of trade and tourism (as a % of GDP) has grown 63.9%, from 3.6% in 2013 to 5.9% in 2018. In the meantime, exports (as a % of GDP) grew by 51.9% (from 5.4% in 2013 to 8.2% in 2018). The tourism increasingly important role was reflected in the hotel industry: the number of hotels grew 49.4% (from 1,039 in 2013 to 1,552 in 2018) and the number of overnight stays in hotels increased 55,1% (from 26,100,320 in 2013 to 40,469,825 in 2018), which fueled a higher competitive rivalry in the industry (INE, 2019).

In face of this, the challenges faced by managers of hospitality organizations are unique and go beyond those faced by other services' executives. Besides the fast pace of the business environment, hospitality businesses are open systems that have all the characteristics of an uncertain and ambiguous environment (Bowen & Ford, 2004; Tzortzaki & Mihiotis, 2012). Hospitality organizations need to increase the degree of processing and analysis of customer information, the co-production element of the tourist experience, and the growing need for service customization (Tzortzaki & Mihiotis, 2012). In fact, the dynamism and agility of hotels is only viable if it is well-matched with an active and complete organizational memory, where knowledge is created, circulates smoothly and grows (Davenport & Prussak, 1998; Fraj, Matute, & Melero, 2015; Veiga, Ambrósio, & Ferreira, 2020). Knowledge is unique and valuable for gaining competitive advantage, and effective knowledge management should be a major concern of business owners (Wu & Lin, 2009). Hotels, which operate in very competitive environments, must implement appropriate knowledge management (KM) strategies to address the strong competitiveness of markets and thereby improve their operational and financial performance (Cheraghalizadeh & Tümer, 2017; Hallin & Marnburg, 2008 Salem, 2014; Veiga et al., 2020).

New opportunities in the hotel industry stem from knowledge-based assets, which is a key resource and driver of value creation, performance, competitiveness and innovative capacity in the hotel industry (Hallin & Marnburg, 2008; Kim, Kim, Park, Lee, & Jee, 2012; Shamim, Cang, & Yu, 2017). Hotel companies can adopt innovations by incorporating in-house expertise (eg, employees knowledge, market research and information systems) or by developing their own innovations based on knowledge gathered from outside agents or events (eg customers, trade shows, and professional conferences) (Nieves & Diaz-Meneses, 2018). Transforming this knowledge into new products and innovations, and gaining competitive advantage, requires the establishment of internal communication channels that foster collaboration, knowledge creation and exchange among employees (Chalkiti, 2012; Nieves & Diaz-Meneses, 2018; Salem, 2014). KM, accelerating knowledge sharing and innovative capabilities are essential business tourism performance.

Organizational knowledge, and hence the knowledge of organizational workers, expressed through procedures, routines, information systems and databases (processes and infrastructures) is an element that effectively develops businesses in a competitive environment (Dahou, Hacini & Burgoyne, 2019; Kmieciak, Michna, Kmieciak, & Michna, 2018; Z. Wang, Wang, Cao, & Ye, 2016). However, in the hospitality industry, KM processes and infrastructures have not reached the same scale of empirical research as in other fields (T. T. Kim & Lee, 2013; Martínez-Martínez, Cegarra-Navarro, & García-Pérez, 2015; Salem, 2014), and a effective KM processes and infrastructures can create more value for the hotels. Therefore, the main goal of this study is to identify the role of organizational KM processes and infrastructures as a determinant of operational and financial performance in the hotel industry. An additional goal is to evaluate how processes of KM, infrastructure of KM and technological infrastructure of KM have an impact on the operational and financial performance of hotels. The methodology of this research is based on a survey that was taken by hotel directors in Portugal, which collected data on KM processes and infrastructures, as well as operational and financial KPIs.

This study entails two relevant contributions. In the first place, it fills a research gap by providing both theoretical and empirical support for the links among KM processes and infrastructures' impact in two types of performance (operational and financial). Secondly, it was applied in the Portuguese hotel industry, a sector with intense competition and growth (50% in the last five years). Another contribution is related to the analysis of the moderating effect of KM IT infrastructure on the impact of KM process and KM infrastructure in operational and financial performance.

The study is organized as follows: section 2 presents the theoretical framework and the development of the research model; section 3 describes the methodology; section 4 presents the results; section 5 discusses the results and conclusions and points out the limitations of the study and potential future lines of research.

THEORETICAL BACKGROUND AND RESEARCH MODEL

Knowledge Management

Due to the high diversity of concepts and terminologies adopted by different authors, the definition of knowledge may have a practical, conceptual or philosophical approach, either specific or wide (Anand, 2011; Latilla et al., 2018; Nonaka & Takeuchi, 1995). For Nonaka and Takeuchi (1995), knowledge is a justified belief in the truth, which is a function of a special position, perspective or intention; it implies an action, being specific to a given context and dependent on unique points of view, personal sensitivity and experience. For Erden, von Krogh and Nonaka (2008), the concept of knowledge is linked to a belief justified in the truth, meaning that individuals need to justify the veracity of their observations based on the observations of the world; it is also the ability to define a problem and act accordingly. Davenport and Prusak (1998) define knowledge as a fluid mix of shared experiences, values, contextual information and the specialized knowledge that provides a framework for its use for incorporation into new experiences and information. For Latilla et al. (2018), knowledge is constituted as a combination of consciousness, perceptions and feelings that define and make products unique, exclusive, innovative and timeless, as well as a source of organizational competitive advantage.

Indeed, KM covers a diversity of processes, systems and practices in organizational environment, which is designed to create value through the application and use of knowledge (Barley, Treem, & Kuhn, 2018; Teece, 1998); it is concentrated on the creation and distribution of knowledge in organizations with technologies such as the internet, intranets and e-mail (Bouncken, 2002; Hallin & Marnburg, 2008;

Kremp & Mairesse, 2004; uit Beijerse, 1999; Yang & Wan, 2004). It refers to the strategic value of knowledge oriented to practices, critical to a firm's chances of success (Barley et al., 2018). For Gloet and Terziovski (2004), KM is a management function that create or locate knowledge, it manages the flow of knowledge and it guarantees that knowledge is effectively and efficiently used for the organization's long-term success.

An appropriate KM strategy enables organizations to create, acquire, access and leverage knowledge in a timely manner, resulting in rapid and effective delivery of products or services to meet customer preferences (Martinez-Conesa, Soto-Acosta, & Carayannis, 2017), ultimately resulting in better organizational performance (Gray & Meister, 2004; TH Kim, Lee, Chun, & Benbasat, 2014). A company's KM planning and structuring should be in line with its business strategy (Anand, 2011; Du Plessis, 2007) and organizational processes, culture and information technologies (Bagnoli & Giachetti, 2014; Choi, Poon, & Davis, 2008; Singh, 2018). An appropriate KM strategy supports competitive advantages and results in better organizational performance (Alavi & Leidner, 2001; Bloodgood, 2019; Cepeda-Carrion & Cillo, 2019; Choi et al., 2008; Gloet & Terziovski, 2004; Halawi, Aronson, & McCarthy, 2005; S. Lee et al., 2012; Ragab & Arisha, 2013; Shankar, Acharia, & Baveja, 2009; Z. Wang et al., 2016).

The hospitality industry is a knowledge-intensive industry as a consequence of the intense use of technology and the essence of product and service (Hallin & Marnburg, 2008; TT Kim, Kim, Park, Lee, & Jee, 2012; Marco-Lajara, Claver-Cortés, Úbeda-García, García-Lillo, & Zaragoza-Sáez, 2019; Martínez-Martínez et al., 2015; Sigala & Chalkiti, 2015; Wilkea, Costa, Freire, & Ferreira, 2019) as well as due to the high interaction between employees and customers (Hallin & Marnburg, 2008; Salem, 2014; Sigala & Chalkiti, 2015; Yang & Wan, 2004). In the hospitality industry, knowledge is embedded into clients and company problem solving, operational work procedures, products and services, and competitors' attitude (Shamim et al., 2017; Shaw & Williams, 2009; Veiga et al., 2020; Yang & Wan, 2004), and KM focuses on systematic planning and guidance for managing the flow of knowledge as an intangible asset to continuously improve employees' knowledge of the specific needs of clients (Bouncken, 2002; Shamim et al., 2017; Shaw & Williams, 2009). KM plays a key role in hotel success and competitive advantage (Hallin & Marnburg, 2008; Salem, 2014; Shamim et al., 2017; Sigala & Chalkiti, 2015; Tzortzaki & Mihiotis, 2012; Yang & Wan, 2004), due to the increase of quality standards (Hallin & Marnburg, 2008; Shamim et al., 2017; Yang, 2015), with the creation of new products/services (Sigala & Chalkiti, 2015) or the promotion of the supply of personalized tourist experiences for clients (Bouncken, 2002; Tzortzaki & Mihiotis, 2012). For effective KM in the hotel industry, processes of knowledge creation, management and transfer are essential, as well as all technological and non-technological infrastructures (TT Kim & Lee, 2013; Martínez-Martínez et al., 2015; Salem, 2014).

Knowledge Management Process

KM process is the systemic management of relevant knowledge flows including creation, discovery, acquisition, transfer, sharing, and diffusion of knowledge in order to support problem-solving, dynamic learning, strategic planning, and decision-making (Anand, 2011; Ferraris, Santoro, & Dezi, 2017; Gonzalez & Meli, 2017; Salem, 2014).

KM processes help hotels acquire, store and use knowledge for problem-solving and decision making, dynamic learning and strategic planning (Salem, 2014). These processes can consist of various components such as knowledge management processes associated with acquiring knowledge about customers, knowledge filtering, knowledge utilization to solve new problems and protection of misuse knowledge within the hotel (Gold, Malhotra, & Segars, 2001; Salem, 2014).

Knowledge Management Infrastructure

KM requires the existence of a specific infrastructure by identifying strategic orientations on technology, structure, culture and human resources in order to increase their efficiencies (H. Lee & Choi, 2003; Salem, 2014). The KM framework alludes to the structural willingness of an organization, particularly the hotel industry, to encourage knowledge-related activities (Gold et al., 2001; Salem, 2014). These structures enable knowledge-related vitalizing interactions as well as allow hotels to adjust to a constantly changing environment (Salem, 2014; Sanchez & Mahoney, 1996). The ability of a hotel structure to facilitate the flow of knowledge is adjusted by the policies, processes and system of rewards and incentives, which determine the channels through which knowledge is accessed and how it flows (Salem, 2014). Moreover, the structure of a hotel is important to leverage its technological architecture (Salem, 2014). KM culture describes the level at which organizational culture supports the view of knowledge as active and valuable resources, and is one of the most important factors for a successful KM strategy (Gold et al., 2001; Salem, 2014). The culture of the organization is central to the hotel's ability to manage its knowledge more efficiently, with the relevant corporate vision and purpose as its relevant component (Salem, 2014). Human resource-based KM strategies, where they are at the centre of knowledge creation, describe the expertise required for employees and the ability to interact with others (Salem, 2014).

Knowledge Management Technological Infrastructure

KM systems refers to a class of information systems applied to organizational KM to support and improve the organizational processes of knowledge creation, storage/retrieval, transfer and application (Alavi & Leidner, 2001; Ferraris et al., 2017; Gloet & Terziovski, 2004; Gonzalez & Meli, 2017). For the implementation of a KM system it is necessary to understand and develop the infrastructure elements necessary to support the acquisition, management and transfer of organizational knowledge, with emphasis placed on people, processes and technology (Halawi et al., 2005; Ragab & Arisha, 2013; Shankar et al., 2009; Singh, 2018; Usoro & Abiagam, 2018). Information and communication technologies are powerful facilitators of organizational knowledge processes and building appropriate IT systems (hardware, software and processes) is an integral part of KM's success (Andreeva & Kianto, 2012) and is applied for two purposes: the creation of knowledge repositories and the creation of corporate knowledge directories (Davenport & Prussak, 1998). KM's technology infrastructure can create more value for the hotel through its interaction with other organizational processes and can lead to a differential value of the organization (Salem, 2014).

Research Model

The literature review reveals that a correct hotel KM strategy based in processes, infrastructure technological and non-technological impact on hotels performance. Figure 1 shows the research model.

Figure 1. Research Model

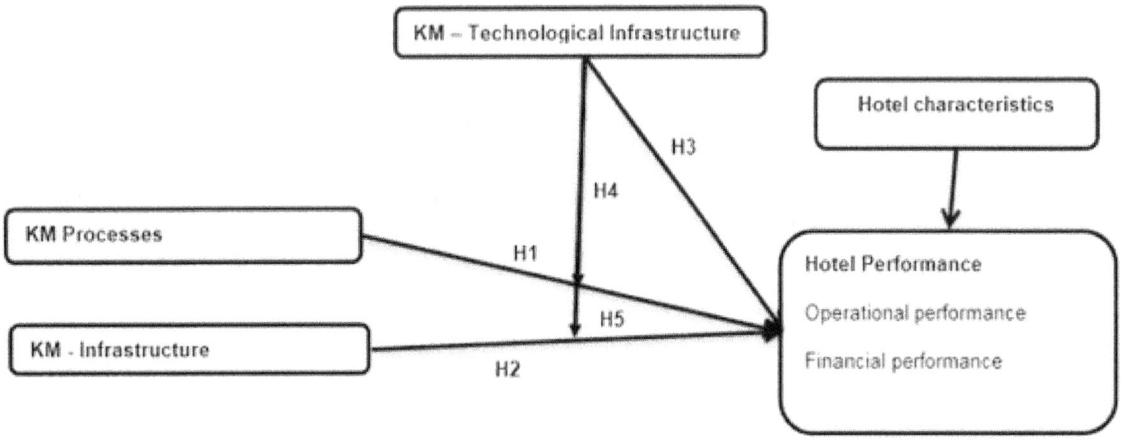

The following hypothesis has been proposed:

1. **Hypothesis 1 –** KM Processes have a significant impact on hotels performance
   Hypothesis 1a – KM Processes have a significant impact on hotels Operational Performance
   Hypothesis 1b – KM Processes have a significant impact on hotels Financial Performance
2. **Hypothesis 2 –** KM Infrastructure share a significant impact on hotels performance
   Hypothesis 2a – KM Infrastructures have a significant impact on hotels Operational Performance
   Hypothesis 2b – KM Infrastructures have a significant impact on hotels Financial Performance
3. **Hypothesis 3 –** KM Technological Infrastructure share a significant impact on hotels performance
   Hypothesis 3a – KM Technological Infrastructures have a significant impact on hotels Operational Performance
   Hypothesis 3b – KM Technological Infrastructures have a significant impact on hotels Financial Performance
4. **Hypothesis 4 –** The interaction between KM and Processes KM Technological Infrastructures have a significant impact on hotels performance
   Hypothesis 4a – The interaction between KM and Processes KM Technological Infrastructures have a significant impact on hotel Operational Performance
   Hypothesis 4b – The interaction between KM and Processes KM Technological Infrastructures have a significant impact on hotel Financial Performance
5. **Hypothesis 5 –** The interaction between KM Non-Technological and Technological Infrastructures have a significant impact on hotels performance
   Hypothesis 5a – The interaction between KM Non-Technological and Technological Infrastructures have a significant impact on hotels Operational Performance
   Hypothesis 5b – The interaction between KM Non-Technological and Technological Infrastructures have a significant impact on hotels hotel Financial Performance

METHODOLOGY

Data Collection

The population of this study includes all Portuguese hotels and the data collection was done through a structured questionnaire answered by the Hotel Managers. In terms of sampling, an electronic data survey was sent to the emails included in a database of 700 Portuguese hotel directors' contacts. Data were collected between February and March 2019; 55 hotel managers answered the questionnaire, producing a response rate of 7.9% and corresponding to 3.5% of Portuguese hotels. The sample had 26 four-star hotels, 19 five-star hotels, eight three-star hotels, one two-star hotel and one one-star hotel.

Measurement

All items in the survey were evaluated with a seven-point Likert scale ranging from 1 to 7.

Measures of KM Processes and KM Infrastructures were based on scales developed and validated by Salem (2014).

Scale of KM Technological Infrastructures was based on measures developed and validated by Andreeva & Kianto (2012).

Finally, scales of operational and financial performance were developed and validated by C. L. Wang et al. (2015), Z. Wang et al. (2016) and Veiga et al. (2020). The full measurement items are displayed in Appendix.

To account for differences between hotels, hotel age, hotel integrated in a group (No; Yes); client origin (% Foreign), location (North, Center, Lisboa, Alentejo, Algarve, Autonomous Region of Madeira; Autonomous Region of the Azores) and category (1 to 5 stars) were included in our research model as a control variable. These variables were selected because of its potential impact on firm performance.

Data Analysis

All data were analyzed using IBM-SPSS software version 25.0 (IBM Corporation, New York, USA). To describe the hotels sample it was computed the descriptive statistics (means and standard deviations) of the variables included in the research and correlations between them. To modeling KM processes, KM infrastructures, KM technological infrastructures and how control variables influence the hotel's operational and financial performance, it was used multiple linear regression. For the estimation of the standard errors of variables coefficients it was computed used robust standard errors, to eliminate possible problems of heteroscedasticity. In all regressions, the existence of variables with potential multicollinearity effects was evaluated by variance inflation factors (VIF). The estimated multiple regression models were as follows:

$$Perf = \beta_o + \beta_i Control \ldots \qquad 1.$$

$$Performance = \beta_o + \beta_i Control + \beta_j Process + \beta_k Infrastructure + \beta_l ICTInfrastructure$$

$$Performance = \beta_o + \beta_j Process + \beta_k Infrastructure + \beta_l ICTInfrastructure$$

$$Performance = \beta_o + \beta_i Control + \beta_j Process * ICTInfrastructure + \beta_k Infrastructure * ICTInfrastructure$$

One for each Performance variables (Operational Performance and Financial Performance)

RESULTS AND DISCUSSION

Table 1 shows the results referring to descriptive statistics (mean and standard deviation), Pearson's correlations between all the variables used in the study and VIF of predictor variables. Regarding the sample, 36% (20) of the hotel were in the Lisbon region, 24% (13) were in the Algarve and 13% (7) in the Madeira Island, the three leading tourist regions in Portugal. The hotels had been operating on average for 20.13 years, 56.4% (31) hotels were integrated into a hotel group and, on average, 66.2% of the guests were foreigners.

Table 1. Mean, Standard Deviation (SD), correlations between the variables and VIF

	Mean	SD	1	2	3	4	5	6	7	8	9	10	11
(1) Operational Performance	4.82	0.93	**NA**										
(2) Financial Performance	4.83	1.21	.786*	**NA**									
(3) Hotel age	20.13	18.21	.265	.193	**1.290**								
(4) Integrated in a group	0.56	0.50	-.011	.091	.018	**1.570**							
(5) Algarve	0.24	0.43	-.071	-.022	.077	-.028	**1.500**						
(6) Lisboa	0.36	0.49	-.059	-.039	-.156	.132	-.421*	**1.850**					
(7) Category (stars)	4.11	0.85	.132	.029	-.045	-.363*	-.021	.037	**1.320**				
(8) International market (%)	66.22	25.14	.051	.036	-.269*	.389*	-.019	.361*	-.023	**1.780**			
(9) KM Process	4.52	1.45	.292*	-.071	.152	.360*	-.153	.110	.019	-.057	**4.092**		
(10) KM Infrastructure	4.49	1.48	.085	.221	.205	.194	-.036	.026	-.124	.008	.527**	**1.879**	
(11) KM ICT Infrastructure	4.76	1.41	.399**	.106	.220	.377*	-.047	.021	.056	-.016	.833**	.662**	**4.882**

Note: * $p < 0.05$; NA – Not Applicable

In relation to the factors that have impact in hotel competitiveness, modeled through multiple linear regression (Table 2), the models I, II and IV shows that hotel's age have a positive significant impact in operational performance ($B = 0.02$; $p < 0.05$) and the model II presents that hotel's age have a significant impact in financial performance ($B = 0.02$; $p < 0.05$). In model IV, % of foreign tourists has a positive significant impact on financial performance ($B = 0.01$; $p < 0.05$).

Regarding strategies oriented to knowledge management, KM processes (Model II: $B = 0.44$; $p < 0.05$; Model III: $B = 0.42$; $p < 0.05$) presents a significant impact in financial performance. KM process helps firms manage relevant knowledge flows including creation, discovery, acquisition, transfer, sharing, and diffusion of knowledge in order to support problem-solving, dynamic learning, strategic planning, and decision-making, contributing to firms performance (Anand, 2011; Ferraris, Santoro, & Dezi, 2017; Gonzalez & Meli, 2017; Salem, 2014).

Table 2. Linear regressions (Estimated coefficients)

	Model 1		Model 2		Model 3		Model 4	
	OP	FP	OP	FP	OP	FP	OP	FP
Hotel age	0.02*	0.01			0,02*	0,02*	0,02*	0,02
Integrated in a group	0.06	0.20			-0,19	-0,43	-0,16	-0,43
Algarve	0.06	0.54			0,11	0,19	0,17	0,23
Lisboa	0.39	0.35			-0,07	-0,86	-0,09	-0,82
Category (stars	0.15	0.10			-0,41	-0,23	-0,64	-0,28
International market (%)	0.01	0.01			0,01	0,00	0,01*	0,00
KM System			-0,11	0,44*	-0,05	0,42*		
KM Infrastructure			0,21*	0,21*	0,18*	0,26*		
KM IT Infrastructure			0,51*	0,32*	0,44*	0,29		
KM System x IT Infrastructure							0,04*	0,04*
KM Infrastructure x IT Infrastructure							-0,01	0,06*
R^2	17.5%	8.4%	22,4%	12,8%	40,2%	23,7%	38,1%	21,8%
Adjusted R^2	7.2%	3.0%	17,5%	7,2%	22,8%	11,6%	26,6%	12,2%

Note: * $p < 0.05$; OP – Operational Performance; FP – Financial Performance

KM infrastructure shows that exist a statistically significant influence in operational (Model II: B = 0.21; $p < 0.05$; Model III: B = 0.18; $p < 0.05$) and financial (Model II: B = 0.21; $p < 0.05$; Model III: B = 0.26; $p < 0.05$) performance. The ability of a hotel infrastructure to facilitate the flow of knowledge supports the view of knowledge as active and valuable resources, and is one of the most important factors for a better performance (Gold et al., 2001; H. Lee & Choi, 2003; Salem, 2014; Sanchez & Mahoney, 1996).

KM IT infrastructure (Model II: B = 0.32; $p < 0.05$) presents a statistically significant impact in operational (Model II: B = 0.51; $p < 0.05$; Model III: B = 0.44; $p < 0.05$) and financial (Model II: B = 0.32; $p < 0.05$) performance. Information and communication technologies are powerful enablers of organizational knowledge processes and building appropriate IT systems (hardware, software and processes) is an integral part of KM's success (Andreeva & Kianto, 2012) and firm performance (Davenport & Prussak, 1998).

Finally, the interaction terms between KM IT infrastructure with KM processes and KM infrastructure shows that exist a statistically significant impact of interaction between KM IT infrastructure and KM processes (Model IV: B = 0.04; $p < 0.05$) in operational and financial performance, there is a multiplier effect on the operational and financial of hotels resulting from the use of combined KM systems and IT infrastructure. Similarly, KM IT infrastructure and KM infrastructure shows that exist a statistically significant impact of interaction between both (Model IV: B = 0.04; $p < 0.05$) in financial performance. There is a multiplier effect on the financial of hotels resulting from the use of combined KM infrastructure and IT infrastructure. KM's technology infrastructure can create more value for the hotel and higher performance through its interaction with other organizational processes and can lead to a differential value of the organization (Salem, 2014); and the structure of a hotel is important to leverage its technological architecture (Salem, 2014).

CONCLUSION

A knowledge management strategy encompasses systems, technological and non-technological infrastructure. Knowledge promotes distinction, which in turn encourages the development of new knowledge. This study results show that the KM strategy of systems, technological and non-technological infrastructure is critical to operational and financial performance of the hotel industry and to obtaining competitive advantages.

These results have several implications for hotel management. If managers want to improve performance and hospitality it is crucial to have a process to acquire knowledge about customers, to filter it and to leverage it to solve problems, while protecting it from inappropriate use inside the hotel. The hotels organizations need to have structure that facilitates the discovery of new knowledge, benefits of sharing knowledge outweigh the costs, and employees needs to be specialists in their own part.

The results show a multiple effect on performance of technological infrastructure, so it is imperative that hotels use technology to retrieve knowledge about its markets and competition. The IT infrastructure should have the capability to retrieve knowledge about its markets and competition, and supporting management decisions and knowledge work, and it is widely accepted, monitored, and updated. This IT architecture needs to be capable of sharing data and information, knowledge and expertise with all stakeholders in the organization's extended value chain and must dimensioned to support the daily work. The hotels managers need to have strategy of implementation intranet and extranet applications with employees' profiles, groups of individuals with shared interests and knowledge, blogs, wikis, forums for employees exchanging ideas, and social network analysis.

The limitations of the study are inevitable. The methodological design cannot be used to reveal possible causalities between process, non-technological and technological infrastructure, so a methodology based on longitudinal data collection should be employed in future research. Secondly, the data used in this study, particularly those of operational and financial performance, were collected from the perceptions of hospitality managers, which may lead to some bias in the results, and future research may collect objective KPIs from the hotels. Finally, the study refers to a specific sample of a country (Portugal), which limits the conclusions. Therefore, future research should be done to test this model in different countries and sectors, making it possible to generalize these results.

This work is financed by National Funds through FCT - Foundation for Science and Technology within the scope of UID/GES/04630/2019.

REFERENCES

Alavi, M., & Leidner, D. E. (2001). Knowledge Management and Knowledge Management Systems: Conceptual Foundations and Research Issues. *Management Information Systems Quarterly*, *25*(1), 107–136. doi:10.2307/3250961

Algieri, B., Aquino, A., & Succurro, M. (2018). International competitive advantages in tourism : An eclectic view. *Tourism Management Perspectives*, *25*, 41–52. doi:10.1016/j.tmp.2017.11.003

Anand, A. (2011). Understanding Knowledge Management: A literature review. *International Journal of Engineering Science and Technology*, *3*(2), 926–939.

Andreeva, T., & Kianto, A. (2012). Does knowledge management really matter? Linking knowledge management practices, competitiveness and economic performance. *Journal of Knowledge Management, 16*(4), 617–636. doi:10.1108/13673271211246185

Bagnoli, C., & Giachetti, C. (2014). Aligning Knowledge Strategy and Competitive Strategy in Small Firms. *Journal of Business Economics and Management, 16*(3), 571–598. doi:10.3846/16111699.2012.707623

Barley, W. C., Treem, J. W., & Kuhn, T. (2018). Valuing multiple trajectories of knowledge: A critical review and agenda for knowledge management research. *The Academy of Management Annals, 12*(1), 278–317. doi:10.5465/annals.2016.0041

Bloodgood, J. M. (2019). Knowledge acquisition and fi rmcompetitiveness: The role of complements and knowledge source. *Journal of Knowledge Management, 23*(1), 46–66. doi:10.1108/JKM-09-2017-0430

Bouncken, R. B. (2002). Knowledge Management for Quality Improvements in Hotels. *Journal of Quality Assurance in Hospitality & Tourism, 3*(3–4), 25–59. doi:10.1300/J162v03n03_03

Bowen, J., & Ford, R. C. (2004). What experts say about managing hospitality service delivery systems. *International Journal of Contemporary Hospitality Management, 16*(7), 394–401. doi:10.1108/09596110410559078

Cepeda-Carrion, G., & Cillo, V. (2019). Tips to use partial least squares structural equation modelling (PLS-SEM) in knowledge management. *Journal of Knowledge Management, 23*(1), 67–89. doi:10.1108/JKM-05-2018-0322

Chen, W.-J. (2013). Factors influencing internal service quality at international tourist hotels. *International Journal of Hospitality Management, 35*, 152–160. doi:10.1016/j.ijhm.2013.06.004

Cheraghalizadeh, R., & Tümer, M. (2017). Journal of Hospitality and Tourism Management The effect of applied resources on competitive advantage in hotels : Mediation and moderation analysis. *Journal of Hospitality and Tourism Management, 31*, 265–272. doi:10.1016/j.jhtm.2017.04.001

Choi, B., Poon, S. K., & Davis, J. G. (2008). Effects of knowledge management strategy on organizational performance: A complementarity theory-based approach. *Omega, 36*(2), 235–251. doi:10.1016/j.omega.2006.06.007

Dahou, K., Hacini, I., & Burgoyne, J. (2019). Knowledge management as a critical success factor in developing international companies' organizational learning capability. *Journal of Workplace Learning, 31*(1), 2–16. doi:10.1108/JWL-12-2017-0118

Davenport, T. H., & Prussak, L. (1998). *Working Knowledge: how organizations manage what they know*. Boston: Harvard Business School Press.

Du Plessis, M. (2007). The role of knowledge management in innovation. *Journal of Knowledge Management, 11*(4), 20–29. doi:10.1108/13673270710762684

Erden, Z., von Krogh, G., & Nonaka, I. (2008). The quality of group tacit knowledge. *The Journal of Strategic Information Systems, 17*(1), 4–18. doi:10.1016/j.jsis.2008.02.002

Ferraris, A., Santoro, G., & Dezi, L. (2017). How MNC's subsidiaries may improve their innovative performance? The role of external sources and knowledge management capabilities. *Journal of Knowledge Management, 21*(3), 540–552. doi:10.1108/JKM-09-2016-0411

Fraj, E., Matute, J., & Melero, I. (2015). Environmental strategies and organizational competitiveness in the hotel industry: The role of learning and innovation as determinants of environmental success. *Tourism Management, 46*, 30–42. doi:10.1016/j.tourman.2014.05.009

Gloet, M., & Terziovski, M. (2004). Exploring the relationship between knowledge management practices and innovation performance. *Journal of Manufacturing Technology Management, 15*(5), 402–409. doi:10.1108/17410380410540390

Gold, A. H., Malhotra, A., & Segars, A. H. (2001). Knowledge management: An organizational capabilities perspective. *Journal of Management Information Systems, 18*(1), 185–214. doi:10.1080/07421222.2001.11045669

Gonzalez, R. V. D., & Meli, T. M. (2017). Linkage between dynamics capability and knowledge management factors: A structural equation model. *Management Decision, 55*(10), 2256–2276. doi:10.1108/MD-03-2017-0180

Gray, P. H., & Meister, D. B. (2004). Knowledge sourcing effectiveness. *Management Science, 50*(6), 821–834. doi:10.1287/mnsc.1030.0192

Halawi, L., Aronson, J., & McCarthy, R. (2005). Resource-based view of knowledge management for competitive advantage. *Electronic Journal of Knowledge Management, 3*(2), 75–86. Retrieved from https://commons.erau.edu/publication/319/

Hallin, C. A., & Marnburg, E. (2008). Knowledge management in the hospitality industry: A review of empirical research. *Tourism Management, 29*(2), 366–381. doi:10.1016/j.tourman.2007.02.019

INE. (2019). *Estatísticas do Turismo 2018*. Lisbon, Portugal: Instituto Nacional de Estatística.

Kim, T. H., Lee, J. N., Chun, J. U., & Benbasat, I. (2014). Understanding the effect of knowledge management strategies on knowledge management performance: A contingency perspective. *Information & Management, 51*(4), 398–416. doi:10.1016/j.im.2014.03.001

Kim, T. T., Kim, W. G., Park, S. S.-S., Lee, G., & Jee, B. (2012). Intellectual Capital and Business Performance: What Structural Relationships Do They Have in Upper-Upscale Hotels? *International Journal of Tourism Research, 14*(4), 391–408. doi:10.1002/jtr.1868

Kim, T. T., & Lee, G. (2013). Hospitality employee knowledge-sharing behaviors in the relationship between goal orientations and service innovative behavior. *International Journal of Hospitality Management, 34*(1), 324–337. doi:10.1016/j.ijhm.2013.04.009

Kmieciak, R., Michna, A., Kmieciak, R., & Michna, A. (2018). Knowledge management orientation, innovativeness, and competitive intensity : evidence from Polish SMEs intensity : evidence from Polish SMEs. *Knowledge Management Research and Practice, 16*(4), 559–572. doi:10.1080/14778238.2018.1514997

Latilla, V. M., Frattini, F., Petruzzelli, A. M., Berner, M., Latilla, V. M., Frattini, F., ... Berner, M. (2018). Knowledge management, knowledge transfer and organizational performance in the arts and crafts industry : A literature review. *Journal of Knowledge Management*, *22*(6), 1310–1331. doi:10.1108/JKM-08-2017-0367

Lee, H., & Choi, B. (2003). Knowledge management enablers, processes, and organizational performance: An integrative view and empirical examination. *Journal of Management Information Systems*, *20*(1), 179–228. doi:10.1080/07421222.2003.11045756

Lee, S., Gon, B., Hoyal, K., Lee, S., Kim, B. G., & Kim, H. (2012). An integrated view of knowledge management for performance. *Journal of Knowledge Management*, *16*(2), 183–203. doi:10.1108/13673271211218807

Marco-Lajara, B., Claver-Cortés, E., Úbeda-García, M., García-Lillo, F., & Zaragoza-Sáez, P. C. (2019). The role of internal knowledge generation and external knowledge acquisition in tourist districts. *Journal of Business Research*, *101*, 767–776. doi:10.1016/j.jbusres.2018.12.045

Martinez-Conesa, I., Soto-Acosta, P., & Carayannis, E. G. (2017). On the path towards open innovation: Assessing the role of knowledge management capability and environmental dynamism in SMEs. *Journal of Knowledge Management*, *21*(3), 553–570. doi:10.1108/JKM-09-2016-0403

Martínez-Martínez, A., Cegarra-Navarro, J. G., & García-Pérez, A. (2015). Environmental knowledge management: A long-term enabler of tourism development. *Tourism Management*, *50*, 281–291. doi:10.1016/j.tourman.2015.03.006

Nieves, J., & Diaz-Meneses, G. (2018). Knowledge sources and innovation in the hotel industry: Empirical analysis on Gran Canaria Island, a mature mass-tourism destination. *International Journal of Contemporary Hospitality Management*, *30*(6), 2537–2561. doi:10.1108/IJCHM-07-2016-0341

Nonaka, I., & Takeuchi, H. (1995). *The knowledge-creating company: how Japaneses companies create the dynamics of innovation*. New York: Oxford University Press, Inc.

Nunes, S., Estevão, C., & Filipe, M. N. (2018). Determinant factors of competitiveness in the hotel sector : The case of Portugal. *Competitiveness Review*, *28*(2), 122–135. doi:10.1108/CR-07-2016-0043

Ragab, M. A. F., & Arisha, A. (2013). Knowledge management and measurement: A critical review. *Journal of Knowledge Management*, *17*(6), 873–901. doi:10.1108/JKM-12-2012-0381

Salem, I. E. B. (2014). Toward better understanding of knowledge management: Correlation to hotel performance and innovation in five-star chain hotels in Egypt. *Tourism and Hospitality Research*, *14*(4), 176–196. doi:10.1177/1467358414542265

Sanchez, R., & Mahoney, J. T. (1996). Modularity, flexibility, and knowledge management in product and organization design. *Strategic Management Journal*, *17*(SI), 63–76. doi:10.1002mj.4250171107

Shamim, S., Cang, S., & Yu, H. (2017). Supervisory orientation, employee goal orientation, and knowledge management among front line hotel employees. *International Journal of Hospitality Management*, *62*, 21–32. doi:10.1016/j.ijhm.2016.11.013

Shankar, R., Acharia, S., & Baveja, A. (2009). Soft-system knowledge management framework for new product development. *Journal of Knowledge Management, 13*(1), 135–153. doi:10.1108/13673270910931224

Shaw, G., & Williams, A. (2009). Knowledge transfer and management in tourism organisations: An emerging research agenda. *Tourism Management, 30*(3), 325–335. doi:10.1016/j.tourman.2008.02.023

Sigala, M., & Chalkiti, K. (2015). Knowledge management, social media and employee creativity. *International Journal of Hospitality Management, 45*, 44–58. doi:10.1016/j.ijhm.2014.11.003

Singh, P. K. (2018). Knowledge strategy, sharing behavior and performance. *Management Research Review, 41*(3), 395–411. doi:10.1108/MRR-01-2017-0001

Te Yang, J. (2015). Effect of internal marketing on knowledge sharing and organisational effectiveness in the hotel industry. *Total Quality Management & Business Excellence, 26*(1–2), 76–92. doi:10.1080/14783363.2012.661131

Te Yang, J., & Wan, C. S. (2004). Advancing organizational effectiveness and knowledge management implementation. *Tourism Management, 25*(5), 593–601. doi:10.1016/j.tourman.2003.08.002

Teece, D. J. (1998). Research Directions for Knowledge Management. *California Management Review, 40*(3), 289–293. doi:10.2307/41165957

Tzortzaki, A. M., & Mihiotis, A. (2012). A three dimensional knowledge management framework for hospitality and tourism. *Foresight, 14*(3), 242–259. doi:10.1108/14636681211239773

Uit Beijerse, R. P. (1999). Questions in knowledge management: defining and conceptualising a phenomenon. *Journal of Knowledge Management, 3*(2), 94–109.

Usoro, A., & Abiagam, B. (2018). Culture effect on knowledge management adoption in Nigerian hospitality industry. *Journal of Information and Knowledge Management Systems, 48*(3), 314–332. doi:10.1108/VJIKMS-11-2017-0080

Veiga, P. M., Ambrósio, F., & Ferreira, R. R. (2020). Competitiveness of Hotel Industry: a Knowledge Management Approach. In S. J. Teixeira & J. J. Ferreira (Eds.), *Multilevel Approach to Competitiveness in the Global Tourism Industry* (pp. 9–25). IGI Global. doi:10.4018/978-1-7998-0365-2.ch002

Wang, C. L., Senaratne, C., & Rafiq, M. (2015). Success traps, dynamic capabilities and firm performance. *British Journal of Management, 26*(1), 26–44. doi:10.1111/1467-8551.12066

Wang, Z., Wang, N., Cao, J., & Ye, X. (2016). The impact of intellectual capital - knowledge management strategy fit on firm performance. *Management Decision, 54*(8), 1861–1885. doi:10.1108/MD-06-2015-0231

Wilkea, E. P., Costa, B. K., Freire, O. B. D. L., & Ferreira, M. P. (2019). Interorganizational cooperation in tourist destination : Building performance in the hotel industry. *Tourism Management, 72*, 340–351. doi:10.1016/j.tourman.2018.12.015

Wu, I.-L., & Lin, H.-C. (2009). A Strategy-Based Process for Implementing Knowledge Management: An Integrative View and Empirical Study. *Journal of the American Society for Information Science and Technology, 60*(4), 789–802. doi:10.1002/asi.20999

KEY TERMS AND DEFINITIONS

Knowledge Management Infrastructure: The specific orientations on technology, structure, culture, and human resources to knowledge acquisition, knowledge conversion and knowledge application.

Knowledge Management Processes: The management of creation, discovery, acquisition, transfer, sharing, and diffusion of knowledge in an organization.

Knowledge Management Technological Infrastructure: The IT technology and systems to support knowledge acquisition, knowledge conversion and knowledge application.

APPENDIX

MEASUREMENT ITEMS

Knowledge Management Process

1. My hotel has processes for acquiring knowledge about our customers.
2. My hotel has processes for filtering knowledge
3. My hotel has processes for using knowledge to solve new problems.
4. My hotel has processes to protect knowledge from inappropriate use inside the hotel

Knowledge Management Infrastructures

1. Our hotel uses technology to retrieve knowledge about its markets and competition.
2. Our hotel structure facilitates the discovery of new knowledge.
3. The benefits of sharing knowledge outweigh the costs.
4. Employees are specialists in their own part.

Knowledge Management Technology Infrastructures

1. Our organization's ICT is capable of supporting management decisions and knowledge work
2. KM systems and tools in our organization are widely accepted, monitored, and updated
3. Our organization's ICT architecture is capable of sharing data and information, knowledge and expertise with all stakeholders in the organization's extended value chain
4. Our organization's current ICT systems are sufficient to support the daily work

Operational performance

1. Customer satisfaction of our company is better than that of key competitors
2. Quality development of our company is better than that of key competitors
3. Cost management of our company is better than that of key competitors
4. Responsiveness of our company is better than that of key competitors
5. Productivity of our company is better than that of key competitors.

Financial performance

1. Return on investment of our company is better than that of key competitors
2. Return on assets of our company is better than that of key competitors
3. Return on sales of our company is better than that of key competitors
4. Average profitability of our company is better than that of key competitors
5. Profit growth of our company is better than that of key competitors
6. Sales growth of our company is better than that of key competitors.

Chapter 11
Barriers and Enablers to Knowledge Management in the Pakistani Hospitality Industry:
An Application of Fuzzy Delphi Method

Nazima Elahi
Foundation University, Islamabad, Pakistan

ABSTRACT

Knowledge has become a fundamental factor for success of organizations in today's dynamic and competitive era. Due to globalization, the survival of organizations rests upon their ability to attain and retain knowledge. Like other assets, knowledge management (KM) has gained considerable attention in the field of management. Present research is a case study carried for a focalized examination of knowledge management practices in hospitality industry of Pakistan and highlighting the antecedents that impede the KM application to this industry. The study conducted a detailed and comprehensive review of literature including both qualitative and quantitative research studies and formed a list of antecedents. Later, the application of fuzzy Delphi method (FDM) concluded that there are factors like bad law and order situation, lack of support from top management, employees' turnover, conflict among employees, lack of formal knowledge on KM, and its application and poor infrastructure facilities are among the top barriers that hinder the application of KM in the hospitality industry of Pakistan.

INTRODUCTION

Knowledge, being the most valuable asset in an organization and a key segregating factor for formulating successful strategies and hence achieving the overall organizational objectives (Snyman and Kruger, 2004). From a strategic management point of view, the organizations generate, code, stock, and leverage both tacit and explicit knowledge, in order to gain competitive gain (Evans *et al.* 2014). Knowledge further is broadly categorized into two broad categories *i.e. Explicit knowledge* (which is easily acquired can be articulated in written and oral form) and *tacit knowledge*

DOI: 10.4018/978-1-7998-3142-6.ch011

Figure 1. The knowledge value chain
Source: (Weggeman 1997)

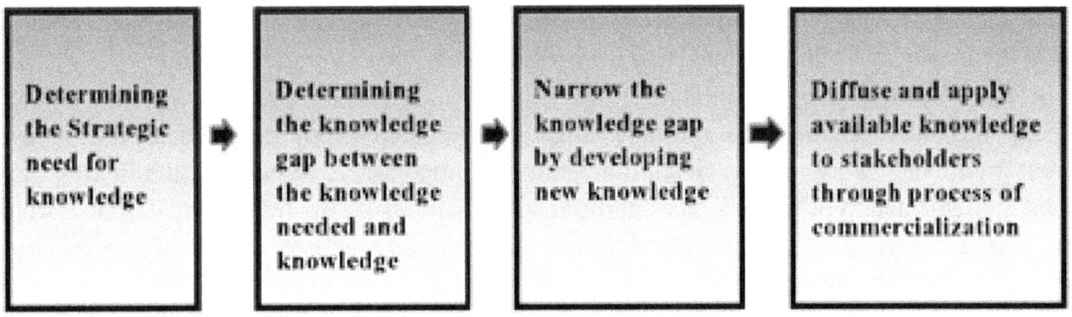

Knowledge management (KM) is a very comprehensive term, the academicians and practitioners define KM from two different perspectives where the former considers it a combination of subjects[1] and disciplines while later define it as an art of creating values from intangible assets. Combining the viewpoint of these two groups, studies define it, as KM is application of knowledge asset to create competitive advantage to organizations (Ahmed *et al.*, 2002) different it is "any process or practice of acquiring, creating, capturing, using knowledge and sharing wherever it resides to improve learning and routine in organizations" (Fukey& Issac, 2014). Yu articulates KM as a set of skills that transforms intellectual capital to business value (Zerega, 2008).

In addition to all this, we are living in an ever-changing business environment there are four major and significant vicissitudes which have changed the business environment in large are globalization, enterprises transformation, the introduction of digital firms and transformations in the fields of tourism and hospitality (Wu & Lin, 2009). These transformations require a dynamic environment and implementation of KM. In this regard, a few studies in literature state that firms following the KM practices perform better than other firms do not follow. More, Brooking (1997) stated that knowledge is a tool, which makes a company powerful in the market and copes up the challenges in an ever-changing environment, moreover, KM plays a critical role in the survival and success of companies.

The Implication of KM in the Hospitality Industry

The hospitality industry occupies a central place in the services sector. It serves to provide catering, accommodation, and numerous supplementary services to visitors and tours nationally and internationally. From an economic point of view, this industry contributes towards sustainable economic development through job creation, investment prospects, and becoming a source of foreign currency. With globalization, entry of big hotel chains in the industry developed a positive market competition and improvement in the quality of services. It ultimately creates a necessity to standardize the quality of services in order to match the competitive standards.

Currently, the hospitality industry works in a knowledge-based economy, where the primary objective of hotels is to achieve sustainable competitive advantage through profit maximization. To gain this, their major focus is on the withholding of customers' satisfaction, reducing operating expenses and employees (Saleem, 2014). The success of this sector lies in the fact that how the customers' needs are fulfilled and how their satisfaction level is catered. Moreover, the hospitality industry is a service provider where the human factor plays a critical role as a determinant of its success. The employees have direct interaction with the customers and hence represent the whole image of the hotel. (Nieves & Segarra-Ciprés, 2015) contended that KM could be helpful to efficient management by using the information on the employees' disposal. Scott and Laws (2006) argued that from a theoretical and empirical viewpoint, better KM holds great promise for tourism and hospitality.

Hospitality and tourism are not the same in nature compared to other organizations as it involves different traditional environments. Given the paramount importance, research in the field of application and implementation of KM practices in the hospitality industry, the research in this field would be useful. Zaei and Zaei (2014) analyzed the tourism and hospitality industry from a KM perspective and argued that in the pursuit of the ever-changing global business environment, it is on the disposal of the organization to utilize limited resources and skills optimally. The study also advocated that Tourism and hospitality is an industry, which needs the enactment of current changes in KM practices. They emphasized the major issues antagonized by this sector in commencing the KM practices.

Significance / Policy Relevance and Specific Objectives of Chapter

Against the importance of the hospitality industry given above, and prominence of KM initiative in this industry, still there remains a research gap in terms of finding the factors which hinder (or serve as an obstacle) to the application of KM practices in this industry. The present study would highlight the significant pitfalls that confront the hospitality industry in its attempt to implement KM practices. The understandings developed through this study would positively contribute towards the body of knowledge that would help to formulate policies for improvement and efficacy of the hospitality industry towards overreaching sustainable development goals (SDGs). The specific objectives of the current study include highlighting the factors, which are enablers or barriers to the KM practices in the hospitality industry of Pakistan. The study also provided a description of challenges faced by the tourism industry in Pakistan and recommended a framework to cope with the issues faced by it.

LITERATURE REVIEW

Literature Review on KM and Hospitality Industry

KM has rapidly emerged and has become the most contended concept in management science. Crnjar (2013) discussed a few principles of the KM approach stating that it is an expensive approach that could prove to be a part of long-term strategy rather than proposing short-term solutions. While achieving organizational goals, it requires top management to be highly involved and supporting. The study explains that KM does not have any direct contribution towards organizational performance; rather it contributes through the indirect channel. Since KM is a broader term, but there are various studies that mentioned the types of knowledge required for the hospitality industry. This includes firstly, *Task-oriented knowledge:* comprise of processes and strategies required fulfilling a particular task (Canon-Bowers and Salas, 2001). The internal coordination of employees can fulfill the purpose without having a direct or indirect interaction extensively; it could be internalized through training. Secondly, *Guest oriented knowledge*: includes the choice of customers (guests), and their future desires in terms of service quality.

A recent study by Voegeli (2019) discussed the trends of KM practices, its limitations and future directions of the hospitality industry. Because of globalization, China allowed several foreign firms to enter their market of the hoteling industry and compete. In this scenario, there arises a potential opportunity to discuss the practices and processes of KM in the hospitality industry of China. In this regard, Wong *et al.* (2017) discussed the issues and challenges faced by the hospitality industry in terms of KM of the Marriot group while entering the Chinese hospitality industry. Saleem (2014) conducted an analysis to check the KM practices in the Egyptian hospitality industry and its relevance to innovation and organizational performance. The study chose 113 five star chain hotels for data collection through questionnaires and empirical analysis. Findings suggested a positive and significant relationship between KM practices, innovation and hotel performance in Egypt.

Dimitrios *et al.* (2018) studied how KM could be used to leverage the tourist organization in Greek. This stressed and explored how KM can be helpful to overcome the crisis and contribute towards innovation and ideas creation. Shah and Baloch (2017) studied the hospitality industry of Pakistan from a customer's satisfaction point of view. In qualitative data analysis, the study found that in order to get a better hospitality industry, customers' satisfaction and service quality is important. Yiu and Law (2013) studied how information and communication technologies (ICT) influences knowledge sharing in the tourism industry of Hong Kong. The study found that ICT has deep roots in the given industry and recommends that top management should focus on KM practices in the given industry to achieve the organizational goal of success.

Avdimiotis *et al.* (2016) argued that the application of tacit knowledge to the hospitality industry is of vital importance and requires more than a positive organizational culture. Cheng (2010) conducted a descriptive analysis and applied comprehensive content analysis to identify current trends in KM research. It included studies conducted between 2006 to 2010 in the field of tourism and hospitality. Out of 1992 searches, only 62 studies relevant to this had been identified Review of literature between this period found that contrary to previous studies, current study found that most of the studies conducted in the context of KM and in the hospitality and tourism industries are empirical in nature and currently the researchers are eager to conduct qualitative studies. Moreover, the combined method to data collection has attracted enormous attention of the researchers. Ozigbo (2015) studied the implementation of KM practices in Nigerian hospitality industries. It tries to develop a model of KM where innovation in the industry could be incorporated. It applied the survey method where data is collected from 240 respondents. Empirical analysis showed that KM is the most important thing that improves the competitive advantage of the organizations. The study also highlighted that there is an important and significant role of top management in legitimizing and empowering the KM practices. Ababneh (2008) studied the nature of the link between organizational innovation and KM and found that organizational learning along with innovation has a positive correlation with KM practices.

Ruhanen and Cooper (2015) presented a view that the KM approach must become a part of future research agendas in order to transform intellectual capital in tourism to achieve sustainability and competitiveness in the industry. Hallin and Marnburg (2008) discussed the theoretical and empirical literature on KM and its link with the tourism and hospitality industry and presented details on static and dynamic perspectives of knowledge management. The study highlighted that future insight into the exploration of empirical research is highly needed in this field. There is also a dire need for a sound and strong research design in this context to rationalize the empirical research. Research also suggested conducting analysis to find the factors, which hinder the KM practices before its implementations. Bouncken (2015) checked the quality management of the hospitality industry and discussed how it can enhance the quality of services by raising the knowledge of employees on how to cater to customers through knowledge. The study also focusses on how employees turn over, employees' rotation, increasing ratio of unskilled workers and seasonal demand on customers' preferences are significant in reducing the quality of performance. The findings provided a framework, where the application of KM practices is required in the presence of these factors. Tzortzaki and Mihiotis (2012) discussed a three-dimensional knowledge management framework for the tourism and hospitality industry. The study focused on how managers of hotels can best use the knowledge of their employees in order to enhance their adaptability and help to develop unique capabilities. Findings highlighted that managers' lack of knowledge on the market and the importance of employees' internal knowledge and its applicability.

The tourism and hospitality industry contributes towards sustainable development and has a direct association with sustainable development goals (SDGs). The SDGs are also termed as Global goals introduced since 2015, have been adopted by all member nations of the United Nations (UN) to end poverty and bring prosperity and peace to everyone by 2030.SDGs broadly address the challenges of poverty, inequality, prosperity, peace, environmental degradation and justice (United Nations, 2018). Improvement and application of KM in this industry positively enhance the real GDP growth through new job creation, attracting foreigners' spending and making payments through taxes. A report in 2017 showed that the hospitality industry holds a prominent place in the jobs market, and it created 313 million jobs worldwide which is 9.9% of total employment and a part of 20% in overall jobs created for one decade. It promotes cultural industries and fine arts fields. The hospitality industry on a wider ground occupies a central place from customers' needs perspectives and its importance for employees.

Literature Review on Enablers and Barriers to Knowledge Management

Currently, the major findings of the studies depict that the hotel industry lacks the application and practices of a knowledge-based economy. As suggested by Lengnick-Hall & Lengnick-Hall (2003), technological growth demands the application of knowledge management to achieve a competitive advantage. Since, in this effort, there is the direct involvement of employees hence the management of human resources i.e. HRM is a part of this process. However, managing people is not sufficient rather there is a need for managing organizational capabilities, managing relationships and managing learning and knowledge (Ulrich, 1997; Saint- Onge, 2001; Coates, 2001; Lengnick-Hall & Lengnick-Hall, 2003)[2]. In the context of the hospitality industry, in order to achieve a sustainable competitive advantage, Oliver and Kandadi (2006) suggested that four keys areas are important i.e. roles, responsibilities, strategic focus and learning.

Rudez (2010) discussed the interrelationship between the KM and hospitality industry of Slovenia, before and after its entry into the European Union (EU). The findings articulated that entry into the EU market did not have any touch on the performance of its hospitality industry. The major endorsements of the analysis included exposing the Slovenian hospitality industry to KM practices, the construction of clearer guidelines on KM in the said industry, and barriers that affect KM in the hospitality industry must be tackled efficiently to get good results. On similar grounds, Mahapa (2013) investigated the influence of KM practices on the organizational performance of the hospitality industry in Zimbabwe. The data collected through structured interviews from 15 managerial employees and 35 non-managerial staff members of large hotels in Zimbabwe showed that organizations are in need of exploiting more knowledge resources and using them to achieve ideal organizational performance. Moreover, the organizations necessitate making KM practices into their overall business strategies. Hawamdeh and Al-edenat (2019) tried to identify the barriers to the hospitality industry of Jordan. The study collected data through questionnaires and utilized it to conduct a quantitative analysis. Findings suggested that there are individual as well as organizational barriers, which affected the hospitality industry of Jordan. The most significant factor of organizational structure has affected largely and it supported introducing and encourage a reward system for all those employees who share knowledge meritoriously. Likewise, discussing the barriers Freydouni and Woods (2010) discussed the factors of knowledge sharing which act as barriers and affect the hoteling industry of Malaysia. The study collected data for empirical analysis through 23 semi-structured interviews of the managers and senior staff members of four-star and five-star hotels across the country. Results suggested that transmission of knowledge within the organization and personal attitudes serve to be a significant barrier to knowledge management in the said hotels of Malaysia. The study comprehended that Malaysia is a tourist country and it possesses the potential to enhance efficacy and organizational capability.

Jackson (2017) studied, what challenges are confronted with the hospitality and tourism industry of Malawi and what are the possible opportunities in this regard. It discussed that industries face certain hindrances such as the development of infrastructure, marketing issues, corrupt practices and lack of ICT technologies. It collected data through a questionnaire survey from 195 respondents and analyzed this information using both qualitative and quantitative methods. It also pointed out that quantitative methods of data analysis are more reliable and provide a better understanding of the insight to the research problem.

By considering the review of the studies given above, the current study may be right in making the argument that the application of the KM approaches to tourism, the hospitality industry will help to increase the pace of opportunities, and the sector will become more profitable, competitive and efficient. Moreover, its operations will be intelligent and characterized by learning. In addition to this, the literature on issues and obstacles to the application of KM practices highlights various critical factors and issues that must be tackled to improve its performance and make this sector in line with the other productive sectors of the economy.

TRENDS OF HOSPITALITY INDUSTRY AND KNOWLEDGE MANAGEMENT

Historical Origin of KM, its Cycle and Processes

Historically, the development of KM consists of three significant phases. Initially (Phase 1, 1960) this concept emerged as an idea in the field of Information Technology (IT), where KM focused on the IT and communication field, using which, organizations made different ways to enhance productivity by means of available information on a large scale. While, during the second phase (the mid-1980s to onward), the businesses experiences shift from production orientation to market orientation. During this phase, the customers were attracted to a service based on service quality and the best deal on goods and services. Few organizations of this time applied KM practices to become leaders in the market. The last phase (current), it shifted to a knowledge economy. Cooper (2006) argued that out of these three phases the current one is rendering best to the tourism and hospitality industry.

[3]Demarest (1997) in Rudez, (2010) defined major areas of KM as construction, embodiment, dissemination, use, and management. The knowledge management cycle (KMC) comprises of seven exclusive stages. These phases are Identifying/creating, storing, sharing, using, learning, improving and creating organizational knowledge (Evans *et al.*, 2014). In terms of achieving sustainable organizational performance, KM implementation plays a significant role through innovations, the role of innovation has been

discussed by Smith and Schurink (2005), who termed KM as an approach that prompts knowledge from well-informed individuals and sharing it with appropriate individuals and putting this to actions in order to increase organizational performance.

There are various KM cycles, one of which has been discussed by Mohajan (2016). It presented four KMCs in his study. Firstly, Meyer and Zack KMC (1996), consisting of five stages; acquisition, refinement, storage/retrieval, distribution, and presentation. Secondly, the Bukowitz and Williams KMC (2000), consisting of seven stages given as getting knowledge, using knowledge, learning, contributing to knowledge, assessing knowledge, building and sustaining knowledge, and divesting knowledge. Thirdly, the McElory KMC (1999) comprising of four steps; individual and group learning, knowledge claim formulation, information acquisition, and knowledge validation and Fourthly, Wiig KMC (2005) made by four consecutive stages i.e. building, holding, pooling, and using knowledge. After a careful review of all the KMCs, Dalkir (2005) proposed an integrated cycle of KM. Figure 2 below highlights the major stages of integrated KMC.

Figure 2. Integrated KMC
Source: (Dalkir, 2005)

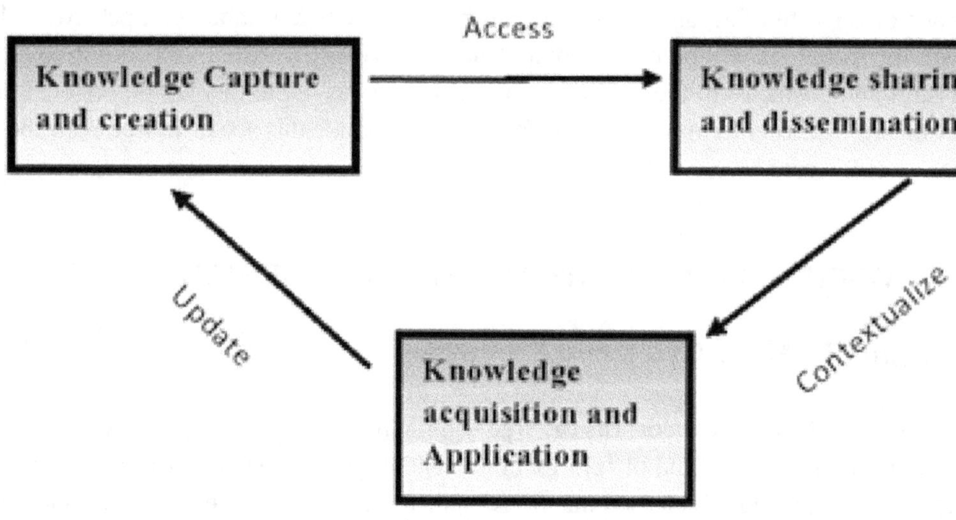

Overview of Hospitality Industry of Pakistan

The hospitality industry is a kind of services sector, which is characterized, by a high level of customer engagement. This industry in Pakistan, despite its promising potential, has been facing a crisis due to prevailing law and order situation. According to (Kandampully & Juwaheer, 2009), Pakistan occupies a position of 47 out of 200 nations in the world travel and tourism council development. The sector could prove to be a significant source of job creation and foreign currency generation. Over time, with continuous support from the Ministry of Commerce, and Pakistan Hotels Association (PHA). PHA was established in1993, under Trade Organization Ordinance (TOO), 1961 and termed as a sole authority, which represents the hotel industry of Pakistan. It is mandatory to all hotels (including all categories) to be a member of this association. The objectives of its establishment were to coordinate and give directions and suggestions on tourism development in Pakistan. Moreover, being a leading sector, it contributes towards national development through payment of taxes and generating revenues. [4]

The hospitality industry of Pakistan, hospitality has remained active since independence; however, it has continued in its infancy for long. During the whole process, it has contributed towards the generation of foreign currency and socio-economic development through job creation and multi-tasks. It is evident that terroristic activities and poor security situation created a gloomy impression of Pakistan and its revenues from tourism and hospitality declined abruptly (Ali, 2010) over the last few decades. As an example, Swat is a famous tourist destination because of its rich cultural significance and civilizations. Due to bad law and order situation and military operation, this valley had a loss of 60 billion and around 40.000 people associated with the hospitality industry remained jobless between 2007 to 2009 (Ali, 2010).

Looking at growth trends, we see that the tourism and hospitality industry has not grown at a faster pace, rather its growth has remained slowed down, particularly the decades of the 1980s and 1990s (Hashim *et al*, 2010). (Prabakaran. 2003) stated that " The total number of hotels in Pakistan has increased from 945 in 1991 to 1250 in 1996, showing an improvement growth of more than 26% in 1993 which reduced to 2.4% in 1995 before again increase in 1996." The rapid urbanization and industrialization resulted in the cities development and people from other areas and foreigners are attracted to tourists' spots. The hotel industry was also disturbed after Marriot's bomb attack. Recently, the government of Pakistan has initiated various measures to improve the hospitality and tourism sector, and this industry is making its way to contribute heavily to socio-economic development.

There is enormous importance of Knowledge and Human resources (HR) *(i.e.* intangible assets) in the hospitality industry. In the framework of Pakistan, the efforts to implement KM in this industry are taking place at a slow pace hence, it is important to identify and discuss the critical factors serving as a barrier to its implementation. Despite the rapidly growing significance of the hospitality industry, there are barriers, which happen to create difficulties in the initiation and implementation of KM practices. The president of Pakistan Dr. Arif Alvi shared in a meeting that "*Pakistan has a lot of potential for hospitality industry, as it has diverse culture, landscape, religious places, to attract more local and international tourists*", More, it has been reported that "In *2018, travel and tourism accounted for over 10% of the world's GDP, created 1 out of 5 new jobs, and grew by 3.7%. The number of international tourists in Pakistan is expected to increase to 1.4 million in 2019.*" (May 27, 2019, The Nation). In this regard, contemporary work done suggested the structural transformation of the Pakistan Tourism Development Corporation (PTDC) and the National Tourism Coordination Board (NTCB). The shaping up of the tourism industry will open new avenues towards economic growth and poverty alleviation (Global Village Space, August 2019). Hashim *et al.* (2016), because of a SWOT analysis, collected data from local and foreign tourists and comprehended that they prefer to stay at a 5-star hotel. It also showed that Pakistan could attract more foreign tourists by providing quality services, creating innovations and avoiding risk.

Writing by Sarfaraz Ali (2016) [5] highlighted that improved law and order situation in Pakistan is a clear determinant of the tourism industry, improvement in this is enhancing tourism and hence ultimately it will advance the hospitality industry. Arshad *et al.* (2017) discussed the problems faced by the Pakistani tourism and hospitality industry and highlighted that there are many factors that are hindering the development of this sector. It suggested that law and order situations, support from media, slow development of infrastructure, and lack of industry-academia linkages are playing a critical role in the development of the tourism industry and hence the hospitality industry has been highly affected. Moreover, it argued that efforts from governmental agencies to formulate policies and code of ethics are also mandatory for development. Jamal Shahid (2019)[6] highlighted a general practice on social media that bloggers usually present a gloomy impression of the Pakistani tourism industry and hence there is negativity observed around. Abuzar Khan Buzdar [7] discussed that efforts from the government on infrastructure development are seriously missing and there is the poor condition of roads and hotels. Sometimes land sliding seriously hits the areas and tourists do not get an alternative for this. There are no sufficient spaces in hotels for tourists (if exceed a number) and no quality checks from the government are made to fix the rates or facilities.

Enablers and Barriers to Knowledge Management in the Hospitality Industry of Pakistan

The literature cites various studies, which highlight the application of KM to different organizations and industries of Pakistan. These studies try to check the impact of KM on organizational performance. A study by Munir and Bibi (2016) studied the impact of KM practices on the performance of the hospitality industry of Pakistan and found that absence KM practices affect the performance of the hospitality industry in Pakistan. Kim and Hencer (2010) discussed the KM application and organizational effectiveness of the restaurant industry and found that leadership influences the KM practices in the restaurant industry. The major factors as highlighted by a number of studies are given as:

Information and Communication Technologies (ICT): The growth in information and communication technologies has had an impact on sustainable competitive advantage and acts as an enabler to facilitate KM application in the hospitality and tourism industry Zvobgo (2015). *Human Resource Management*: HRM acts as a facilitator and enabler to enhance KM practices in the tourism and hotel industry, as mentioned by various studies including The primitive focus on managing people has been broadened to managing organizational capabilities, managing relationships and managing learning and knowledge (Lengnick-Hall & Lengnick-Hall, 2013. They argue that under traditional HRM it is not sufficient to manage people only rather it needs broadening of concept and must include managerial capabilities, managing relations, and learning.

Crnjar (2013) divided these factors into four broad categories *i.e. methodology*, *technology*, *human capital,* and *organization*. It argued that *organizational culture* is a very important factor and determines the efficacy of implementation of KM if it is encouraging of knowledge sharing. *Motivation and employee education* is important and comes under the human capital head. It also included the organizational support and adaptability of employees to accept knowledge management practices. Moreover, according to this study, *organizational structure* and movement of employees across the organization also matter for the best implementation of knowledge management. Studies highlighted that *motivation* from managers/ peers and their skills to motivate employees in order to utilize their knowledge in implementing KM practices are of critical importance. The *support from management* helps to achieve this goal of KM practices implementation in industries in a short span of time (Jafari, 2006). Moreover, it also suggested that culture, *strategic planning* also act as an enabler to facilitate quick and effective KM implementation. In addition to these factors, *uncertainty*, *law and order* situation, *protection laws* also serve as factors of importance.

METHODS AND FINDINGS

Here for analysis purposes, we used qualitative methods to analyze the barriers, which impede the application of KM in the hospitality industry of Pakistan. We applied the Fuzzy Delphi Method (FDM), where experts from the field helped us to narrow down the barriers, which are most relevant to the study. The FDM has long been used in various studies and it is a method consisting of a group of experts who review a particular issue and form a consensus. The study is characterized by the strength that it can be helpful to conduct a study and make a decision in a short time and it could help to finalize on a point by one round only (Kamarulzaman *et al.*, 2015).

Lezama *et al.* (2014) argued that large set of variables (identified after careful review) may complicate the model and further policy implications, therefore FDM helps to narrow down the set, and this method has three strengths i.e. anonymous responses, interaction and controlled feedback, as well as a statistical group response. We followed a three steps methodology of FDM. In step 1, we solicited viewpoints of a panel of experts on a particular issue through an open-ended survey tool. In step 2, we ask experts to rate the statements according to their expertise while in step 3, we ask them to reassess the factors based on results in step 2.

In the first round, we have carried out a detailed and exhaustive review of the literature and screened out a list of various factors, which affect the KM application to the hospitality industry in Pakistan. A detailed list of barriers is given below in Table 1 below.

Table 1. Barriers to KM practices, Hospitality Industry of Pakistan

No.	Barriers faced in the context of Knowledge Management Implementation
1.	Conflicts among employees
2.	Uncertainty due to bad law and order situation
3.	Lack of support from top managers
4.	Employees turnover
5.	Lack of teamwork
6.	Confrontation to Change
7.	Protection laws for seas and northern areas
8.	Lack of investment for infrastructure development
9.	Lack of support from academia
10.	Lack of skill development in the hospitality industry
11.	Lack of media support, particularly social media
12.	Lack of professional education

Source: Detailed review of literature, personal interviews, and questionnaires

In the second round, we discussed these barriers to a group of experts in order to smooth further data analysis. FDM helps to decide on experts according to our convenience. We selected a group of seven experts, because of their expertise and interest in KM. A variety of experts group was selected including three from academia (university professionals), two from the hotels sector (Hotel owners and managers), and two from tourism and development corporations. We asked the question "What factors, you think are impeding the application of KM in the hospitality industry of Pakistan?" We recorded their responses and made a matrix of factors. At the final stage, we identified the most important factors, which are actually serving as a hindrance to KM application in this industry. The barriers, which have been identified, include employees' turnover, lack of support from top management, conflict among employees, uncertainty arising due to bad law and order situation and lack of teamwork.

CONCLUSION

Knowledge Management has been attracting the attention of academicians and policymakers since the last decade. Its application to different industries is regarded as a driver to success and achieving sustainable competitive advantage. By following the growing importance of the hospitality industry, which contributes to national development, we tried to find out the factors that act as a barrier to KM application. This case study applied a qualitative data analysis approach to meet the specific objectives. We started by presenting the concept of KM for a layman, its different stages, cycle, and KM processes. Further, we presented an extensive literature review that provided evidence on knowledge management and its link with the hospitality industry along with highlighting the barriers, which hindered the KM application to the hospitality industry in general and particularly to Pakistan. In the next step, we formed a group of experts to apply the Delphi method and reach the narrowed list of factors.

The findings highlighted that among many factors, there are a few critical elements, which are a true hindrance to KM application in the hospitality industry. This includes, firstly, *employees turnover*, as supported by (Nieves & Segarra-Cipres, 2015), which argued that employees get a piece of knowledge on a sector and with their rapid turnover, new employees enough time to settle in a new environment, and hence process of KM application and knowledge sharing gets halted. Moreover, the employees do not have formal education, training, and result in the slow process. Secondly, *bad law and order situation*, this has been supported by Ali (2010) and (Kandampully & Juwaheer, 2009), weakens the process of KM application in this industry. Overall tourism and hospitality industry have been affected due to (particular examples include) terroristic activities (a bomb blast in Marriot hotel, Islamabad) and military operation in Swat two years back. More, on social media, bloggers also present a poor impression of the tourism and hospitality industry in order to get them viral. Thirdly, support from top management, in case of Pakistan, since there is no proper code of ethics and top managers are not very much aware of KM importance in this industry hence they do not support their employees for this. Fourthly, infrastructure development and lack of quality checks on hotels. Lastly, the lack of teamwork, which is a critical factor to support KM application. Based on the analysis conducted we may be right in developing the consensus that:

a) Cooperation from International Hotels and Restaurants Association (IR&HA) is needed and in this regard efforts from the government are required to arrange awareness seminars with international hoteliers to share their practices and process for the whole course of action.
b) Human Resource Management is a critical factor and there is a need to go beyond the only people management and moving towards managing capabilities, enhancing training facilities in the long run.

c) There is a dire need from media and particularly from social media to present a positive impression of the tourism and hospitality industry of Pakistan. For this, again initiatives are invited from public sector organizations.
d) Since the law and order situation in the country is controlled, still there is a need to ensure the security of tourists through various measures.
e) Infrastructure development is a critical factor in this regard. Public-private investment collaborations in this regard could be helpful to improve the infrastructure and facilities at the hotels.
f) A linkage between academia and industry seems missing in the case of Pakistan. To develop, strong hospitality and tourism industry practicing KM practices we need to devise a framework for the development of this linkage.

LIMITATIONS AND FUTURE DIRECTIONS

While looking at the current literature on KM and its implementation, the data availability posed to be the major problem and further research and insight requires more rigorous and reliable data and data analysis strategies. More research is needed in the areas of the actual industrial sector. A linkage between research communities and the hotel industry is mandatory to explore the issues related to this area of research and finding further ways for its improvements. In the future, more studies with a mixed-method approach and quantitative data analysis approach are required. In this context, comparative analysis between big food chains operated domestically and international hotel chains could help to find more and more effective strategies.

APPENDIX: ACRONYMS

EU European Union
FDM Fuzzy Delphi Method
HR Human Resources
HRM Human Resource Management
ICT Information and Communication Technologies
IT Information Technology
IR&HA International Restaurants and Hotels Association
KM Knowledge Management
KMC Knowledge Management Cycle
KMP Knowledge Management Process
NTCB National Tourism Coordination Board
PHA Pakistan Hotels Association
PTDC Pakistan Tourism Development Cooperation
SDGs Sustainable Development Goals
SWOT Strengths, Weaknesses, Opportunities & Threat
TOO Trade Organization Ordinance

REFERENCES

Ababach, R. J. (2008) The Impact of knowledge management and organizational learning on organizational innovation. Proceedings of the knowledge management international conference on Transferring managing, maintaining knowledge for National Capacity Development.

Al Hawamdeh, N., & Al-edenat, M. (2019). Determinants of Barriers to Knowledge Sharing in the Jordanian Hospitality Industry. *International Business Research, 12*(7), 121–132. doi:10.5539/ibr.v12n7p121

Ali, A. (2010). Socio-Economic cost of terrorism: A case study of Pakistan. *Strategic Studies, 57*, 1–14.

Arshad, M. I., Iqbal, M. A., & Shahbaz, M. (2017). Pakistan tourism industry and challenges: A review. *Asia Pacific Journal of Tourism Research*. doi:10.1080/10941665.2017.1410192

Avdimiotis, S. (2016). Tacit knowledge management within hospitality establishments: Revealing the body of the iceberg. *International Journal of Knowledge Management, 12*(3), 15–29. doi:10.4018/IJKM.2016070102

Bouncken, R. B. (2002). Knowledge management for quality improvements in hotels. *Journal of Quality Assurance in Hospitality & Tourism, 3*(3-4), 25–59. doi:10.1300/J162v03n03_03

Brooking, A. (1996). *Intellectual Capital: Core Asset for the Third Millennium Enterprise*. Thomson Business Press.

Cooper, C. (2006). Knowledge Management and Tourism. *Annals of Tourism Research, 33*(1), 47–64. doi:10.1016/j.annals.2005.04.005

Crnjar, K. (2013). *Drivers and Barriers to Introducing Knowledge Management Initiatives in Hotel Enterprises in Croatia*. Academic Press.

Demarest, M. (1997). Understanding knowledge management. *Long Range Planning, 30*(3), 374–384. doi:10.1016/S0024-6301(97)90250-8

Dimitrios, B., Ioannis, R., Efstathios, V., Christos, A., Dimitrios, T., & Labros, S. (2018). Successful and Efficient Knowledge Management in the Greek Hospitality Industry: Change the Perspective! *Academic Journal of Interdisciplinary Studies, 7*(1), 185–191. doi:10.2478/ajis-2018-0019

Evans, M., Dalkir, K., & Bidian, C. (2014). A holistic view of the knowledge life cycle: The knowledge management cycle (KMC) model. *Electronic Journal of Knowledge Management, 12*(2), 85–97.

Freydouni, S. G., & Woods, P. (2010). Barriers of knowledge sharing in Malaysian hotels. In *5th Knowledge Management International Conference (KMICe)* (pp. 461-466). Academic Press.

Fukey, L. N., & Issac, S. S. (2014). Connect among green, sustainability and hotel industry: A prospective simulation study. *Energy Conservation, 6*, 8.

Hallin, C. A., & Marnburg, E. (2008). Knowledge management in the hospitality industry: A review of empirical research. *Tourism Management, 29*(2), 366–381. doi:10.1016/j.tourman.2007.02.019

Hu, H. H., Kandampully, J., & Juwaheer, T. D. (2009). Relationships and impacts of service quality, perceived value, customer satisfaction, and image: An empirical study. *Service Industries Journal, 29*(2), 111–125. doi:10.1080/02642060802292932

Jackson, S. W. (2017). *A case study of hospitality and tourism challenges and opportunities in Malawi.* Academic Press.

Joong Kim, Y., & Hancer, M. (2010). The effect of knowledge management resource inputs on organizational effectiveness in the restaurant industry. *Journal of Hospitality and Tourism Technology, 1*(2), 174–189. doi:10.1108/17579881011065065

Kamarulzaman, N., Jomhari, N., Raus, N., & Yusoff, M. Z. M. (2015). Applying the Fuzzy Delphi Method to analyze the user Requirement for user Centered Design Process in Order to Create Learning Applications. *Indian Journal of Science and Technology, 8*(32), 1–7. doi:10.17485/ijst/2015/v8i32/92146

Khan, I. A., & Khan, M. S. (2018). *Developing Sustainable Agriculture in Pakistan.* CRC Press. doi:10.1201/9781351208239

Mahapa, M. (2013). Impact of knowledge management strategies on organizational performance in the hospitality industry of Zimbabwe. *Public Administration Review, 2*(1), 76.

McIver, D., Lengnick-Hall, C. A., Lengnick-Hall, M. L., & Ramachandran, I. (2013). Understanding work and knowledge management from a knowledge-in-practice perspective. *Academy of Management Review, 38*(4), 597–620. doi:10.5465/amr.2011.0266

Mohajan, H. (2016). *A Comprehensive Analysis of Knowledge Management Cycles.* Academic Press.

Nieves, J., & Segarra-Ciprés, M. (2015). Management innovation in the hotel industry. *Tourism Management, 46*, 51–58. doi:10.1016/j.tourman.2014.06.002

Ozigbo, N. C. (2015). Knowledge Management Implementation in the Nigerian Hospitality Industry. *Review Pub Administration Manage, 3*(147), 2. doi:10.4172/2315-7844.1000148

Peyman, A. M. J. (2006). Critical issues for knowledge management implementation at a national level. *Vine, 36*(1), 52–66. doi:10.1108/03055720610667372

Prabakaran, R. (2003). Good practices in planning and management of integrated commercial poultry production in South Asia. Food and Agriculture Organization.

Rudez, H. (2010). Knowledge management in the hotel industry before and after the entry in the EU: The case of Slovenia. *Organizacija, 43*(4), 197–201. doi:10.2478/v10051-010-0019-9

Salem, I. E. B. (2014). Toward better understanding of knowledge management: Correlation to hotel performance and innovation in five-star chain hotels in Egypt. *Tourism and Hospitality Research, 14*(4), 176–196. doi:10.1177/1467358414542265

Sanchez-Lezama, A. P., Cavazos-Arroyo, J., & Albavera-Hernandez, C. (2014). Applying the Fuzzy Delphi Method for determining socio-ecological factors that influence adherence to mammography screening in rural areas of Mexico. *Cadernos de Saude Publica, 30*(2), 245–258. doi:10.1590/0102-311X00025113 PMID:24627054

Scott, N., & Laws, E. (2006). Knowledge sharing in tourism and hospitality. *Journal of Quality Assurance in Hospitality & Tourism*, *7*(1-2), 1–12. doi:10.1300/J162v07n01_01

Shah, s. N. U., & Baloch, D. Q. B. (2017). *Comparative Analysis on Service Quality & Customer Satisfaction: Empirical Evidence from Pakistan Hotel Industry*. Academic Press.

Smith, H., & Schurink, W. (2005). The interface between knowledge management and human resources: A qualitative study. *SA Journal of Human Resource Management*, *3*(1), 6–13. doi:10.4102ajhrm.v3i1.58

Snyman, R., & Kruger, C. J. (2004). The interdependency between strategic management and strategic knowledge management. *Journal of Knowledge Management*, *8*(1), 5–19. doi:10.1108/13673270410523871

Soliman, M., & Shou, M. A. A. (2016). Knowledge Management and Tourism: An Exploratory Study Applied to Travel Agents in Egypt. *Environments*, *22*, 31.

United Nations. (2018). *About the Sustainable Development Goals*. Retrieved from https://bit.ly/2jHjQmD

Voegeli, C. (2019). Understanding the Current State and Future Directions of Hotel Sector Knowledge Management: A Literature Review. *Business and Economic Review*, *9*(2), 118–145. doi:10.5296/ber.v9i2.14778

Weggeman, M. (1997). *Kennismanagement – inrichting en besturing van kennisintensieve organizations*. Scriptum Management.

Wong, T., French, L., & Wickham, M. (2017). An examination of the knowledge management process in the emerging Chinese hotel industry. *Knowledge Management Strategies and Applications*, 239.

Wu, C., & Lin, C. (2009). Case study of knowledge creation facilitated by Six Sigma. *International Journal of Quality & Reliability Management*, *26*(9), 911–932. doi:10.1108/02656710910995091

Xu, C. (2010). *A Systematic Review of Knowledge Management Research in the Hospitality and Tourism Industry* (Dissertation). University of Nevada Las Vegas.

Yiu, M., & Law, R. (2013). Review and application of knowledge management and knowledge sharing in tourism. *Asia Pacific Journal of Tourism Research*, *19*(7), 737–759. doi:10.1080/10941665.2013.812128

Zaei, M. E., & Zaei, M. E. (2014). Knowledge management in hospitality and tourism industry: A KM research perspective. Information and Knowledge Management, 4(9), 114-122.

Zerega, B. (2008). *Art of knowledge management*. http://www.accessmylibrary.com/article-1G1-20967956/art-knowledge-managementexperts

Zvobgo, T. C. (2015). *A comparative study into the effectiveness of business rescue strategies in South Africa and the judicial management system in Zimbabwe* (Doctoral dissertation). University of Pretoria.

KEY TERMS AND DEFINITIONS

Employee Turnover: Employee turnover refers to the number or percentage of workers who leave an organization and are replaced by new employees. Measuring employee turnover can be helpful to employers that want to examine reasons for turnover or estimate the cost-to-hire for budget purposes.

Fuzzy Delphi Method: A combination of the traditional Delphi Method with Fuzzy Set Theory in order to address some of the ambiguity of the Delphi panel consensus (Ishikawa et al., 1993). The Fuzzy Delphi is a more advanced version of the Delphi Method in that it utilizes triangulation statistics to determine the distance between the levels of consensus within the expert panel.

Hospitality Industry: A broader term which is referred to a services industry which serves the process of lodging, food provision, event plans, transportation, parks and amusement activities along with provision of cruise line and other services to tourism industry.

Human Resource Management: A strategic approach to achieve effective people management at a work environment in any organization. The human resource management serves as a source to achieve comparative advantage and optimize employee performance.

Human Resources: The people who make a work force for an organization is called human resource. Another term of 'human capital' is also used to replace the term.

Information and Communication Technologies: This is referred to technologies that provide access to information through telecommunications. It is similar to Information Technology (IT), but focuses primarily on communication technologies. This includes the Internet, wireless networks, cell phones, and other communication mediums.

Knowledge: A term used in the context of awareness, familiarity, and understanding of something including information, description, or skills acquired either through process of learning or by discovering, perception discovering, or learning. Knowledge can refer to a theoretical or practical understanding of a subject.

Knowledge Management: A process that creates, shares, uses and manages the information or knowledge of a particular organization. This usually comprise of a multidisciplinary approach to achieve the organizational targets and utilizes the knowledge to its best.

Knowledge Management Process: A process of knowledge collection, analysis, and its management is termed as knowledge management process. The business information is disseminated across the organization.

Pakistan Tourism Development Cooperation: Pakistan Tourism Development Corporation is an organization of the government of Pakistan governed by the Board of Directors and responsible for providing transportation services to various tourist attractions. It is also running the Motels across the attractive places of countries, and was established in 1970.

Sustainable Development Goals: A collection of 17 global goals, which are particularly formulated, to be a blueprint to attain sustainable future for everyone in a society is Sustainable Development Goals (SDG). United Nation General Assemble sets these SDGs in 2015.

ENDNOTES

1 Management, Information Technology, Psychology and organizational behavior.

2 https://kmmgt6660sept2015.wordpress.com/2015/11/01/how-can-knowledge-management-improve-customer-service-in-hotel-industry/

3 The details on historical origin of KM has been taken from Cheng (2010).

4 The description and purpose of this association has been taken from www.pha.org.pk

5 Why Hospitality Industry has bright future in Pakistan, sep 26- 2016- published in Daily Pakistan Global).

6 April 22, 2019: Foreign tourists face several problems, concede tour operators: Dawn News).

7 The Nation, (June 17, 2019), Tourism Industry: Potential Challenges and Misplaced Focus of Government.

Chapter 12
A Survey on Recent Recommendation Systems for the Tourism Industry

S. Ranjith
Vignan's Foundation for Science Technology and Research, India

P. Victer Paul
Indian Institute of Information Technology, Kottayam, India

ABSTRACT

Data mining is an important field that derives insights from the data and recommendation systems. Recommendation systems have become common in recent years in the field of tourism. These are widely used as a tool that can input various selection criteria and user preferences and yields travel recommendations to tourists. User's style and preferences should be constructed accurately so as to supply most relevant suggestions. Researchers proposed various types of tourism recommendation systems (TRS) in order to improve the accuracy and user satisfaction. In this chapter, the authors studied the current state of tourism recommendation system models and discussed their preference criteria. As a part of that, the authors studied various important preference factors in TRS and categorized them based on their likeness. This chapter reports TRS model future directions and compiles a comprehensive reference list to assist researchers.

INTRODUCTION

Data Mining (DM) (Dunham, 2002) is used to discover information from the company's various databases and re-construct it for uses other than the databases which were initially planned for data mining implementation is different for variant organizations depending upon the nature of data and organization. It specifies to juice or mining insight from enormous data, mining knowledge from data is called data mining. In the process of discovering knowledge data Mining plays an essential step,

DOI: 10.4018/978-1-7998-3142-6.ch012

Figure 1. Organization of the work

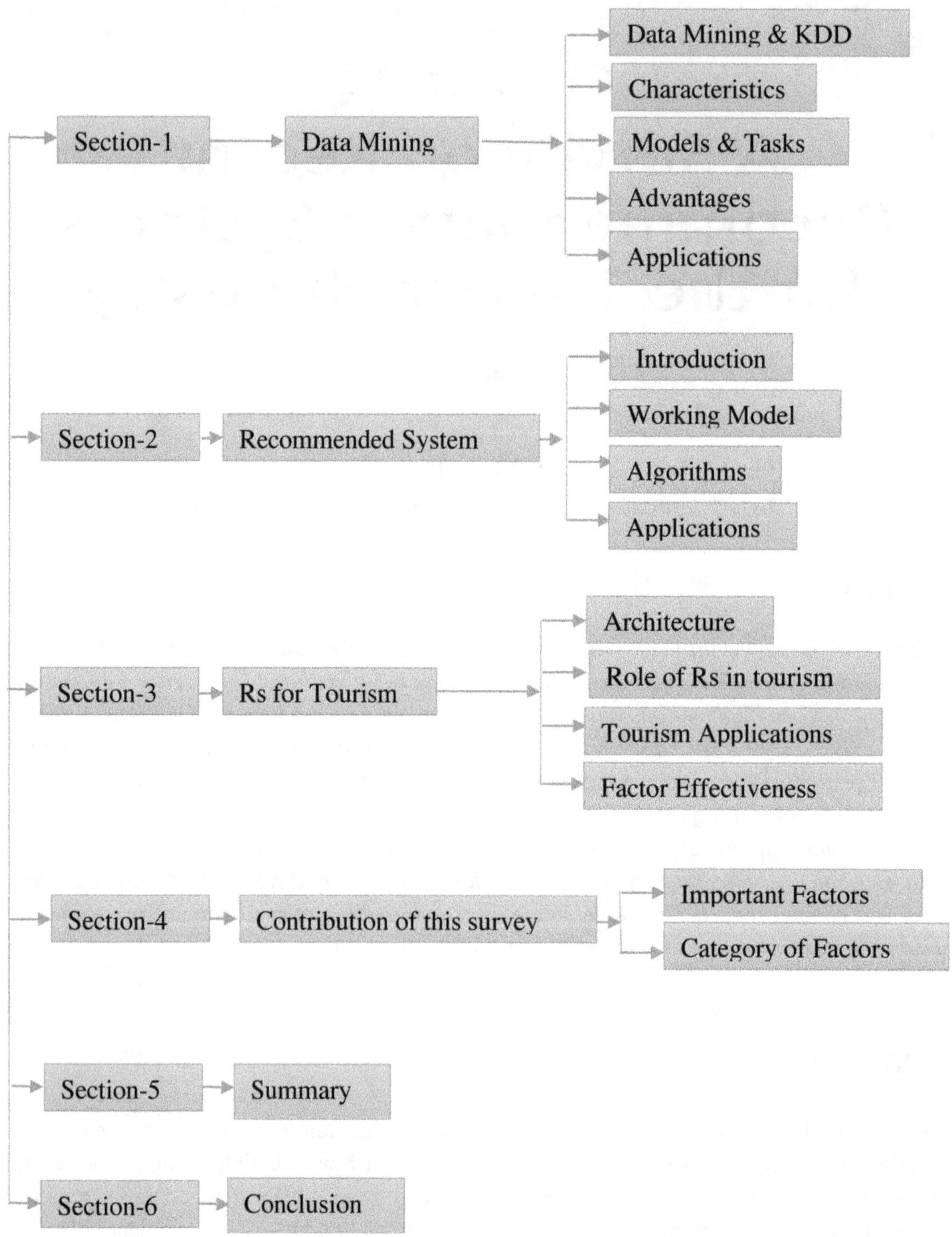

Figure 2. Data mining knowledge discovery process

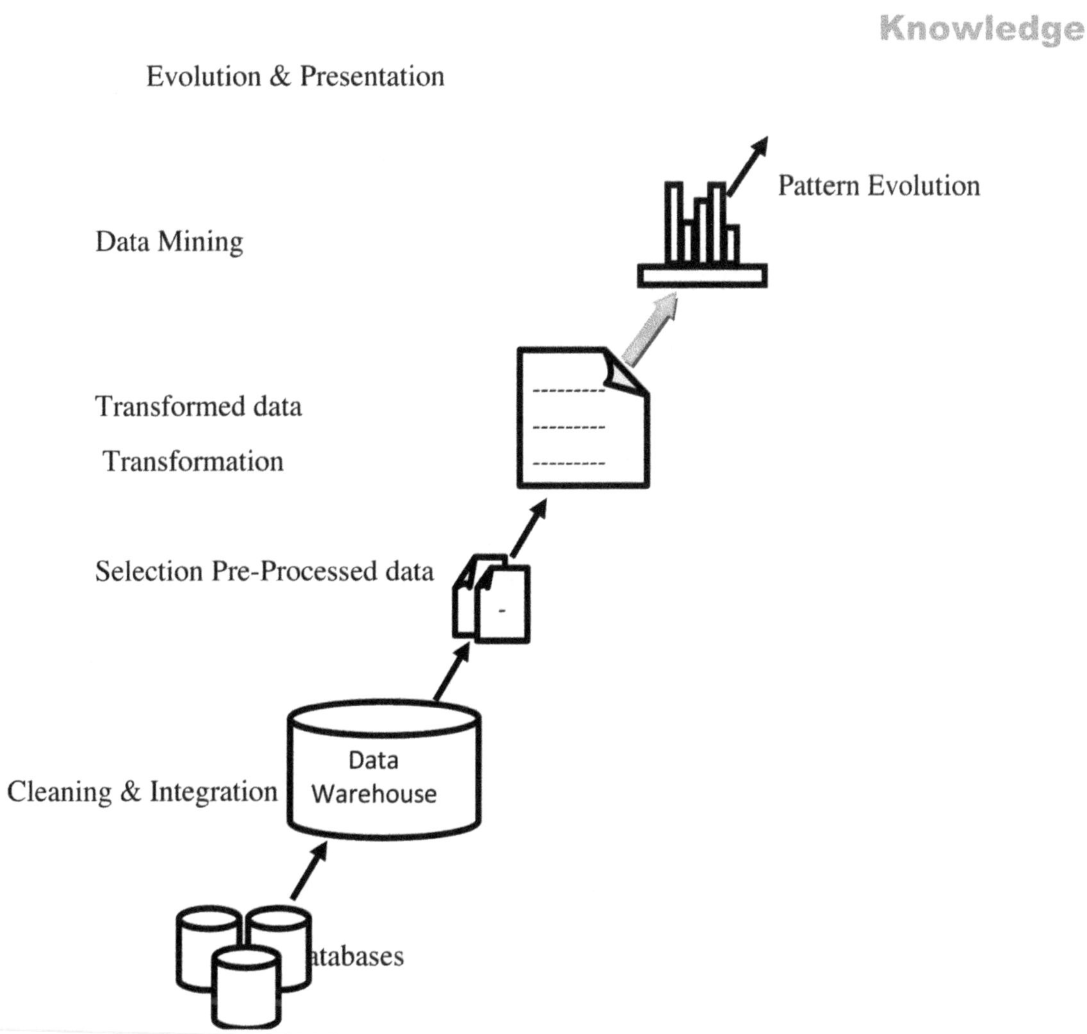

- **Data Cleaning**: Multiple resources are combined to make consistent data integration data and remove noise.
- **Data Selection:** Data relevant to analysis will be retrieved.
- **Data Transformation:** we can convert the data from one form to another form by applying operations like aggregation.
- **Data mining intelligence**: To get interesting data patterns in the dataset data mining methods are used.
- **Pattern Evaluation**: Real interesting patterns are identified by applying interesting measures to data patterns
- **Knowledge Representation**: User will get a visualization of mined knowledge and knowledge representation inappropriate kinds

Data mining is having a lot of characteristics in that some are following are.

- The volume of data so great it has to figure out by implicit techniques, e.g.stakeholders' information, fraud detection credit cards, phone cards, etc.
- It is applicable for a wide variety of information, for example, relational database, data warehouses, object-oriented and object-oriented relational databases, www, etc
- Tools of data mining are used to extract the buried information
- The environment of data mining is client/server architecture
- Sometimes necessary to use parallel processing for data mining because of a large amount of data.
- Clusters, classifications, sequences, forecasting, and associations are five types of functions in the data mining system
- Spreadsheets and other end software tools are easily combined with data mining tools.

By implementing an algorithm on data, the mining model is created. However, it is more than an algorithm. To generate predictions and make inferences about relationships we need to apply statistics and patterns on new data. The data mining algorithm analyzes the data of the mining model, which gets the data from the mining structure. Both mining structure and model are various objects structure stores information that describes data source, model stores information derived from the statistical processing of the data. After a mining model has been processed, it contains metadata, results.

Figure 3. Classification based on data mining types

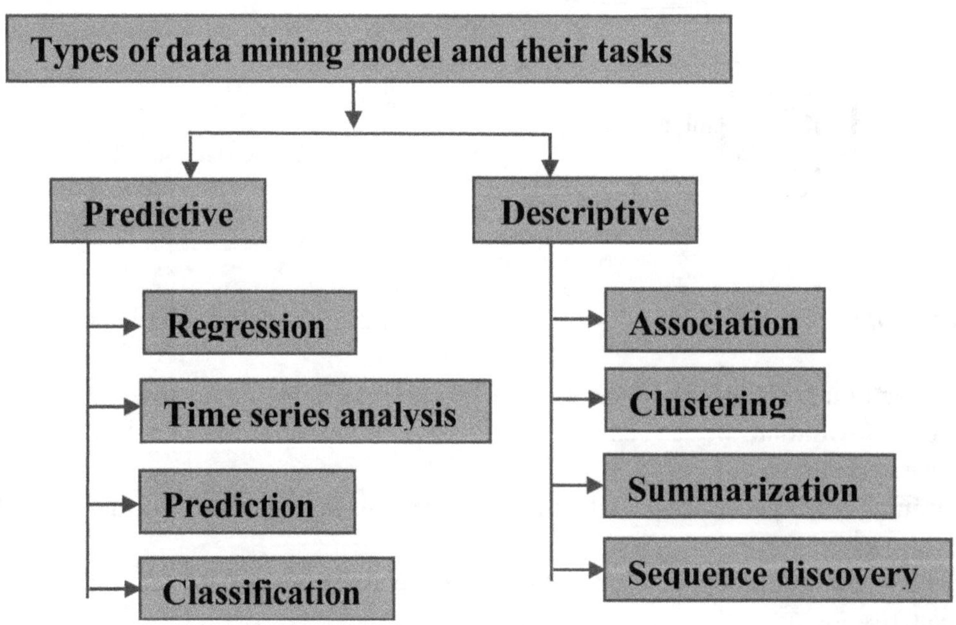

Predictive

To predict the potential behavior predictive modeling used, it is a common technique of statistical. In data mining technology (Kamber, 2007) it is one of the modeling solutions, it works by Dissecting (analyzing) verifiable information, current information and producing a model is utilized to foresee future results by the predictive model. Tasks of the prescient (predictive) model are.

- **Regression:** To comprehend which among the autonomous factors are related to the reliant factors, and regression analysis is used to investigate the forms of these connections. For instance by utilizing a relapse (regression) model conceivable to predict kids' tallness, given their age, weight and different elements.
- **Time-series Analysis:** It is used to try to extract all meaningful knowledge from the shape of data. Previously observed values are used to predict future values by using it
- **Prediction:** It will give the results based on current and past data. It uses many techniques from data mining, AI (Artificial Intelligence), statistics.
- **Classification:** It is an act of discovering a function either model that defines and differentiates data concepts or classes for future prediction. Example: classify bikes based on petrol mileage.

Descriptive

Its most basic form is a concept description. The universal properties of the information in the database are characterized by using descriptive. It describes data in a comparative manner and data in brief; descriptive data mining will present data which is having interesting general properties.

Tasks of the Descriptive model are.

- **Association rules:** Used to find relationships between items
- **Clustering:** discovering groups of similar items
- **Summarization:** Develop delegate information
- **Sequence Discovery:** Used to find sequence patterns

Advantages of Data Mining

1. Possible to analyze a volume of data, get abstract and useful knowledge from mining (data mining) techniques like artificial intelligence (AI), statistical and machine learning.
2. It plays a vital role in the lead marketing organization, customer profiling, credit risk analysis, fraud detection.
3. Possible to get meaningful patterns by processing data or executing the query in a data warehouse, but data mining will do.
4. It saves time, money for organizations in creating new business.
5. Improves customer satisfaction/service and increases profitability

Applications of Data Mining

Some applications of data mining are shown in Figure 4.

Figure 4. Common domains of Data Mining application

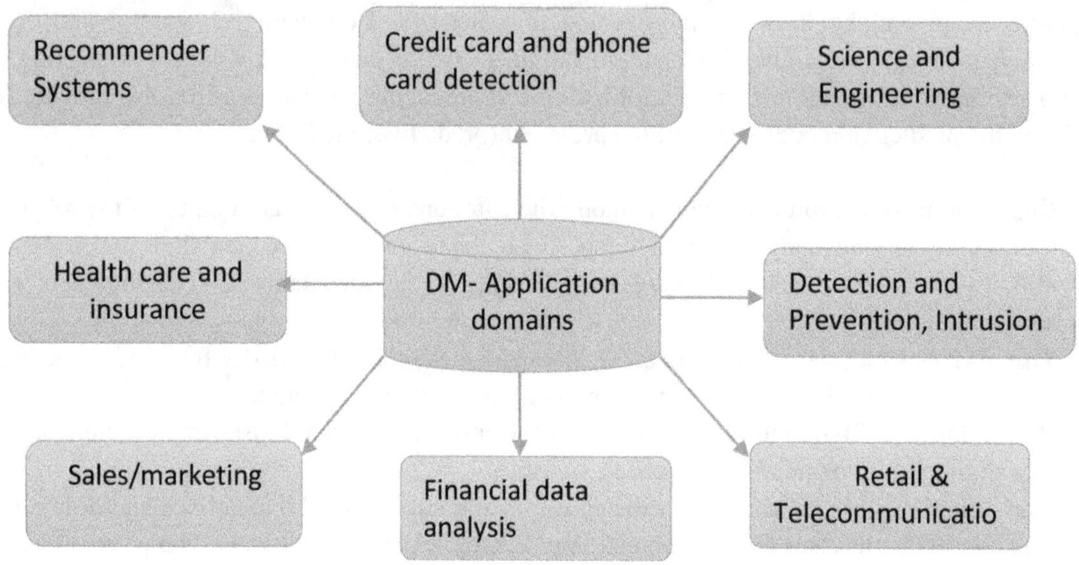

Mining application domains are many, but in this paper, we are mainly focusing on Recommender system application

RECOMMENDER SYSTEM

The system which can recommend something you are maybe interested in that you haven't a try is called Recommender Systems. For example, if you bought a book about Artificial Intelligence it gives a recommendation list including some books about pattern recognition, data mining, etc. Recommendations such as things to be useful to a client are offered by Rs software tools and techniques.

Recommendation System – Working Model

The number of choices is vast on the internet; the filter is required, important and efficiently delivers appropriate data in order to reduce the issue of data overwhelm, which has made a future complication to numerous online clients. Recommender systems figure out this problem by finding through a big amount of aggressively developed data to provide users with personalized info and services.

Working off the recommendation system in five steps, the following are

Step 1: Data collection

Collecting data is an initial phase in making a proposal engine. The type of information can be either implicit or explicit data. Information like ratings and comments which are given by the user on products is treated as explicit knowledge. Request history and return history, Cart events, Page sees, Click through and find log treated as implicit data. When every user visits the site this data set will be created.

Step 2: Storing

We can get good recommendations by providing more data to our algorithms. In a recommendation project if data is more quickly turn into a big data project. The database tool is decided by the type of data that is used to create recommendations. Examples of database tools area standard SQL database, NoSQL database or even some sort of item repository.

Step 3: Analyzing

How will we notice things that have the same user engagement information? So as to try to do, therefore, we have a tendency to clean the info by utilizing totally variant analysis ways. If you wish to produce emergence suggestions to the client as they're seeing the item then you may want an additional quick thinking kind of analysis.

Some of the ways by which we can break down the information are:

- **Real-time systems:** Created data can process by it. Processing and analyzing streams of events tools usually involve in a real-time system. This system would be needed to provide in the bit recommendations.
- **Batch analysis:** Processing the data periodically is a demand for batch analysis. This concern suggests enough information needs to be created in order to form the analysis closely, such as daily sales capacity. This system might work well to send an e-mail on the next date.
- **Near-real-time analysis:** you can refresh the analytics for every few minutes or seconds if you collect information quickly. Working in the near-real-time system is best in giving recommendations during the clone (same) browsing period.

Algorithm

For each item in the product catalog, I1

For every client C who purchased I1

For every item I2 purchased by client C

A record that a client purchased I1 and I2

For every item I2

Calculate the similarity between I1 and I2

Step 4: Filtering

The final step is data filtering it is used to bring consistent information need to give suggestions to the client. We have to select a method which is better suit for proposal motor (engine).

The following are the algorithms.

- **Content-based:** A famous, suggested product has the same features as what a consumer likes or views.
- **Cluster**: Recommended items go together fine, it doesn't care about what other users have done.
- **Collaborative**: Other customers, who like the similar item as another customer views or likes, will also call a recommended product.

Data about communications or ratings is representing as a group of a rectangular grid of numbers (matrices), dimensions are items and clients. Suppose that the below 2 matrices are the same, but then we reduce the 2nd from the 1st by following old appraisal with the digit 1 and no ratings by the digit 0. The appraisal matrix is a fact table where a digit 1 represents a communication by consumers with items.

Figure 5. Interaction between user and items

Ref: https://cloud.google.com/solutions/recommendations-using-machine-learning-on-compute-engine#storing_the_data

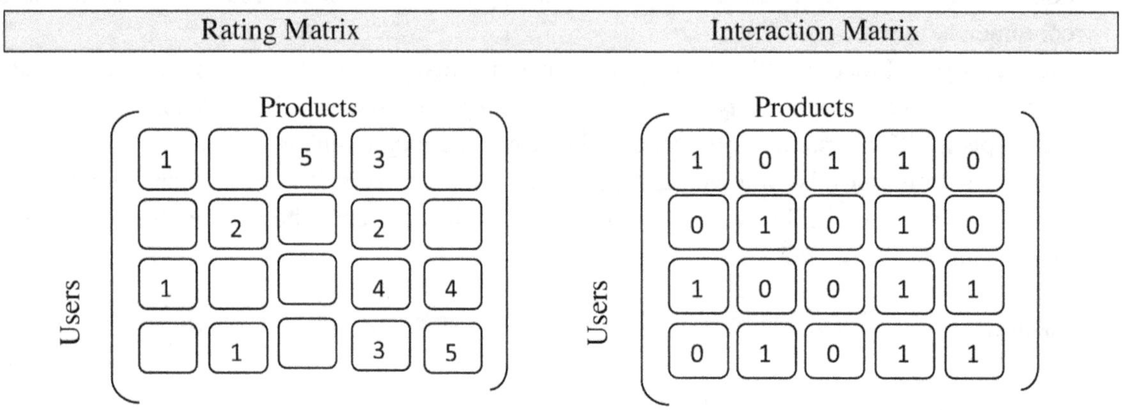

To better describe the information groups of users for proposing based on the appraisal or item use Dijkstra's algorithm, cosine similarity K-Nearest algorithm, Jaccard's coefficient.

Figure 6. Graph shows the work of cluster filtering

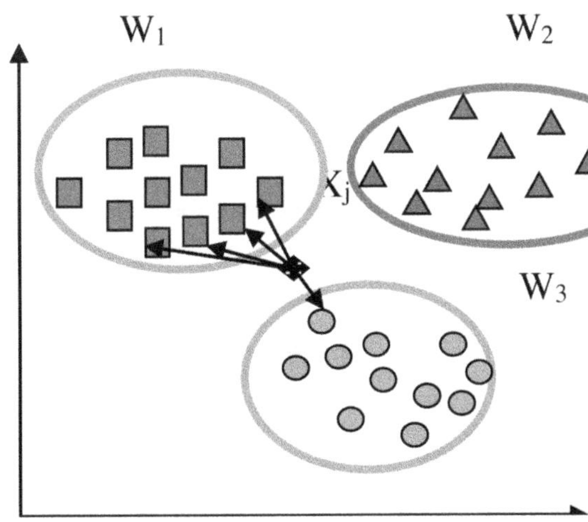

W_1, W_2, W_3 are three classes, and class label finding is a goal for the unknown example X_j. For this situation, we utilize the Euclidean separation and estimation of k=5 neighbors. Of the 5 closest neighbors, 4 have a place with W1 and 1 has a place with W3, so Xj is allotted to W1, the dominating class. The choice of *k* is a hyper-parameter that can be tuned or set heuristically. After filtering and using the algorithm result will get finally, based on the auspiciousness of the sort of proposal user will get suggestions. Whether sending an email or real-time recommendation later after the tiny period.

Recommendation System Algorithms

In the recommendation system three main filter techniques, these techniques contain

Algorithms techniques like content, collaborative, hybrid-based techniques. Collaborative technique divided into two categories: model-based and memory-based

Model-Based Techniques

The description model develops from the database and an active user will get predictions by this model. Model building processes done by variant pursuing algorithms such as the following are.

Figure 7. Classification based on recommendation algorithms

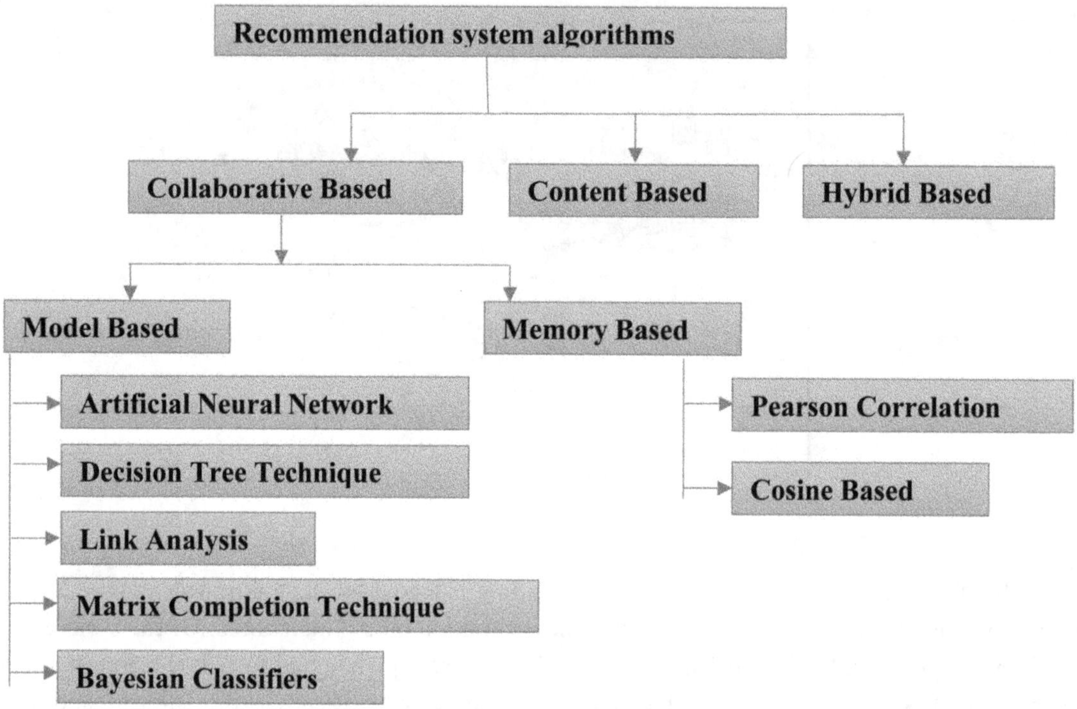

- **Artificial Neural Network (ANN):** Structure of numerous associated neurons that are organized in layers in efficient ways is called ANN. The associations between neurons have weights related to them relying upon the measure of impact one neuron has on another. In a few special problem situations we can use neural networks there are having some advantages. For instance, because of the way that it contains numerous neurons and furthermore assigned weight to every association, unnatural neural systems are very strong in dealing with the disturbance and mistaken Data sets (John, 2005). Suppose nonlinear capacities and catching difficult connections in information groups can also be done by artificial neural networks, they can be proficient and even work if part of the system fails. Finding the perfect system topology for a given issue is the major disadvantage of ANN and it acts as lower headed for the group mistake, once topology is decided.
- **Decision Tree Technique (DTT):** Decision Tree Induction. It is a structure that incorporates a root hub, branches, and leaf hubs. Each interior hub indicates a test on quality, each branch means the result of a test, and each leaf hub holds a class mark. A decision tree is built by breaking down an arrangement of preparing cases for which the class names are known and it depends on the philosophy of tree diagram. They are then connected to characterize already unseen instances. If trained on top quality information, they have the ability to make exceptionally exact predictions (Caruana, 2006). Compare to SVM (support vector machine) and ANN (artificial neural network) classifiers decision trees are more interpretable since they join basic inquiries regarding information in an understandable way. In handling products with a mixture of the real-valued and

categorical options decision tree is very flexible and also in handling objects that have few specific missing options

- **Link Analysis**: Connection Analysis is the way toward working up systems of interconnected items in order to investigate patterns and trends (Linoff, 1997). It is used to improve the accomplishment of web searches by presenting its great potentials. HITS algorithms and page rank also present in link analysis, a website page as a solo hub in the web graph (Deng, 2004) handle by mostly link analysis algorithms
- **Matrix Completion Techniques**: Within the user-item matrices to predict unknown values we can use the matrix completion technique. In collaborative filtering recommendation systems (38) Correlation-based K-nearest neighbor is one of the significant methods utilized. They lean on widely on the chronicled rating information of clients on things. Many Items which are represented within the matrix (Caruana, 2006) are not given rate by the user because more often than not rating grid is enormous and sparse. This issue dependably prompts the failure of the framework to give solid and precise suggestions to clients. In matrix completion for practice variations different of low-rank models have been utilized especially in collaborative filtering toward application is used.
- **Bayesian Classifiers**: Classification problems can solve by the probabilistic framework. It depends on the definition of Bayes theorem and conditional probability. Class label and each attribute are considered as random variables by Bayesian classifiers (Friedman, 1997). The primary advantages of Naive Bayes classifiers are that they are powerful to isolated commotion points and insignificant properties, and they handle missing qualities by disregarding the instance during likelihood evaluation computations.

$$P(H|X) = \frac{P(X \mid H)P(H)}{P(X)}$$

P = Probablity
H = Posterior Probability of a Hypothesis
X = training data

Memory Based Technique

Prediction generation can do by the entire client item database; we can find the neighbor using statistical techniques, it is also known as the nearest neighbor. Through item, user depends skills can achieve Memory-based CF (collaborative filtering). By looking at client evaluations on a similar thing client based collaborative filtering techniques calculates likeness between clients, and it then calculates the predicted rating for a thing by the dynamic client as a weighted normal of the evaluations of the thing by clients like the dynamic client where weights are the likenesses of these clients with the objective thing. To calculate predictions in item-based filtering techniques use similarity between things but not clients. It assembles a model of thing likenesses by recovering all things appraised by a functioning client from the client thing network (matrix); to calculate the similarity between item/user many kinds of likeness metrics are utilized. Pearson correlation coefficient and cosine are the 2 famous likeness metrics.

- **Pearson Correlation Coefficient:** It will take a range (r) of values from +1 to -1. No association between the two variables is indicated with 0. If the no is greater than 0 is called positive association, if it is less than 0 is called negative association. In positive association value of 1 variable increment other variable value increments but in negative association variable value increments another variable value reduces

Table 1.

$$r = \frac{N(\sum XY) - (\sum X)(\sum Y)}{\sqrt{[N(\sum x^2) - (\sum x)^2][N(\sum y^2) - (\sum y)^2]}}$$

N = number of pairs of scores

$\sum XY$ = sum of the products of paired scores

$\sum x$ = sum of X scores

$\sum y$ = sum of Y scores

$\sum x^2$ = sum of squared X scores

$\sum y^2$ = sum of squared Y scores

- **Cosine Based:** Compare to Pearson-based measure cosine similarity is different in that it is a vector-space show which depends on linear algebra-based math rather than a statistical approach. Between 2 n-dimensional vectors depend on the angle between them relativeness measured by it. Text mining (TM) to compare 2 text files and fields of information retrieval uses the cosine based measure, in this case, vectors of terms represents the documents. The comparability between two things closeness measure is also alluded to as similitude metric, and their techniques used to figure the scores that express how comparable clients or things are to each other. Thing or Client based suggestion developing use scores as a foundation. Based on the condition of use, distance measure or correlation metrics can also be alluded to as similarity measures

There are many domains using recommendation systems, for example, tourism and travel, e-commerce, movie and music, health and Sensex, etc. This paper reviews about recommendation system in the tourism domain.

Figure 8. System architecture for tourism product recommendation framework

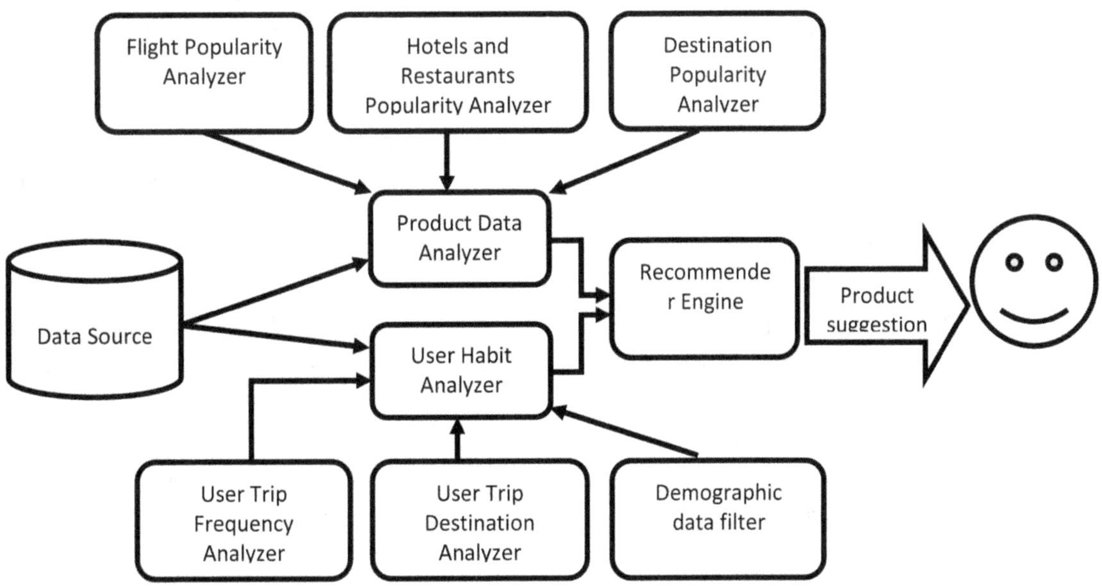

RECOMMENDATION SYSTEM FOR TOURISM

For recommendation research tourism is an interesting domain. More choices of target and desire are provided to travelers by this system. Different kinds of attractions are Vacation, business, adventure, cultural and historical. There are many cities and places all over the world where these variant features may suit originally. With such more number of choices, tourists need to have suggestions about what to see, where to go, how to visit, etc. Commonly, travel agents help tourists who have some knowledge and capabilities to provide such guidelines. Moreover, human factors restrict recommendations from these agents like less memory, finite knowledge about society and their tourist options which may not match a tourist's requirements saved in a DB (database). A few times the last conclusion depends upon the tourist agent. Getting a signal about what choices are well suited in an important case for a particular person or work in the social process, these two activities done by recommender system software. In a tourism domain systems cities are indicated by recommendations (by human or system) to reach targets, tourist plans, hotel facilities, beauty to view, way maps, the program to participate, air companies, etc. It is not a definitive work since mostly client visits the tourist agent with no idea. A recommender system can help the tourists to plan their trips. Using the software, the user will get recommendations, in a Web site, an ATM device, as in a multimedia totem or by the travel agent. Decision support gives detailed information to travel agents.

A Recommender Engine Architecture for Tourism Applications

Many filtering methods are used to build a recommender engine. Each and every method contains the pros and cons. Filtering methods possible dis-advantages can be avoided by grouping them to form a hybrid solution. User needs, requirements, time constraints had been collected by most researchers to focus on generating personalized recommendations in previous recommender engine studies on the tourism domain. Budget limitations and clients' travel practices like (targets depend on seasonality and trip frequencies) will play a crucial role in suggesting better recommendations to the user. To extract such valuable information, for this existing travel portal extension is a recommender system. Depends on client rating suggesting an item is a too simple approach. This approach can be upgraded by utilizing a travel site database that contains clients' past interests, traveling information and habits. As illustrated in the below figure, the information contains destinations, restaurants, hotels, airways and past customer activities, personalized rules can generate by applying data mining algorithms on items

- **Product Data Analyzer**: Based on the season's popularity analysis of flights, hotels, travel destinations and restaurants are performed by this product analyzer. The famous of these things are analyzed with respect to the consumer profile produced from the customer practice analyzer module.
- **User Habit Analyzer**: Based on users', travel destinations and seasons, demographic data, previous travel frequencies creating user profiles is an aim of this module. One important attribute for this module is user income.
- **Recommender Engine**: Data analyzer and user habit module outputs will be processed in this recommender engine. The travel destination is associated with the target user. After selecting the destination, flights and hotel choices will be recommended as an output of this module.
- **Demographic Filtering**: Suggestions are dependent on customers' demographic data, profiles are developed by analyzing product view total versus every customer's demographic data such as gender, age, nation, etc versus view counts of items are utilized to generate demographic profiles. Marketing applications will use more this type of filter.

For example, some tourists are traveling newly they don't know anything about, cities to go, road maps, travel plans, attractions to see these all solved by the recommender system in tourism. No need to search for all tourism places like museums, zoo, etc in search engines is a very complex and time-consuming process. Recommender System will provide a list of places automatically based on the selection of options. Travelers or tourists can share knowledge about hotels, restaurants, attraction places, and cities so that future people can discover new options. Without a travel agent, any complications in tourism we can plan just by selecting options in the e-tourism recommender system, mobile recommender applications, etc.

Recommendation Factors for Tourism

Some are the effects of the important factors tourism recommendation systems are following are.

Table 2. Effects of important factors tourism recommendation systems are following

Factor	Items under Factor	Way of Effecting Explanation
Cost	Cost of Accommodation, Transportation, Food and Beverages and other expenses	It is a very important factor in the tourism recommendation system. A tourist mainly considers the cost of accommodation, transport, and food; it will depend upon the season, visit places, attractions and cities also. It recommends the places based on past visited places and current selection options.
Punctuality	Time to visit, leave, to spend, etc	Time plays a vital role in selecting tourism place using recommendation system because tourist depends on holiday time or multiday plan, they want to visit more places in a short span of time, no need to waste more time at one place or in traveling.
Environment	Natural Resources, Air Quality, Climate condition, Personal security, and safety.	This factor depends upon season; climate conditions, and such as cleanliness, air quality, and feelings towards destination are environmental conditions this total can be defined as an environment. During the season system recommends good environmental places
Hospitality	Accommodation, Food, Infrastructure(electricity, beverages, public lavatory, and safety, etc)	Hospitality is a special factor and only factor effects more to the tourism recommendation system. For example a welcoming atmosphere, variant food, cleanliness. The system recommends famous hospitality places also based on user selection.
Attraction	Cultural, Natural, Recreation, and Activities	This factor plays a decisive role because based on attracting, variety, exciting places the only tourist will choose the place. If already visited, the recommendation system suggests different one based on visitor options
Accessibility	Basic Medical First Aid/Hospitals Transportation Authorities	Last but not least is accessibility factor for emergencies, ability to provide travel throughout the destination and appropriate user access into a destination
Feedback	User satisfaction,	Factor plays a vital role in future recommendation results and recommends places based on past tourist user, agent, guide feedback on overall tourism visited places.

SURVEY ON RECOMMENDATION SYSTEM FOR TOURISM

This paper mainly contributing to the recommendation system in the tourism domain with different categories with variant factors. The first three sections discussed data mining, a recommendation system and the role of the recommendation system in the tourism domain. In the final section's main objective discussed, it is about different categories of factors with sub-factors in the tourism domain using a recommendation system that we studied and find from different articles. First, we have three main categories; they are recommendation approach, recommendation engine model and type of data interpretation and these three will have subfactors. In the first category, we provided preference criteria like single, multi-criteria and type of interfaces like by human or travel agent, by the system. In the second category provided with user accesses methods like web, mobile, hybrid and recommendation technique types like recommendation filtering methods, artificial intelligence (AI) techniques. Finally ends with data collection models like single, group, hybrid and dynamic decision approach like adaptive.

The following are the vital factors can be considered for recommendation.

- **Data Collection Model:** Giving recommendations based on user type. Users may be individual or group.
- **Recommendation Approach:** Criterion may be single or multi means recommendations giving to single (individual) or multi (individual and group).
- **User Accesses Method:** Interface may be a website, mobile app or both
- **Recommendation Technique Types**: Number of techniques for recommendations but mostly prefer to recommendation filtering techniques or AI (Artificial Intelligence) techniques.
- **Dynamic Decision Approach**: we can also change our decision in the middle of tourism is called adaptive.
- **Type of Interface**: Recommendations can be provided by a human (intermediatory or travel agent) or by the system through the website or by a mobile app to tourists.

Classification of Tourism Recommendation Criteria

The literature is classified according to its content into the following major categories; recommendation approach, recommendation engine model, type of data interpretation with some different sub-factors.

Figure 9.

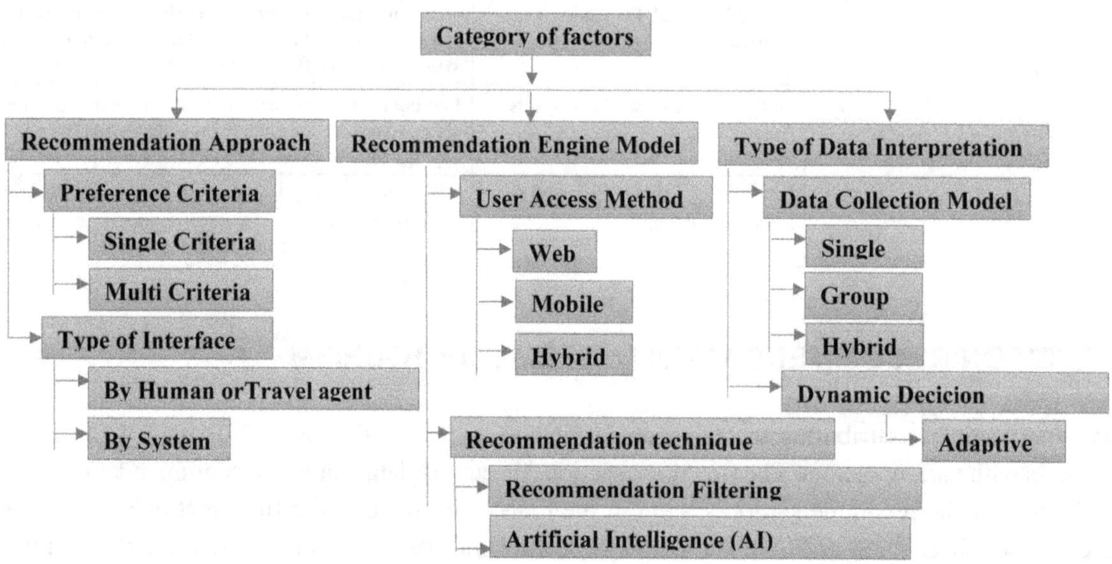

Recommendation Approach

Nowadays, if people want to plan a tour or date out initially search on the internet about tourism information but it will take a lot of time, complicated, confusion to select the places. So the recommendation system helps tourists in selecting tourism places based on demographic profiles, past visited places it recommends the places to tourists with clear information. The system offers suggestions for the solo user or set of clients (with friends and family members etc)

Figure 10. System of E-Tourism for Individual User

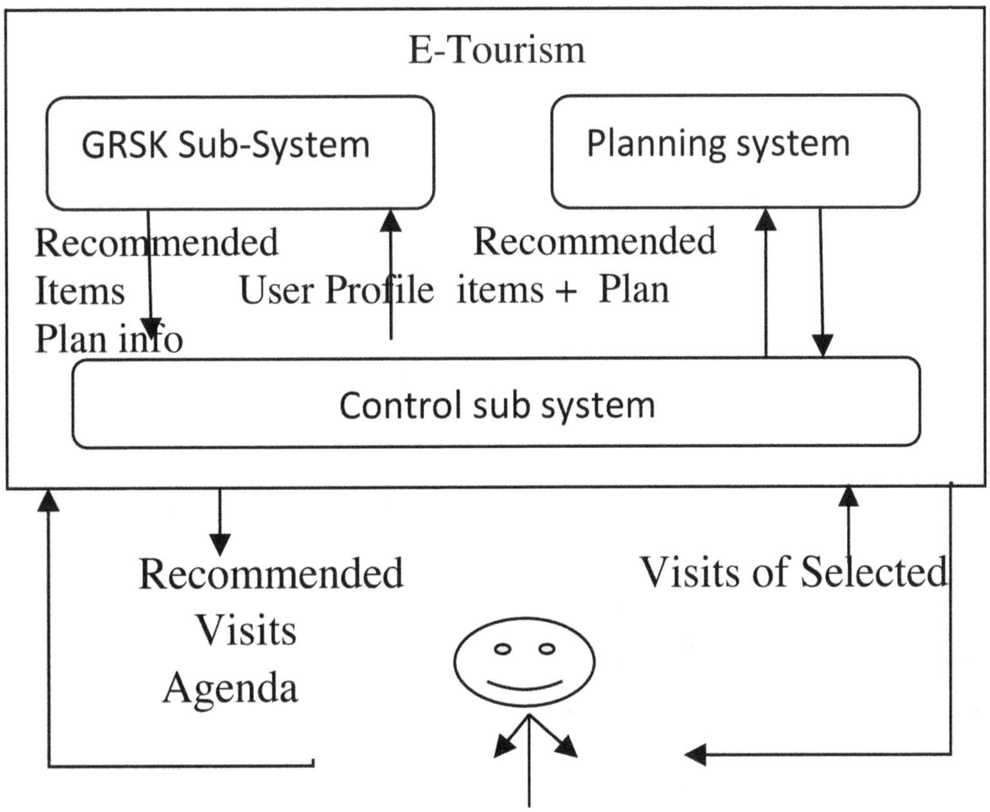

Preference Criteria

In this section, we will know about the criteria of the recommended system. We have two criteria first one single criterion and second criteria are about multi-criteria

Single Criteria (Single recommended proposed system)

Depends upon demographic simplification this system recommends about tourism, based on past visited places information and interest of users, this system offers suggestions for solo user or group of users.

E-Tourism (Sebastia, 2009) is an example of a tourist web application recommender system and individual recommended system that helps the client on making customized travel plans or schedules for anyone of metropolitan. The tourism website relieves the work of handing out a great quantity of data, deciding on the more favorable actions for a specific client. Moreover, it permits for understanding and planning of user actions by managing special origins of data as time spent on the visit, space between the visiting places or opening hours of places, in this method we make customized tourist plans for the suggested plans. The section mainly focuses on the interest of a particular user on the recommendation.

Figure 11. Hybrid E-Tourism System

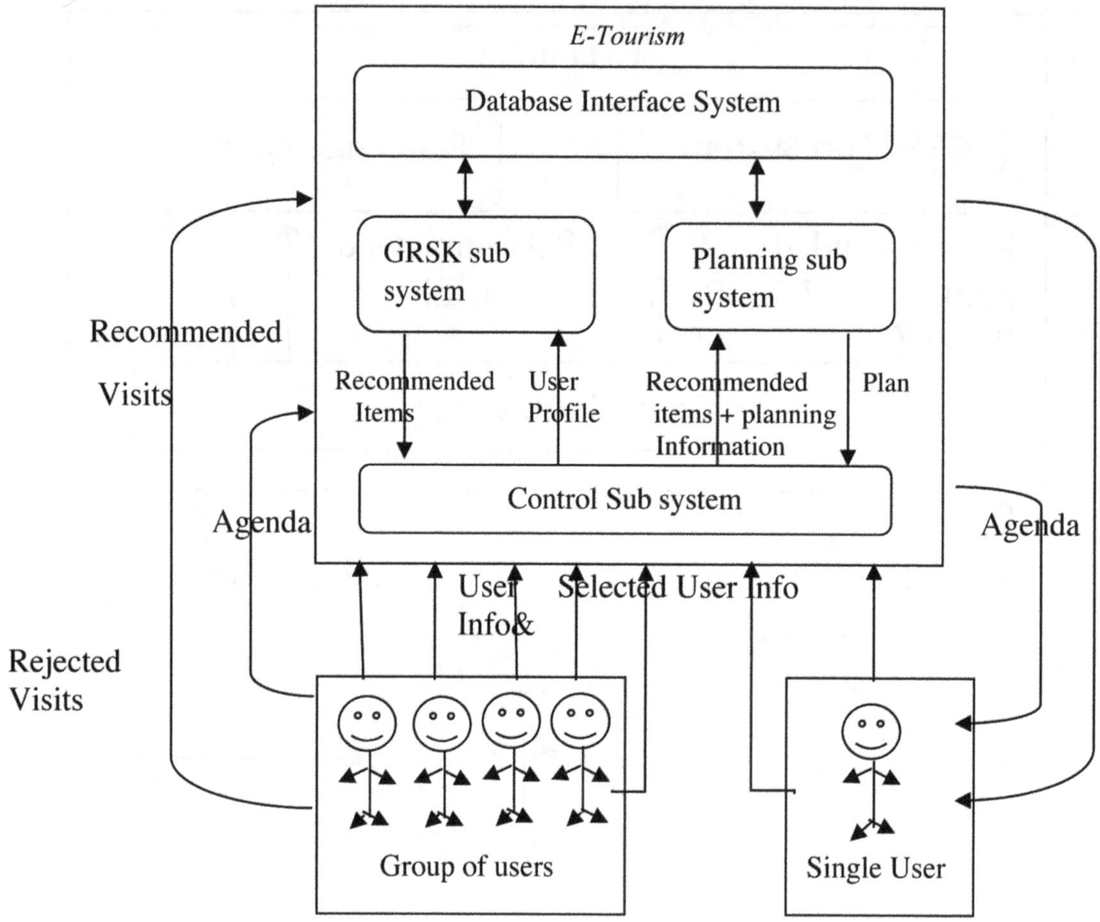

The framework shown in Fig.10 does not take care of the issue of setting out to a particular place yet it centers on suggesting a rundown of the exercises that a visitor may appreciate in a city. It also believes that timetables, exercises, and spaces between the actions to calculate a relaxation and traveler agenda. E-Tourism is arranged for three subtasks. Fig.7 shows the GRSK (Generalist recommender system kernel)the control sub-framework, and the planning subsystem. The workflow of the e-Tourism system in steps wise, initially, the user profile is built for the first time as the user utilizes e-Tourism; he should enlist and enter his private data and general inclinations. As general inclinations, the client stops over Science Museums. With this data the system builds a first user sketch that will be modernized appropriately with the significance response attained from the client; i.e. the exercises which have been at long last performed by the client

The module responsible for building the client profile is the control subsystem, the center of e-Tourism. It functions as a UI, starts the execution of alternate subsystems and unifies the trading of data. The second step is to create an arrangement of exercises that are likely important to the client. This rundown is figured by the GRSK subsystem, whose information is the client's general and particular inclinations. The GRSK figures the suggestions as indicated by the present client's profile and other clients' profiles (contingent upon the proposal method). The computed proposals constitute the rundown of proposed things to visit.

The third step is to figure the visitor plan with the chose exercises of the past advance. At this stage, the framework plans the exercises as per the time confinements of the client and nature. The module accountable for figuring the arrangement is the arranging subsystem. The contribution of this module is the arrangement of exercises emphatically chose by the client, the exercises left as in-various and different inclinations essential for arranging like the client accessible time. The outcome is a vacationer motivation containing the arranged exercises together with the time when the exercises should begin and the evaluated term of every action. Arranging subsystem (Sebastia, 2009) computation

The last step is to process the client's input. At the point when the client logs again in the framework, he is requested to rate the exercises in the last prescribed arrangement. The data acquired from these evaluations are utilized to enhance the client profile and acquire appropriate suggestions

Multi-Criteria System

The system which will support recommendations for both single and group is called multi-criteria recommender system. This system initiates a technique for providing suggestions of client activities to set of clients (family or friends) and single user (individual or couples). This system builds suggestions depends upon the clients' population classification, visited locations and their tastes in the previous tour. The set suggestion is figured from single and private suggestions during the need of systems like intersection, collection or incremental junction, both implementations are done through one system only.

Figure 11 shows the architecture of e-tourism. Generalist Recommender System Kernel (GRSK), control, planning, and database interface are the subsystems of tourism architecture. The control subsystem starts the execution of alternate subsystems and unifies the trading of data and it works as a user interface. It incorporates exchanging clients' demand into an appropriate suggestion inquiry and displays the chain of suggested activities and trip planning, i.e. the client schedule. The storage mediator operates the inquiries approaching from whatever remains of segments in the framework. 2 more essential components in architecture are the planning and GRSK subsystems. GRSK is a common useful recommender framework in the blame for making the list of suggested exercises and elected activities like building the tourist plan (Garcia, 2008) plans by planning subsystem.

E-Tourism (Garcia, 2011) mechanism working, section one is to develop a single client report. The client signup in the framework and gives his private information and common interests. In the case of gathering, it is fundamental to give the clients that shape the gathering, who must be already enrolled in the framework. Section two executes by GRSK is to make a chain of exercises that are the interest of attention to the client or set of client. The final section is to plan the chosen options and revisit the user plan as shown in Fig. 8, GRSK architecture (Garcia, 2011)

Type of Interface

Recommendations can be an offer by the system or through a travel agent. By System website or mobile app can get tourism recommendations and in some systems may get through a travel agent.

By Human Intermediatory or Travel Agent

Recommendations for tourism domain suggested by a travel agent or by latest technology (website or mobile app) may assist to reach the destination, programs to participate, airways, tour plans, way maps, desire to watch, etc. Here decision support framework gives information to the travel agent. Travel agent provides options to the customer with the help of the recommender system. The system shows relevant user features to travel agent but recommendations will not send directly to the user because the system acts as a decision support system. The communications are provided by the system between the user and tourist agent via online chat, where information is interchanged via machines associated with the network. TM (Text mining) is used by the software methods to find fascinating regions in the messages sent (i.e., objectives of the movement, sorts of attractions intriguing to the client, seasons of intrigue, and so on.). Later that, the database is checked by the system and retrieves options for the tourists (like urban communities and attractions) arranged in these intriguing territories

In the tourism domain, knowledge is shared by the people about cities and attractions, so that new options can be discovered by potential travelers or tourists. The public database stores these self-experiences of tourist options. Moreover, the travel agent uses collaborations to understand the user's interests, likes, needs, and targets. Ricci et al. (2002) In order to recognize the users' intrigue, post the test of connecting with the client through a discourse. This communication with users may include text information, utilizing methodological sources (e.g., online chat (Stanley, 2004). Once information is interchanged, a discussion is analyzed by an automated system to find subjects in the messages.

By System (Website or with mobile app)

A recommender framework is programming to help in the public procedure of demonstrating or accepting signs about what choices are more qualified in a unique case for a certain individual Resnick and Varian, (2002). Suggestions might be built about books, hotels, CDs (compact discs), tourism, data, etc. Domain recommendations in tourism can give by system or human may help tourists, visiting places, desirability to see and more are there. This is certifiably not an unimportant work because many times a client visits a travel organization without a proper agenda of user plan. The recommended framework can take care of this work by corresponding the user's choice and wishes against every single accessible element and facilities and after that helping the client to choose his/her tour design. The system can be used by the user to plan the tourism by choosing available alternatives in the system may websites or mobile applications a lot of websites and mobile applications refer to table no 1 and 2.

Recommendation Engine Model

This factor categorized into three sub-factors; model interface type, model techniques, model architecture.

User Accesses Method

This subfactor categorized into three sub-sub-factors; website, mobile app, and hybrid.

Web site or Web Application

A web-based interface is a very general choice is chosen by e-tourism (Sebastia, 2009) proposals. Interfaces of this kind will let users watch further for data in a flexible way. Clients really have the best cooperation with the framework utilizing a wide screen which licenses showing an enormous measure of information developed with maps, pictures or significantly higher quality videos. Moreover, the PC (personal computer) can be cooperated with the mouse and travel through maps, perform zoom activities, select things or even intuitive them. It is very valuable for users when they are in the phase of arranging their excursions. In any case, Web sites are really not intended to be utilized amid the stay since for the most part the users won't have simple access to PCs with the network. In spite of the fact that an expanding number of users having versatile handsets or tablets with network association, recommend demonstrates the data ridden Web sites which can't be effortlessly perused or controlled on such little screens. Based on this alarm portion segment we remark some great highlights abused in Web-based interfaces to refresh the collaboration with the clients.

Table 3. list of recent studies about web-based recommender system applications in tourism

Year	Title	Application Name
2014	Multi-agent recommender system for e-tourism using reputation-based collaborative filtering: MARST	MARST
2012	Tourism recommendation system: Empirical investigation	MyTravel Pal
2009	On the design of individual and group recommender systems for tourism	e-Tourism system
2010	A web-based pervasive recommendation system for mobile tourist guides	MTRS
2013	Hybrid recommendation system for tourism	Trip Advisor
2000	Deep map and challenging IT research in the framework of a tourist information system	DEEP MAP
2003	A context-aware tourist information system	CATIS

Mobile App

Mobile interfaces offered by the systems have grown considerably in the past some decades since many of the users having versatile handsets with net connectivity are many or recently the well-known smart gadgets. Electronic handsets are tiny and their network connectivity is generally slow; due to this amount of data that can be displayed in these handsets cannot be distinguished with a Web site. Due to this cause recommenders have to make an effort to give only the required information to the user, and it should be well structured to display exactly in short displays. Moreover, the system communicates with the user is finite, web-based interfaces have some basic information (moving, building text) which are not that flexible. However, the latest android phones provide better interaction with big screens. Moreover, the merits of the versatile handset are that the framework can be utilized anywhere anytime with the network connections so that clients may get required data, re-built locations or change their plans during living. In recent days, maximum electronic handsets are having GPS and the recommender can know the present area of the client and it may offer geo-referenced data, counsel or suggestions in light of the information

In Yu and Chang it is reported as the initial concern in the area which has used mobile systems (Yu CC, 2009). This system, designed for PDAs, in order to support personalized tour planning, the system offers location-based recommendation services. Recommendation depends on tourists' preferences, location and time.

Table 4. list of recent studies about mobile-based recommender system applications in tourism

Year	Title	Application name
2004	Personalized multi-criteria decision strategies in location-based decision support	LBS
2006	Proximo, Location-aware collaborative recommender	Proximo
2003	Developing a location-based tourist guide application	Tourist Guide
2006	IT framework for comprehensive and integrative travel guidance	planner

Hybrid

A combination of website and mobile application becomes hybrid. Depends upon user interest can use any one interface, which they feel flexible and comfortable.

Recommendation Technique Types

This subfactor categorized into two sub-sub-factors; recommendation filter techniques, AI techniques

Recommendation Filtering Techniques

Efficient and accurate recommendation techniques are used to offer fine and needful recommends to single clients. This clarifies the significance of understanding the possibilities of various suggestion methods and features different recommendation techniques are like collaborative, content and hybrid-based techniques.

Content-based filtering technique (CBFT): This method is a space subordinate algorithm and it focuses more on the examination of the traits of things keeping in mind the end goal to create forecasts. Content-based sifting method is the best when archives, for example, site pages, productions, and news are to be prescribed; the recommendation is made based on the client profiles utilizing highlights removed from the substance of the things the client has assessed in the past in content-based sifting strategy

Recommended to user

Figure 12. Content-based filtering technique

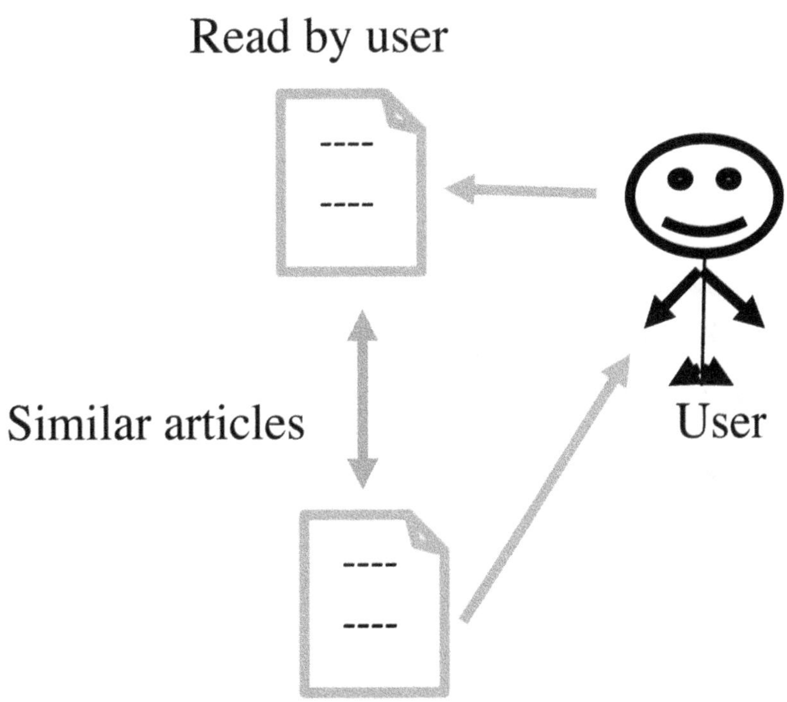

Advantages

CBFT has beaten the tests of Collaborative Filtering (CF). Even if there are no ratings by users they have the capability to propose fresh things. Actually, user preferences are stored in the database. So regardless of whether the storage isn't containing client inclinations, proposal exactness isn't influenced. Likewise, if inclinations of client change, in a short period it has the ability to modify its suggestions. Situations can be managed where the same items cannot be changed by different users, but only indistinguishable things as indicated by their natural highlights. Without sharing their profile clients can get proposals, and this guarantees security

Figure 13. Collaborative based filtering technique

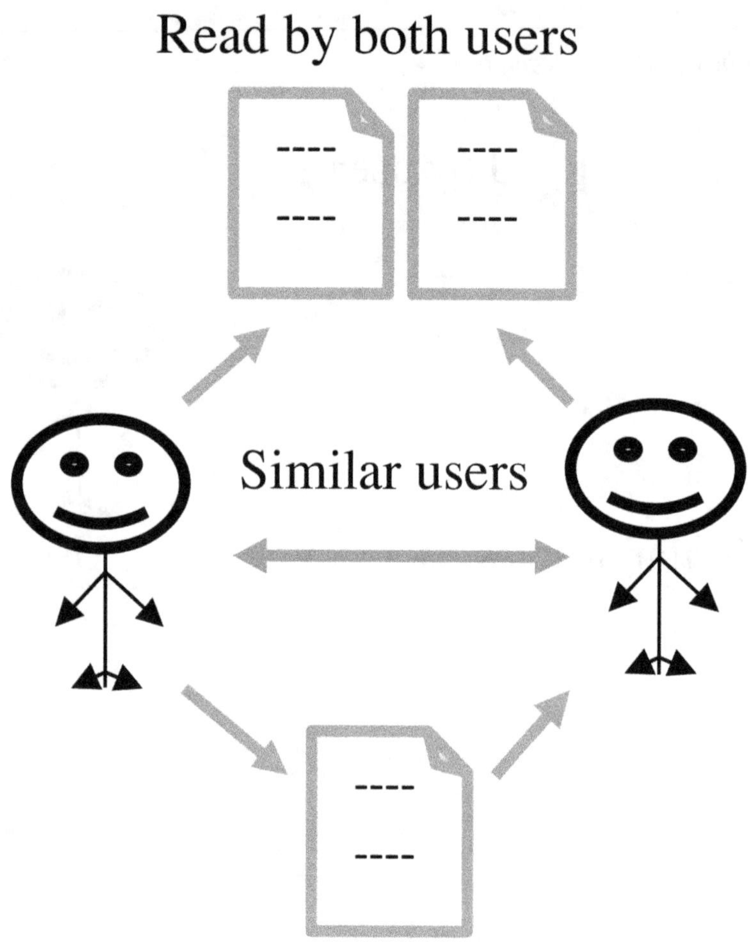

Disadvantages

The procedure requires an inside and out learning and depiction of the highlights of the things in the profile which is the big problem. Filtering techniques which are based on content are based on things' set of data. That is, the item description is wealthy and must have a well-managed client profile before suggestions can be done to clients. This is called limited content analysis. If there is descriptive data then effective CBF is available. Overspecialization in the content is another important issue of the CBF method. Items that are already defined in the profiles by the users are restricted to get recommendations

Collaborative Based Filtering Technique

Independent technique which is a domain for content that can't be effectively and sufficiently depicted by metadata, for example, motion pictures and music. By building a database collaborative filtering technique works of inclinations for things by clients. It at that point matches clients with significant intrigue and inclinations by computing likenesses between their profiles to make suggestions. Association called community is built by such users. Truly rated by clients in his community by the users are recommended to the user even though he has not rated. Recommendations can be of either prediction or suggestions that are developed by collaborative filtering.

Read by her recommended to him

Hybrid Filtering Technique

Hybrid filtering method consolidates distinctive proposal strategies with a specific end goal to increase better framework enhancement to stay away from a few constraints and issues of unadulterated suggestion frameworks.

Figure 14. Hybrid Filtering Technique

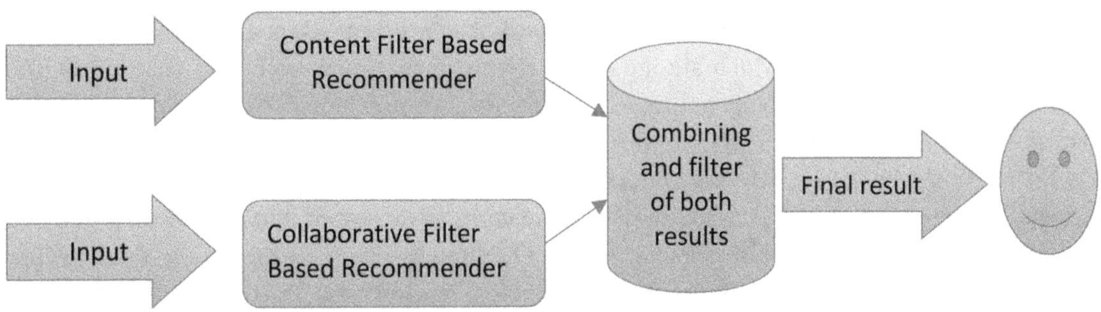

Artificial Intelligence (AI) Techniques

AI techniques like Optimization techniques, automatic clustering, multi-agent system, knowledge representation are described in this section.

Optimization Techniques

The optimization technique produces the best possible result under given circumstances. In designing, constructing, maintaining, decisions are taken by engineers. The objective of every such choice is either to limit exertion or to boost benefit. The design variables function is the effort or benefit. The procedure of finding the conditions that give the greatest or the base estimation of capacity is called the optimization technique.

Tainan (Lee, 2009) is the best example of an agent-based travel route recommender that uses a technique called AOT (ant colony optimization techniques). In these strategies, an arrangement of self-sufficient substances (which speak to the ants) coordinates through pheromone-intervened backhanded and worldwide association to discover the best answer for the voyaging sales representative issue (for this situation, to design a course that experiences diverse purposes of enthusiasm around the city).To visit a city genetic algorithm is utilized by the CT-Planner-4. In each progression of a cyclic procedure it picks the populace of various conceivable plans, which are checked by their utility for the client; changed and recombined by means of hybrid to produce another populace for the following emphasis are the best ones. After a specific number of cycles, the best arrangement is at long last chosen. VISIT system authors are planning to made suggestions adopted based on the mindset of client, which is a group of various elements (weather, area, and sentiments of social media, time and preferences of the client)

Automatic Clustering

CL (Cluster) filtering is a base for many tourism recommenders employ techniques, in which system users are divided into sets that share few general functions. The fundamental plan of these functions is that those products that have been positive feedback given by the same user can be recommended to the user. Democratic information is the base for the idea of comparability utilized to assemble clients, on the common inclinations of the clients over various kinds of touristic exercises, or on the express evaluations of individual exercises. In any case, AI built up a device called automatic clustering which might be effectively used to characterize the vacationers. A straightforward method for partner another member with comparative past clients of the framework is to utilize the k-nearest neighbors' approach (Dasarathy, 1991), computing which are the k-past clients of the framework who were more like the present one (Dasarathy, 1991) (Noguera, 2012)

The best option is to set the clients into variant to use the k-means algorithm (Gavalas, 2011) (Pinho, 2011) the starting seeds of the k wanted bunches are built up in some application-subordinate way. At that point, there is an iterative procedure in which, the items are arranged into the closest group in each progression, and the bunch models are recalculated. The strategy diverts to an answer when the protest has a place with similar bunches in two continuous cycles. In Moreno et al. (Lucas, 2013) the k-means algorithm is connected with three unique purposes: to get an arrangement of starting vacationer fragments to acquire classes of clients with comparative statistic qualities, and to characterize clients as per the unequivocal appraisals they have given.

Multi-agent System

It is a mechanized framework that contains numerous communicating astute specialists. Issues that are troublesome or outlandish for an individual operator or a solid framework to solve can be solved by a multi-agent system.

Independent and proactive programming substances fit for acquiring data from their condition and acting in a keen path upon it keeping in mind the end goal to endeavor to achieve an arrangement of objectives or destinations are operators. Multi-operator frameworks are gatherings of specialists that interface between themselves to share data and assets, arrange their exercises and cooperate in the joint effective answer of a circulated problem (Wooldridge, 2009). Tourist@ (Batet, 2012) is an agent-based framework that gives customized proposals on social exercises.

Every sort of social movement has one specialist, a little database is kept up with the occasions of that write accessible in the city (historical centers are the exemption, as there is one particular operator for every exhibition hall in the city). The client speaks with the framework through a graphical interface gave by a User-Agent. A Broker Agent goes about as an arbiter for correspondence between the User

Agents and the social exercise specialists. The client can enquire, can check an action that she has gone to, or can request a customized proposal. The basic of Tourist@ is the agent of recommender; client data is maintained for every user. Some basic information is initialized in profile with on abnormal state social interests given by the client when she/he utilizes the framework for the starting time. The proposal Agent progressively and consequently alters this underlying information about the client inclinations by predicting down the client's questions and assessments. Proactive recommendations can also be provided by the user because it knows the situation of the client in the city and can give recommendations on social exercises that fit the client's inclinations and are situated in the region. CB (content-based) and CL (cluster) recommendation methods are used by the system

Knowledge Representation

In e-Tourism recommender systems are needed, like any learning-based shrewd framework, an approach to speak to in a proficient way the area information, with the goal that it can be utilized as a part of their thinking forms. Systems created in Ai are satisfactory devices for this reason. Using ontology's domain knowledge can be represented. This use of Philosophy depicts a mutual and unequivocal formal conceptualization of a given space. Segments are classes (speaking to ideas, typically sorted out in some sort of various leveled structure), non-taxonomical and taxonomical relationships Sánchez and Moreno (2008), axioms (Batet, 2012) and instances (representing specific objects). There are a few tourism recommenders that utilize metaphysics' to formalize space learning. As of late most frameworks having nonexclusive cosmology's that store data about various angles that must be considered in the suggestion of social exercises. For instance, the framework portrayed in Wang et al. (2011) incorporates nonspecific travel ontology (with data about convenience, eateries, transport, shopping, culture, and so on.) and a client cosmology in which the statistic qualities of the sightseers and their inclinations are demonstrated. other systems like GeOasis (Martínez-Santiago, 2012), SAMAP (Castillo, 2008) SMART MUSEUM (Ruotsalo, 2013) and the one recommended in Alonso et al. (2012) also contain ontology's to model the various types of touristic exercises and to have the capacity to reason on them in a semantic manner. These frameworks utilize philosophy based comparability measures to derive if two sorts of exercises are comparative and this learning may likewise be utilized to register the similitude amongst clients and give suggestions in view of CL filtering methods.

Type of Data Interpretation

This factor categorized into two sub-factors; data collection or user type, decision change.

Data Collection Model

This subfactor is categorized into three sub-sub-factors; single, group, hybrid.

Individual

The individual recommended system is nothing but providing the recommendations to a particular user based on user interest. A lot of internet sites that help a holidaymaker client on the developing of customized holidaymaker arrange or agenda of the business (city or places to visit) e-tourism (Sebastia,

2009) eases the task of the process an outsized quantity of knowledge and choosing the foremost desirable activities for a specific user. To boot, it additionally permits for the arrangement and calendar of the holidaymaker exercises by taking care of totally variant forms of learning like whole long periods of spots, space between the spots to go to or the time spent on the visit. This fashion we have a tendency to produce a customized agenda for the holidaymaker with the counseled activities. Not only e-tourism (Sebastia, 2009) mobile recommended system (Gavalas, 2014) like applications are providing the recommended system

Group

As an alike individual, the group recommended system also present. It is specially designed for group (family members or friends) travel plan or tourism agenda. Here recommendations ought to contact the interests of most of the set users. This technique makes suggestions boost the gathering tastes, their statistic characterization and furthermore the spots went to by the clients in previous visits. The group suggestion is processed from singular individual proposals through the usage of procedures appreciate accumulation, convergence or dynamic intersection (Garcia, 2011) This methodology is enforced as an addition of the e-tourism tool (Sebastia, 2009), that may be a client adapted business and free website, whose important element is that the GRSK (Generalist recommender system kernel) (Garcia, 2011), an independent domain taxonomy-driven computer program that manages the group recommendation. The principal assignment of this sort of framework is to recognize the individual inclinations thus understand a tradeoff that is acknowledged by all the bunch individuals. This can be the critical reason in an extremely bunch RS because of anyway singular inclinations are figured out how to return up with group inclinations can affirm the achievement of the council.

Hybrid Recommender System

A combination of individuals and groups is called a hybrid recommender system. This is useful for both individual and group, Based on the user type system will recommend. In some applications, the hybrid technique module (Garcia, 2011) is used it gathers along with the three lists things came back by the item chooser and develops one list that represents the ultimate set or client suggested things. These three arrangements of proposed things registered by the things Selector square measure at that point handled by the Hybrid method that applies a mixed hybrid suggestion (Sebastia, 2009) and returns the arrangement of hierarchal things. By controlling these arrangements of things several, we have a tendency to offer far more flexibility to the GRSK (Garcia, 2011), as a result of the other hybrid technique will be utilized by merely exchanging one part by another.

Dynamic Decision Approach

Adaptive: Proposal frameworks and adjustive frameworks are acquainted in tourism applications with help the voyagers in their basic leadership forms. A few frameworks should answer to the astonishing changes all through the tour. Current holidaymaker systems aren't ready to gather all tourists and tourism product data from completely variant resources. Additionally, they did not give a succession of counsel on the different travel items (i.e., primary goal, wanted spending plan, length of the tour, settlement, transportation, action, and eatery) upheld by some individual request inclinations and travel organized. These frameworks don't guide any customization. As an example, skilled somebody or system admin couldn't modification the advice rule settings any longer. They can even tackle user needs by exploiting an adjustive holidaymaker suggested framework (ATRS) (Etaati, 2015) which can modify its behavior to take to the fresh state of affairs. This technique gathers data regarding somebody and therefore the travel product from completely variant assets. Within another stage, it provides customized suggestions that supported the gathered data. ATRS (Etaati, 2015) takes the explorer's phase of movement into thought and recommends the movement item bolstered that. Furthermore, ATRS gathers the voyager's inclinations in regards to the movement item proposal's request inside the underlying spot. At that point, all through the movement, the framework takes after the arrangements. As an example, some individual likes to incite suggestion on the action than on the convenience. At long last, ATRS gives an office to the administrator of the framework or skilled somebody to switch the advice rule factors.

CONCLUSION

This article reported on the current state of data mining and recommendation system in the tourism domain. Characteristics, model, advantages, applications, algorithms of data mining and recommended system. We also discussed a novel recommendation system working flow with one example. Example about the interaction of the user with product using k-means algorithm implementation, types of recommendation algorithms and various types of domains are using recommendation system but this paper focused on recommendation system in tourism domain and its importance in tourism, like giving recommendations to tourist or travel agent based on preference options with demographic profiles. Types of tourism applications can develop using recommendation system is also explained with some examples like e-tourism, LBS, etc, next section contains about factors affecting tourism and importance of factors in tourism domain listed out in table with name of factor, item of factor and way of effecting tourism domain with sub-factors explained clearly in table.

After factor effecting topic, related review of the paper discussed then started the main objective of the paper important factors and categorized into three main factors are a recommendation, building model and type of user these three main factors are categorized into respective subfactors, sub-factors are categorized into sub-sub-factors. Sub factors under recommendation are recommendation criteria and recommendation interface type under building proposed model sub-factors are model interface type, model techniques, and model architecture. Finally under the type of user factor sub-factors are data collection and adaptive. In factors, we explained the importance and functionality of factor advantages and dis-advantages with respective factors and sub-sub- factors. Listed two tables for example reference for tourism web-based applications and mobile applications.

REFERENCES

Batet, M., Moreno, A., Sánchez, D., Isern, D., & Valls, A. (2012). Turist@: Agent-based personalized recommendation of touristic activities. *Expert Systems with Applications*, *39*(8), 7319–7329. doi:10.1016/j.eswa.2012.01.086

Batet, M., Moreno, A., Sánchez, D., Isern, D., & Valls, A. (2012). Turist@: Agent-based personalized recommendation of touristic activities. *Expert Systems with Applications, 39*(8), 7319–7329. doi:10.1016/j.eswa.2012.01.086

Bedi, P., Agarwal, S. K., & Jindal, V. (2014). MARST: Multi-Agent recommender system for e-tourism using reputation-based collaborative filtering. In Proceedings of databases in networked information system (Vol. 8381, pp. 189-201), Berlin: Springer.

Berry, M. J. A. (1997). *Linoff, G. Data mining techniques: for marketing, sales, and customer support.* New York: Wiley Computer Publishing.

Bojnordi, E., & Moradi, P. (2012). A novel collaborative filtering model based on the combination of the correlation method with matrix completion technique. In *16th CSI international symposium on artificial intelligence and signal processing (AISP)*. IEEE. 10.1109/AISP.2012.6313742

Caruana, R., & Niculescu-Mizil, A. (2003). An empirical comparison of supervised learning algorithms. In *Proceedings of the twenty-third international conference*. ACM.

Castillo, L., Armengol, E., Onaindia, E., Sebastia, L., González-Boticario, J., Rodriguez, A., ... Borrajo, D. (2008). Samap: A user-oriented adaptive system for planning tourist visits. *Expert Systems with Applications, 34*(2), 1318–1332. doi:10.1016/j.eswa.2006.12.029

Claus, R., & Martin, R. (2004) Personalized multi-criteria decision strategies in location-based decision support. In *Proceedings of a journal of the association of Chinese professionals in geographic information systems* (Vol. 10, pp. 149-156). 10.1080/10824000409480666

Dasarathy, B. (1991). *Nearest neighbor (NN) norms: NN pattern classification techniques*. IEEE Computer Society Press.

Deng, C., Xiaofe, H., Ji-Rong, W., & Wei-Ying, M. (2004). Block-level link analysis. *Proceedings of the 27th annual international ACM SIGIR conference on research and development in information retrieval*, 440–7.

Dunham, M. (2002). *Data mining: Introductory and advanced topics*. Prentice Hall.

Etaati, L., & Sundaram, D. (2015). *Adaptive tourist recommendation system: conceptual frameworks and implementations. In Proceedings of journal of computer science* (Vol. 2, pp. 95–107). Springer.

Friedman, N., Geiger, D., & Goldszmidt, M. (1997). Bayesian network classifiers. *Mach Learn, 29*(2-3), 131–63.

Garcia, I., Sebastia, L., & Onaindia, E. (2011). *On the design of Individual and Group Recommender Systems for Tourism. In Proceedings of expert systems with applications* (Vol. 38, pp. 7683–7692). Elsevier.

Garcia, I., Sebastia, L., Onaindia, E., & Guzman, C. (2009). A group recommender system for tourist activities. In Proceedings of an international conference on electronic commerce and web technologies (Vol. 5692, pp. 26-37). Springer.

Gavalas, D., & Kenteris, M. (2011). A web-based pervasive recommendation system for mobile tourist guides. *Personal and Ubiquitous Computing, 15*(7), 759–770. doi:10.100700779-011-0389-x

Gavalas, D., Konstantopoulos, C., Mastakas, K., & Pantziou, G. (2014). *Mobile recommender system in tourism. In Proceedings of the journal of network and computer applications* (Vol. 39, pp. 319–333). Elsevier.

Han & Kamber. (2007). *Data Mining Concepts & Techniques*. Morgan Kaufmann Publishers.

Isinkaye, F. O., Folajimi, Y. O., & Ojokoh, B. A. (2015). *Recommendation systems: Principles, methods, and evaluation. Egyptian Informatics Journal* , 16, 261–273.

Koren, Y., Bell, R., & Kolinsky, C. (2009). *Matrix factorization techniques for recommender systems* (Vol. 8). IEEE Comput.

Kurata, Y., & Hara, T. (2013). CT-planner4: Toward a more user-friendly interactive day-tour planner. In *Information and communication technologies in tourism 2014* (pp. 73–86). Springer International Publishing. doi:10.1007/978-3-319-03973-2_6

Larose, T. D. (2005). *Discovering knowledge in data*. Hoboken, NJ: John Wiley.

Lee, C. S., Chang, Y. C., & Wang, M. H. (2009). Ontological recommendation multi-agent for Tainan city travel. *Expert Systems with Applications, 36*(3), 6740–6753. doi:10.1016/j.eswa.2008.08.016

Loh, Lorenzi, Saldaña, & Licthnow. (2004). A tourism recommender system based on collaboration and text analysis. In *Proceedings of cognizant comm. corp.* (Vol. 6, pp. 157-165). Academic Press.

Lucas, J., Luz, N., Moreno, M., Anacleto, R., Figueiredo, A., & Martins, C. (2013). A hybrid recommendation approach for a tourism system. *Expert Systems with Applications, 40*(9), 3532–3550. doi:10.1016/j.eswa.2012.12.061

Martínez, L., Rodriguez, R. M., & Espinilla, M. (2009). REJA: A georeferenced hybrid recommender system for restaurants. In *Proceedings of IEEE/WIC/ACM international conference on web intelligence and intelligent agent technology* (pp. 187–190). 10.1109/WI-IAT.2009.259

Martínez-Santiago, F., Ariza-López, F., Montejo-Ráez, A., & Ureña-López, A. (2012). GeOasis: A knowledge-based geo-referenced tourist assistant. *Expert Systems with Applications, 39*(14), 11737–11745. doi:10.1016/j.eswa.2012.04.080

Meehan, K., Lunney, T., Curran, K., & McCaughey, A. (2013). Context-aware intelligent recommendation system for tourism. In *Proceedings of the 11th IEEE international conference on pervasive computing and communications* (pp. 328–331). IEEE.

Noguera, J. M., Barranco, M. J., Segura, R. J., & Martínez, L. (2012). A mobile 3D-GIS hybrid recommender system for tourism. *Information Sciences, 215*, 37–52. doi:10.1016/j.ins.2012.05.010

Pinho, J., da Silva, B., Moreno, M., de Almeida, A. M., & Lopes, C. (2011). *Applying recommender methodologies in the Tourism sector. In Highlights in practical applications of agents and multi-agent systems in advances in Intelligent and Soft Computing* (pp. 101–108). Springer.

Ricci, F., & Werthner, H. (2002). Case base querying for travel planning recommendation. *Information Technology & Tourism, 4*(3), 215–226. doi:10.3727/109830501108751001

Ruotsalo, T., Haav, K., Stoyanov, A., Rochee, S., Fanid, E., & Deliaic, R. (2013). SMART MUSEUM: A mobile recommender system for the web of data. *Journal of Web Semantics*, *20*, 50–67. doi:10.1016/j.websem.2013.03.001

Sánchez, D., & Moreno, A. (2008a). Pattern-based automatic taxonomy learning from the web. *AI Communications*, *21*, 27–48.

Sánchez, D., & Moreno, A. (2008b). Learning non-taxonomic relationships from web documents for domain ontology construction. *Data and Knowledge Engineering Journal*, *64*(3), 600–623. doi:10.1016/j.datak.2007.10.001

Sebastia, L., Garcia, I., Onaindia, E., & Guzman, C. (2009). E-Tourism: A Tourist recommendation and planning application. *International Journal on Artificial Intelligence Tools, 18*, 717-738.

Sebastia, L., Garcia, I., Onaindia, E., & Guzman, C. (2008). e-Tourism: a tourist recommendation and planning application. *20th IEEE International Conference on Tools with Artificial Intelligence (ICTAI)*. 10.1109/ICTAI.2008.18

Suanmali, S. (2014). Factors Affecting Tourist Satisfaction: An Empirical Study in the Northern Part of Thailand. In *Proceedings of 4th international conference on tourism research SHS web of the conference* (Vol. 12). EDP Sciences.

Ucar, T., & Karahoca, A. (2015). Personalizing trip recommendations: A framework proposal. *Global Journal of Computer Sciences*, *05*, 30–35.

Wang, W., Zeng, G., & Tang, D. (2011). Bayesian intelligent semantic mash-up for tourism. *Concurrency and Computation*, *23*(8), 850–862. doi:10.1002/cpe.1676

Wooldridge, M. (2009). An introduction to multi-agent systems. John Wiley and Sons Yager, R.(1988). On ordered weighted averaging operators. Operators in multi-criteria decision making. *IEEE Transactions on Systems, Man, and Cybernetics*, *18*, 183–190.

Yu, C. C., & Chang, H. P. (2009). Personalized location-based recommendation services for tour planning in mobile tourism applications. In *Proceedings of the 10th international conference on e-commerce and web technologies (EC-Web'09)* (pp. 38–49). 10.1007/978-3-642-03964-5_5

KEY TERMS AND DEFINITIONS

Association: It is used to find relationships between items.

Classification: It is an act of discovering a function either model that defines and differentiates data concepts or classes for future prediction.

Clustering: It is used to discover groups of similar items.

Collaborative: Other customers, who like the similar item as another customer views or likes, will also call a recommended product.

Content-Based: A famous, suggested product has the same features as what a consumer likes or views.

Data Mining: Data mining is the process of extracting useful knowledge from data or big data, for example, information stockrooms, for productive examination, information mining calculations, and encouraging business.

Descriptive: It is the elucidation of past information to more readily comprehend changes that have happened in a business.

Memory-Based Technique: Prediction generation can do by the entire client item database; we can find the neighbor using statistical techniques, it is also known as the nearest neighbor. Through item, user depends skills can achieve Memory-based CF (collaborative filtering). to calculate the similarity between item/user many kinds of likeness metrics are utilized. Pearson correlation coefficient and cosine are the 2 famous likeness metrics.

Model-Based Techniques: The description model develops from the database and an active user will get predictions by this model. Model building processes done by variant pursuing machine learning algorithms, such as decision tree, artificial neural network, etc.

Multi-Agent System: It is an automated framework that contains numerous cooperating smart agents. Issues that are troublesome or outlandish for an individual operator to solve can be solved by a multi-agent system.

Multi-Criteria System: The system that will support recommendations based on both single and group of users.

Optimization Techniques: The optimization technique produces the best/optimal result from the collection of a possible solution set under given circumstances.

Predictive: It is a type of cutting-edge investigation that utilizations both new and authentic information to figure movement, conduct, and patterns. It includes applying measurable examination procedures, expository inquiries, and mechanized AI calculations to informational collections to make prescient models that spot a numerical worth - or score - on the probability of a specific occasion occurring.

Recommended System: Recommender frameworks intend to anticipate clients' inclinations and prescribe things that very likely are intriguing for them. They are among the most dominant AI frameworks that online retailers execute so as to drive deals.

Recommended System in Tourism: Traveler suggestion to help clients on the association of a recreation and vacationer motivation. Initially, a recommender framework offers the client a rundown of the city puts that is likely important to the client. This rundown considers the client statistic grouping, the client enjoys in previous excursions and the inclinations for the present visit.

Regression: To comprehend which among the autonomous factors are related to the reliant factors, and regression analysis is used to investigate the forms of these connections. For instance, by utilizing a relapse (regression) model conceivable to predict kids' tallness, given their age, weight and different elements.

Chapter 13
Signifying Business Tourism Through Knowledge Sharing and Innovation Processes

Nabila Tanvir
National Textile University, Faisalabad, Pakistan

Sajjad Ahmed Baig
National Textile University, Faisalabad, Pakistan

ABSTRACT

The main objective of this chapter is to highlight the significance of business tourism in the global village where the advanced technologies are utilizing for knowledge sharing and implementing innovation to attract the potential customers effectively and enhance the profitability of tourism industry. It also reviews how the professionals are developed by business schools, which are offering degree programs in tourism. With the growing popularity of tourism industry, there is a need for producing graduates in tourism who may perform excellently in launching innovations through effective knowledge sharing processes. It also discusses how knowledge sharing and innovation play an instrumental role in meeting the challenging demands of the potential clients of the tourism industry.

INTRODUCTION OF BUSINESS TOURISM

There are hundreds of people who are traveling around the globe for their business tours. A large number of business conferences and exhibitions have been held all over the world where the business organizers and managers come to attend such festivity. The millions of people around the globe are engaged in the different types of business tourism and this industry has been evolving rapidly with great prospects. The business tourism is known as the big industry, opening new horizons towards the leading destinations for the business people who have to embark all the time for the business trips. In the globalized world, the people are interacting with each other to have good business deals and the business tours are arranged by the tourism firms to facilitate businessmen to make their business tours more successful. To attract

DOI: 10.4018/978-1-7998-3142-6.ch013

potential clients, business tourism firms are evolving day by day to ensure the quality of products and services.

Figure 1.

HOW TO DEFINE BUSINESS TOURISM?

There is a great difference between these two terms, business tourism and business travel as the business tourism is defined in broader sense, encompassing the wide-ranged experiences of the business travelers or business tourists who have to stay away from the native land or homes for one night or more while business travel is related with the movement of the business travelers from one place to another place. Business Tourism relates to traveling of people who have to travel for the business, job-related work and trade.

According to the World Tourism Organization (WTO), tourism is defined as the individuals are involved in traveling and staying out of the native homes or places for some specific period for some recreational or business activities (Tribe, 1999). While in the viewpoints of Smith, the tourist is a wandering person who visits different places for enjoying life to experience a change (Smith, 1989). According to Mathieson and Wall, tourism is defined as the short-term traveling towards the destinations out of the typical residence and workplace, the activities, performed during the stay and the facilities, developed for catering to the needs of the tourists (Mathieson & Wall, 1982).

In the context of tourism, the destination is termed as the very conception of place, undergoing the process of epistemological and historical changes. According to the literature of tourism, destinations are known as the places and territories in terms of administrative and spatial hierarchy (Framke, 2002). In the view of Mill and Morrison, the destination is a significant part of the tourism system, incorporation of interdependent elements to entertain the tourists with the satisfying vacation experience. In other words, the destination is defined as the unification of manifold elements which include infrastructure, attractions, transportation, facilities and hospitality (Mill & Morrison, 1985).

The concept of destination is based on economic and job-oriented factors that offer all-inclusive services from the perspective of a sales pitch (Jensen, Hansen, & Metz, 1993). The destination can be elaborated as a dynamic whole, comprising three main resource bases which include the attraction base, demand base and facility base (Jensen et al., 1993). This resource-based conception of destination develops understanding about the changing demands and supply of destinations according to the specific event management, having adaptive responsiveness to the new trends and demands and innovative development of the attractive products and services for the tourists (Framke, 2002).

TYPES OF BUSINESS TOURISM

The multi-faceted types of business tourism have broadened the scope of business tourism in the modern era where the people are much concerned about the development of businesses more successfully out of the boundaries of the native lands. The tourism firms should develop awareness about the changing demands of the customers and organize business trips or events according to the different types of business tourism. It would help to target the potential customers specifically and obtain productive results in the form of generation of revenues massively.

John Swarbrooke and Susan Horner describe the different forms of business tourism in the book, *"Business Travel and Tourism"* which include local, regional and national meetings and conferences, associations corporate, major international congresses and conventions, training courses, exhibitions and trade fairs, product launches, incentive travel, short-term migration for employment, student and teacher exchanges, taking goods to markets, delivering goods to customers, military service at places other than one's normal base, aid charities/ NGOs working on projects away from their main base, government employees traveling in the service of the state, e.g. diplomats, individual general business trips, daily commuting to work outside one's home area (Swarbrooke & Horner, 2012)

In other words, the business tourism encompasses all types of business travels which are related with the business meetings, business deals with clients, training sessions, negotiation for new business deals, trade fairs or exhibitions, quality control management by visiting suppliers and contractors (Beaverstock, Derudder, & Faulconbridge, 2016).

SIGNIFICANCE OF BUSINESS TOURISM

The business tourism is a new terminology in the advanced tourism industry but it is also true that for many decades and centuries, the people are used to travel for business or job-related works. The researchers and academicians should pay attention to highlight the chronological development of business tourism as this business sector has been growing rapidly since its inception. It is an old phenomenon of the primitive ages but the very specific forms of business tourism have been developed in the last few decades, emerging as the wider global industry.

Figure 2.

To develop the business tourism industry, tourism organizations have to operate within the ostensible business environment which can be categorized into the micro-environment as well as macro-environment. In the micro-environment, the organizations are influenced by the internal factors in terms of physical or organizational ones e.g. intermediaries and suppliers while the business organizations are influenced by the external factors in terms of broader global ones e.g. international and national conditions. The factors of macro-environment should be identified and anticipated with a proactive approach to minimize the impacts of these factors within the domain of the business tourism industry.

With the advent of the advanced technologies, the growth rate of business tourism has been declining in the current times where the advanced technologies like satellite, webinar and video conferencing have been playing the vital role to connect the business people from the remote areas and they don't feel any need of traveling in this virtual world of internet era. With the development of Information and Communication Technologies (ICTs), there is need of adaptive approaches for the survival of the organizations how the economy and business can be tackled by the global paradigm shift as the internet plays an influential role for value creation of the organizations (Piller & Reichwald, 2009).

The business tourism is a wide-ranged and multi-faceted phenomenon, encompassing all-inclusive business trips for attending meetings, exhibitions, conferences and incentive travels (Biletska, 2011). Over the last few decades, the massive growth rate in the domain of business tourism has been observed on both international as well as the domestic level (Ladkin, 2006). The business tourism plays an influential role in promoting the worldwide tourism economy as millions of people are involved in the sponsorship of this industry (Beaverstock & Budd, 2013). However, the industry of tourism is the biggest, having saturated markets that result in the continuous reduction of profit margins along with the rapidly changing demands of the customers (Lohmann, 2004). The expansion of the business tourism industry has reached the boundaries of its environmental, economic and social where there is no more expansion. Tourism organizations are small and medium-sized enterprises whose products and services are very comparable and exchangeable (Buhalis, 1998).

The phenomenon of business tourism is assistive fundamentally in developing the social network as well as knowledge-based economies by sharing and transferring knowledge so resourcefully through the arrangement of conferences and meetings (T. Rogers, 2013). To meet the needs of the business travelers, the huge chains of business hotels have been developed in multiple geographical locations, manifesting the worldwide economy networks and the positive economic flows (Rogerson, 2015). To meet the challenging conditions in terms of climatic change and economic recessions, the key players of this business tourism industry should anticipate and implement the proactive approach to reduce the impacts of such unpredictable changes. There is need of redesigning such tourism training programs which would be useful for redefining the structures of the tourism organizations on the basis of skills and knowledge, bringing about the innovativeness and competitiveness in the field of business tourism.

PRACTICAL APPROACH OF TOURISM FOR CULTURAL AND SOCIAL DEVELOPMENT

The tourism is known as the vehicle of social development and community development, having multi-dimensional impacts on the cultural, social and economic growth (Altinay, Sigala, & Waligo, 2016). For instance, when the tourists plan tours around the world by choosing the different destinations which have entirely the different cultures, social, political and environmental variances, they share their specific norms and values according to the specific culture and society. As recently, Prince William and Kate Middleton, the Duke and Duchess of Cambridge planned royal tour of five days to Pakistan which was fruitful for sharing culture and developing economy of the developing Pakistan tourism society ("In pictures: Prince William, Kate bid farewell to Pakistan after the action-packed 5-day tour," 2019). Such kind of tour is exemplary for knowledge sharing among the people of two different societies and cultures who visited many important destinations of Pakistan with full coverage of social, cultural and political values and norms.

INNOVATION AND BUSINESS TOURISM

Since the 1980s, the tourism companies have been encountering a lot of challenges for sustaining the global competitive edge which is indispensable without innovation (Ottenbacher, 2007). In light of the literature, there is a persistent issue of the tourism firms which acknowledge the potent force of innovation for the enhancement of the competitiveness (Hjalager, 2010). The tourism industry is very responsive and proactive to embrace the technological innovations for serving the customers and launching the marketing campaign for the product development. Many tourism companies have searched out the multi-faceted applications, developed by the information and communication technologies for implementing in the back-office (to tackle the day-to-day operational tasks) as well as in the front office (deployment of the customer relationship management (CRM), web-based services (Web 2.0 tools or application of geo-localization technologies through GPS on phones) and constructing the hotels with the use of clean technologies to be environmentally effectual (Miralles, 2010).

Theory of Innovation

Innovation implies the capacity of generating knowledge, implementing and combining the innovative technologies for introducing the innovative organizational structures, marketplaces, operational processes, and products & services. Innovation can be defined as the process of transforming knowledge and ideas into value (Dvir & Pasher, 2004) and knowledge-based innovation plays the most significant role in producing the effective outcomes of knowledge management in the domain of business tourism (Quintane, Mitch Casselman, Sebastian Reiche, & Nylund, 2011).

Innovation implies for generating unique ideas with a problem-solving and proactive approach which tend towards restructuring, reorganizing and redesigning of the organizational and operational systems to improve the quality of products and communicative skills, reducing the high costs to a minimum level with the new budgetary plans and exclusive management of the teams. In other words, innovation stands for the capacity to adapt and change through the creation, recognition, and implementation of the innovative ideas for improving the processes, services, and products effectively (Hall & Williams, 2008).

Types of Innovations

The rejuvenation and expansion of the wide-ranged products & services, implementation of innovative techniques for producing, supplying and distributing the products in the associated markets, enforcement of rules and regulations for the workforce and revolutionary changes in the organizational structure, such kind of development take place in the organization when the employees and management are well-prepared to bring innovation in the organizations whether it is technological innovation or administrative innovation. There are three major types of innovation in the tourism industry which are elaborated here:

- **Incremental Innovation:** In this type of innovation, there are no radical changes through a revolutionary approach in the markets and technology but just minor improvements in the existing products or services. For instance, introducing such an advanced system to reduce waste in the kitchen of hotels or arrangement of larger pickup trucks for speedy luggage handling of the tourists.
- **Distinctive Innovation:** This type of innovation relates to the changing demands of the customer behavior as well as the organizational behavior where the adaptive approach is implemented to obtain productive outcomes. For instance, advance online buying of discounted travel tickets and in-flight telephone communication services.
- **Breakthrough Innovation:** This type of innovation is entirely related to the radical changes by applying the innovative approach in the organizational structure and system, advanced technology and consumer behavior. For instance, E-Ticketing system and automatic check-in services in the hotels (Hall & Williams, 2008).

The Drivers of Tourism Innovation

The innovation shouldn't be viewed in the context of individualistic entrepreneurs or innovators but it should be viewed in broader prospects of the economic and political conditions and cultural trends (Sundbo, 1998). There are some specific drivers of innovation in business tourism which are discussed here:

- **Competition:** Competition is known as the most powerful driver of innovation, generally and specifically in the domain of business tourism. The tourism companies always are busy to outshine each other by offering attractive incentives to potential clients by reducing travel costs and barriers to minimum levels (Hall & Williams, 2008). In this competition-intensive era, some of the airlines of Europe and the USA have been ruined financially amidst the race of competition for offering the exclusive rates of air travel (Graham & Vowles, 2006). In such an environment of competitiveness, innovation is an influential driver for the survival of the business firms and there is a need for the sustainable innovations in the organizational processes rather than products so that the other competitors wouldn't imitate them.
- **Economic Performance:** In the viewpoints of Schumpeter, economic performance is also an effective driver of innovation. In the context of productivity, there are impacts of introducing budget flights, automatic check-in services and freeze or cook food. There is a need for considerable financial resources for launching radical innovations. Nevertheless, in the industry of tourism, there is much scope of incremental and process innovations rather than technological innovations. The firms can get benefits from the opportunities for sustaining the bilateral relationship between economic performance and innovations (Cainelli, Evangelista, & Savona, 2005).
- **Demand-Led Innovation:** It is the general perception of innovation that it may be recognized as demand pulled (Schmookler, 1966). The innovations apply to the growing demands of valuable customers in the field of tourism (Shaw, Williams, & Nelson, 1995). If there is a growing number of tourists, there would be growing demands for considerable innovation as there would be exclusive flows of tourists which ultimately boosts up the demands of more flights, package holidays and hotels (Hall & Williams, 2008).
- **Technology:** Technology is one of the most significant drivers of innovation either it is an in-house development of technology or outsourced technology. For instance, the online systems of providing information and boosting up sales with the use of the internet are the productive and supportive tools for flourishing business tourism. In the age of the internet, there is a need for innovation for rejuvenating the workforce skills, advanced structures of organizations and innovative services which can be named E-tourism (Weiermair, 2006). In the field of business tourism, the technological innovation plays a remarkable role through innovatory changes in the transportation, material science, advanced accounting systems of hotels, renovation of the kitchen as well as bathroom accessories, use of credit cards and jet technology for promoting the business tourism on the international level.

The different kinds of innovation systems are introduced by the tourism companies within the tourism destinations on the national as well as the regional level (Hall & Williams, 2008). In the industry of tourism, innovation is embedded inherently in correspondence with the locality, depending on the available resources and relating to the particular locations which cannot be reproduced elsewhere (Longhi & Keeble, 2000). Transferring tacit knowledge through the individuals who help to circulate the information within the organizations, it is the most significant mechanism of the knowledge spillovers. Nevertheless, tacit knowledge is linked with the internal innovation which is highly sensitive in the case of losing the skilled people when there is high ratio of employee turnover or bankruptcy of the company while the other organizational structures tend to preserve the codification which takes place frequently in the networks of franchises (Darr, Argote, & Epple, 1995).

There are several factors which slow down the degree of innovation in the small as well as medium-sized enterprises which include the inadequate collaborative and managerial systems, absence of the tough competitiveness and significance of mass tourism and inadequate absorptive capacity of adopting the latest technologies by the tourism companies and destinations (Mayer, 2009). The tourism industry has been undergoing the metamorphosis process due to its changing nature, changing demands of the customers, and changing global environment. In other words, business tourism has to encounter changes in the markets and behavior of consumers which are based on the competitiveness, use of technology, work design and production processes (Coakes, Willis, & Clarke, 2001).

Resultantly, the operational processes and products & services are constantly under the process of modifications with the speedy growth rate. This modifying sector of tourism is interrelated with the intricate pattern of innovation in the different segments of the tourism system, including arts & entertainment, transportation, and hospitality. The innovation in tourism is pervading in every corner of the world where the local restaurants attract the customers by professional websites and innovative delicious dishes within the tourism market. There is a need to understand how the tourism industry is subject to the changes in the broader prospects, economically, politically and socially(Baumol, 2002).

It is commonly considered that tourism is a passive recipient of innovation to some extent but on the other hand, it is known as a powerful player for conveying and transferring innovative ideas and other kinds of innovations. In the industry of tourism and hospitality, the innovation is embedded specifically in the retailing and offering the fantastic cuisines according to the different food cultures. The chefs and cooks are moving around the globe in search of a good idea and the people also come back home with the cultural impacts of the territories where they travel. In the modern era of technologies, the innovation-intensive culture has been promoting rapidly due to the mounting competitiveness (Hjalager, 1996).

KNOWLEDGE CREATION AND BUSINESS TOURISM

Knowledge is defined as the most significant resource which is possessed by the organizations with value-addition characteristics (Grant, 1996). The difference between information and knowledge can be pinpointed on basis of three main characteristics, firstly, knowledge is functional to demonstrate the intention or standpoint of the individuals in the specific perspectives, secondly, knowledge gives meaningful and relational perspectives to accelerate the actions, thirdly, knowledge is about the specific actions and ends (Ikujiro Nonaka & Takeuchi, 1996). Hence, Knowledge is an amalgamation of experiences, contextual information, expert insights, and values, providing a framework for the evaluation and incorporation of the valuable information and new experiences (Davenport & Prusak, 1998).

In the perspectives of resource-based viewpoint (RBV), the firms have a repository to accommodate the resources, knowledge, skills to be responsive to internal and external systems for endless learning. Within the firms, the process of knowledge creation is developed by R & D systems with the tendency of problem-solving (Grant, 2000). The process of knowledge creation on the organizational level is dependent on the capabilities, qualifications, experience, expertise, skills, and training of employees and it is acquired by having interaction with other key partners within the workplace environment (Ikujiro Nonaka & Takeuchi, 1995). This knowledge creation and acquisition always emphasizes the development of informal social networks and self-managed teams as if there would be more empowerment or autonomy, the employees tend towards more complex learning which would be supportive in generating the innovative ideas and models (Ikujirō Nonaka & Takeuchi, 2007). According to I. Nonaka and H.

Takeuchi, the creation, organization, and dissemination of knowledge are valuable for the good future of the business firms (Ikujiro Nonaka & Takeuchi, 1995).

The main purpose of knowledge management or knowledge sharing is to develop an exclusive understanding of the needs of the customers; it can be anticipated by having interaction with the customers as well as end-users, a powerful knowledge channel is developed during the process of product development. In the context, business tourism can be flourished by developing a powerful knowledge channel to communicate with the customers for the anticipation of the changing demands and generation of new ideas to meet these challenging demands effectively. The organizations must empower individuals and create a learning environment at the workplace, it would be helpful for replenishment and development of organizational knowledge bank in the form of human capital as well as the structural capital (Skyrme & Amidon, 1997).

It is very important to develop an understanding of the operational process of innovation and implement such mechanisms that help to generate ideas and inspire the entrepreneurs to go beyond the knowledge creation or ideas generating factories to exploit knowledge for the development of innovative products and services. To accelerate the knowledge flow, the effectual technological infrastructure develops connectivity with the knowledge workers for having accessibility to the knowledge and making the flow of knowledge speedily from the people who know those who are in search of the valuable knowledge (Skyrme & Amidon, 1997). If there is a good flow of knowledge, it would provide leverage to the organizations to develop a useful and flexible system of knowledge creation by overcoming all pitfalls of the organizations internally (Grant & Baden-Fuller, 2004). The mechanism of acquisition, assimilation, transformation, and application of the external knowledge is dependent on the absorptive capacity of the organization as it is explained by the conventional three-dimensional model, developed by Cohen and Levinthal in 1989-1990 (Cohen & Levinthal, 1990). Consequently, innovation is developed by these two constituents which are the main source of knowledge creation internally and knowledge assimilation externally (Zahra & George, 2002).

There are many kinds of knowledge, distinctively two main kinds, tacit and explicit knowledge, introduced by Polanyi in 1996 (Polanyi, 1966). The explicit knowledge is related to the manuals, databases, and technology which are more significant in the industry of tourism as the tourism companies and hotel management have adeptness in organizing and managing the databases of the customers. The role of eCommerce is also significant in business tourism where online reservations, up-to-date information, and sales systems are based on the advanced technology of E-business. The explicit knowledge has some issues of protecting the patent rights against the imitation of the rivals while tacit knowledge is a good source of innovation as it is inexpressible and inimitable (Blackler, 2002).

The knowledge repository of organization implies for the public goods which play a contributory role in sharing ideas with the individuals who gain benefits from the assortments of knowledge (Connolly, Thorn, & Heminger, 1992). When the costs of knowledge sharing are linked with sharing, it would be less prospective while the rewards of knowledge sharing are related to sharing, it would have great prospects. When individuals are motivated to contribute effectively by sharing knowledge, their confidence and competence would be greater in the context that they can perform particular tasks (Bandura, 1997). According to the social exchange theory (Blau, 1964), the individuals are interested to have interactions with each other in the view of benefits and costs. The benefits of incentives don't need to be tangible; the individuals are engaged in interacting with each other based on expected reciprocity of relationship in the future (Gouldner, 1960). In this context, knowledge sharing has positive influences on the individuals when they build trust that their contribution would be reciprocated with some degree of benefits in the future (Davenport & Prusak, 1998).

In the knowledge era, business competence depends on the ability to create and sharing organizational knowledge. To nurture continuous knowledge flows, it is very crucial to introduce the people management practices (Cabrera & Cabrera, 2005). In the modern era, the significance of knowledge management has been increased in the promising knowledge economy which provides a set of guidelines on how to behave, perform, operate processes and add values to the products and customers (Buhalis & Costa, 2006).

In the advanced technological era, the organizations tend to outperform their competitors by developing innovations in the products and services rather than by maintaining the quality of products or services in the minimum costs. It is acknowledged significantly how knowledge sharing and transferring to the tourism industry are supportive of implementing innovation in tourism which includes the trading system, infrastructural system, technology system and regulation system (Hjalager, 2002). The knowledge management is the supportive tools for the organizations in the different perspectives which are discussed here:

- It assists the organizations for operating in an innovative and challenging environment.
- It helps the organizations to undergo the evolutionary process for developing new structures and management systems.
- It helps to activate and develop people in inventive manners.
- It enables the organizations to encounter new challenges and set priorities by developing shared values and focusing on the customer experiences and building relationship management systems.

The knowledge sharing is the most significant phenomenon in the tourism industry which helps to accomplish the assigned task dedication (Shaw & Williams, 2009). Therefore, tourism organizations have interaction with different organizations for sharing valuable information and developing innovations (Denicolai, Cioccarelli, & Zucchella, 2010). In the tourism industry, the accurate, pertinent and timely information should be shared to obtain productive outcomes. Through knowledge sharing with key stakeholders and partners, the communication level can be enhanced among tourism firms with mutual trust and commitment (Medina-Muñoz & García-Falcón, 2000). The knowledge sharing shows the active and successful partnership, participating in the decision-making process and setting goals and enhancing the performance of the partners (Beritelli, 2011).

The significance of learning and knowledge has been increased for the development of the organizational resources effectively. From the viewpoint of Spender, the establishment of knowledge-based organization is the source of sustainable competitive advantage as such an organization has a capacity of developing unique and valuable knowledge through the process of learning and involving the customers in the process of innovation (Spender, 2008). With the advent of the internet era, the universe has become the global village where everyone is connected so closely to each other through internet connectivity. The authoritative and decisive power is shifting out of the corporate headquarters to the individualistic business units which are liable to delegate this power to the self-managed teams, located in any corner of this global village. It is important to learn how to embrace global diversity and how to work within the different cultures which have cross-cultural and cross-national impacts due to the evolution of digital technology around the globe (Allee, 2009).

In the digital and technological revolution, the advanced technologies and digital networks are supportive in creating, assimilating, disseminating and implementing the most accessible knowledge for the accomplishment of the tasks. In the digital world, the social webs and knowledge networks play an instrumental role in creating, diffusing and giving leverage to innovation and knowledge. Organizations must develop a good relationship through the multifarious business networks by interacting and engaging with each other. The industry of tourism is considered one of the most lucrative industries around the globe, having gigantic shares in the employment and growth of GDP (Gross Domestic Product). Ultimately, the tourism industry has been contributing greatly in boosting up the global economy through the value-added services and value chains. Globally, business tourism has been developed as the biggest economic sector with constant and comparatively fast growth in business travel (Dutta & Mia, 2011).

China has been growing this global business by investing massively in the industry of tourism with productive outcomes in its territory. In 1982, the establishment of The China National Tourism Administration was good and proactive governmental planning and strategies to flourish this business tourism sector with innovation (Scott & Ding, 2008). In the perspectives of the economic growth rate, the innovation is also highlighted by the experts as the most significant success factor for the tourism and other services sector (Hall & Williams, 2008). To meet the challenging demands of the tourism sector, there is much space for innovation than others (Hjalager, 2010). In the industry of business tourism, the customers are the key partners in adding value to the services of tourism through their experience which may be entertaining, educational, escapist and aesthetic ones for enhancing and sustaining the performance of the tourism firms through the production of value-added products. Additionally, the engrossment of the customers is very crucial for sharing the valuable knowledge about the needs and wants of the customers which helps to sponsor the product innovation in the tourism industry according to the needs of the customers.

ROLE OF SOCIAL MEDIA OR VIRTUAL COMMUNITIES IN BUSINESS TOURISM

The social media can be termed as internet-based applications which are the advanced digital tool to generate impressions by the customers and share information-based experiences with other vulnerable internet users (Blackshaw, 2004). The internet has brought a revolution in every sphere of life and primarily remodeled the tourism industry on how to share information among the people who tend to travel frequently (Buhalis & Law, 2008). The social media comprises consumer-generated content (CGC) like websites, blogs, forums, virtual communities, reciprocal tagging, videos on Flicker and YouTube, playing a substantial role to attract the massive number of travelers around the globe. By using social media, people tend to share and post their travel-related experiences and knowledge which is the source of information for potential customers (Gretzel & Yoo, 2008). In the internet era, search engines like Google are the most influential interface which works as a gateway to disseminate travel-related information among potential customers. The renowned marketing channels approach potential customers to share all possible information about the destinations and tourism experience (Wang & Fesenmaier, 2003). The industry of tourism is a knowledge-intensive industry and it is very critical to promote the tourism industry by estimating consumer behavior due to the modified technologies, having accessibility to travel-related information (P. J. Sheldon, 2007).

Figure 3.

The involvement of the customers helps foster the learning and create innovative products that add value to the customers. In the Internet Era, it is easy to access and share information about the changing demands of the customers as the customers may have interaction with the companies and other clients by reducing the asymmetrical information to be more empowered. In this context, the virtual or online communities play a vital role in enhancing product innovation in the tourism industry with the customer support relationship. According to the framework of Von Hippel, the customers are also the main source of innovation who share and transfer their inventiveness and expertness to the organizations and other communities too (Von Hippel, 1986).

Figure 4.

The people share their experiences, interests, needs, services, and information, regarding the social and cultural traditions, protocols and languages. The tourism firms are also engaged in developing innovative products through learning from the online or virtual communities where potential customers are involved in sharing their knowledge, expertise, experiences, and needs (Jeppesen & Frederiksen, 2006). The role of social media is also crucial in sponsoring business tourism as the social media tools enable the people to be involved in self-organizing social networks, social feedback, and informal interactions through the connectivity, collaboration and communication within the social media community (Hemsley & Mason, 2012). To attain a sustainable competitive advantage, knowledge sharing and transferring to the key partners, stakeholders and employees are very significant. In the view of Nonaka and Takeuchi, to some degree, every individual is the knowledge worker (Ikujiro Nonaka & Takeuchi, 1995) and all workers have the potentiality to influence the organizational bottom line (Hui, 2003).

The virtual or online communities present the challenging opportunities for the tourism firms by providing the right information and building close relationships within the like-minded people which help to boost up the processes of social learning (Miller, Fabian, & Lin, 2009). For instance, the tourist online communities share the information and experience by using the advanced features of the website of the tourism firms which offer services of booking hotels with other interactive tools like navigation tools and maps. There should be a reasonable layout of the virtual or online tourism community which focuses on developing the understanding of the consumer's needs and behaviors socially and psychologically how they react and interact for the fulfillment of their needs. Such communities are a good source of providing reliable and timely information to active members rather than other traditional means. The tourism companies should formulate and implement such strategies that may facilitate the process of learning by creating, nurturing, assimilating and disseminating knowledge within the community.

In the current times, the knowledge, creativity, and innovation are widely acknowledged as the most critical competitive factors which significantly assist and nurture the adaptive capacity of the enterprises for sustaining its exceptional performance (Palacios Marqués & José Garrigós Simón, 2006). The reciprocity of the relationship between these three factors depends on the organizational capacity of generating and innovating ideas and the capacity of learning, expanding knowledge by sharing with people (Teece, 2007). In the context of rapid environmental changes, tourism is undergoing environmental changes dynamically due to its vulnerability to the different environmental factors. To meet the challenging environmental conditions, there is a need for collecting, sharing and processing a massive volume of information for the enhancement of creativity and innovation process constantly. The knowledge flows can be accelerated by providing the effective technological infrastructure, connecting all knowledge workers.

In the business tourism industry, knowledge sharing accelerates the growth rate of this industry by communicating with the employees, stakeholders, and customers. Being a valuable resource of tourism companies, knowledge helps to generate new ideas and develop innovation successfully. When there is the adequate knowledge flow, circulating the most reliable and valuable information among the key stakeholders, customers and employees, there would be more chances of creativeness and innovativeness to upgrade all systems of tourism industry so that the potential customers should be attracted for choosing the appropriate destinations, airlines, hotels, and reservations of tickets. The knowledge management in the industry of tourism assists to sustain the competitive advantage of the firms. Therefore, tourism firms must develop a knowledge-sharing culture for enhancing organizational performance and competitiveness.

ROLE OF BUSINESS SCHOOLS IN THE BUSINESS TOURISM

The business schools have been playing an important role by introducing bachelor degrees in Tourism and hospitality to improve the critical and professional characteristics in the field of tourism. In the perspectives of the growing popularity and profitability of the tourism industry, it is necessary to provide guidelines through educational institutes like the business schools of the different universities around the globe. The business schools develop the professional skills of the undergraduates in tourism by providing valuable training (Dredge et al., 2011). In the view of the future of the flourishing tourism industry, the business schools are producing graduates in tourism so that professional tourism managers would be developed to flourish the industry of tourism on a large scale. When the students are trained at the educational levels, they would perform excellently to grow business tourism. It is the responsibility of the institutes to prepare such curriculums for the bachelor degree programs in tourism which may help the students to enhance their professional knowledge and skills. In the view of the best tourism education, there is a need for reshaping, redesigning and refiguring the curricula of business tourism at the educational levels (P. Sheldon, Fesenmaier, & Tribe, 2009). If the good curriculum is designed for the graduates of tourism in the business schools of different universities, the students would be more receptive to the learning of the conceptual and professional paradigm of tourism (Greenspoon, 2008).

With the fantastic development of tourism at a higher educational level, many degree programs in tourism are introduced under the disciplines of anthropology, sociology, geography and business management of multiple universities of Europe and the USA (Inui, Wheeler, & Lankford, 2006). The scholastic approach should be implemented by the curriculum developers to give importance to the subject of tourism as the major discipline or university study (McKercher, 2002); and to develop such theoretical framework which would provide a set of guidelines for the students of tourism programs at the institutional levels (Fidgeon, 2010). The main objective of business schools is to develop graduates who are capable to develop critical and creative thinking which would be transformed into efficient and innovative business practices in the domain of tourism (Benn, Bubna-Litic, & Eckstein, 2001). Resultantly, the vocational skills are observed dominantly in getting the education of business tourism (Airey & Johnson, 1999). Tourism education should have an adaptive and transformative approach to develop market-driven graduates with exclusive professional insight and expertness (James, 2002). Furthermore, it is very important to evaluate how the institutions are operating, how the teachers of business schools are teaching and how the students are responsive to the learning environment at the university level (Brookfield & Holst, 2010).

The academicians and researchers should explore how to develop a curriculum of the tourism programs at the university level to meet the future needs of the young generations in this competitive world by taking an overview of the social, ecological and cultural aspects of this planet (WCED, 1987). The tourism education should have a capacity of developing more innovative, creative and proactive leaders in the tourism industry who may act and think critically to promote the business in the advanced era of technology and climatic change (Glasser, Calder, & Fadeeva, 2005). The institutions and universities are a good source of developing the graduates of Tourism at the business schools which gives leverage to the business industry with the development of intellectual capital and sustainability of the competitive advantage. The professionalism and rationalism are two main factors of nurturing the business tourism graduates who would implement their knowledge, skills, and abilities for establishing a business of tourism according to the modern trends of tourists. The different courses of the bachelor degree programs in tourism provide all-inclusive knowledge about the historical backgrounds of all popular destinations and hotels all over the world; how the business firms are developing the tourism industry by launching multi-faceted innovations and advanced system of knowledge management.

The business strategies and tactics are introduced on an institutional level for the tourism students on how business tourism can be flourished more successfully on a large scale. It would enlighten the vision of the future managers and entrepreneurs in the field of tourism who are educated at universities to groom their abilities for managing the tourism firms with the advanced features. The expertise of the professional tourism organizers is shifted towards the students of the tourism business for upgrading the tourism industry through the inter-linked approaches of innovation, knowledge sharing, and knowledge management. In a nutshell, business tourism is a broader subject which is taught at the university level to produce such graduates who play a crucial role in society by investing and promoting the business of tourism. The business schools are producing professional individuals, related to the tourism industry for establishing and promoting business tourism based on their expertise, innovative aptitude, and knowledge, etc.

The business schools provide practical knowledge in the specific profession or business, sponsoring the prosperity and productivity of the tourism industry in the age of advanced technology and globalization. The individuals who are interested to establish tourism business, they should get an education in tourism from the business schools so that they may boost up their professional skills how to target the potential customers through social networking and marketing channels and how to offer multi-dimensional products and services to satisfy the customer effectively with innovative strategies. If the individuals are equipped with knowledge and skills, they would perform excellently to promote the tourism business on national and international levels by capturing the attention of potential customers through internet interfaces. There is a need for substantial research work in the domain of business tourism to enable the professionals to innovate the tourism industry following the new trends and demands of the tourists.

CONCLUSION

With the advent of globalization era in 1980, the role of knowledge and knowledge management has been increased due to the growing changes in the social, cultural, economic and technological fields (Baggio & Cooper, 2010) and knowledge sharing is comparatively evaluated in terms of the tourism sector (Baerenholdt & Haldrup, 2006). Nowadays, the knowledge is considered the source of competitiveness of the tourism enterprises and destinations as it is the most significant to convey the necessary information to the concerned people at the right time (Chris Cooper, 2015). Innovation in tourism is assistive in providing the innovative solutions to all kinds of social, economic and environmental problems, occurred at the different destinations by propagating innovative ideas following the nature of social agreements, resources and capacities which is significant for the sustainability of social and economic transformation (P. J. Sheldon, Pollock, & Daniele, 2017).

Innovation is indispensable, mainly for destinations to attain a competitive advantage in the tourism industry (Quandt, Ferraresi, Kudlawicz, Martins, & Machado, 2017). The tourism social entrepreneurship is linked with creativity and innovation by creating jobs and livelihood, boosting up the financial status through the generation of massive revenues and patronizing the suppliers of the local community (Franzidis, 2018). In marginalized localities, there is not much good orientation for developing the tourism industry and there is a need of implementing top-down tourism development approaches rather than community-based tourism strategies (Porter, Orams, & Lück, 2018). The economic disparities within the nations can be alleviated through the collaboration of the professionals of tourism industry and local communities and the main agenda of flourishing the tourism industry in the marginalized or developing societies is to uplift the financial status as well as well-being of the communities of different destinations (Aquino, Lück, & Schänzel, 2018).

Knowledge is defined as amalgamation of multifarious experiences, contextual information, values and professional insight which is supportive for the evaluation and incorporation of new experiences and information (Czernek, 2017). According to Cooper (2006), the very idea of knowledge sharing can be comprehended in terms of the theory of innovation diffusion, postulated by E. M. Rogers (1995) who elaborated how an innovative idea can be conveyed through the proper channels among the members of the social system over time. In other words, knowledge sharing is a process of transferring knowledge from one entity to another, having diverse kinds of flows within and between the organizations of private and public sector, between the stakeholders and customers (tourists), between tourists and local communities and between the local communities and tourism firms (Weidenfeld, Williams, & Butler, 2010). The knowledge flow is maintained through the different channels which include seminars, trade press, labor mobility, observations and social media networks (Weidenfeld & Hall, 2014).

There is great competitiveness based on tacit knowledge of the workforce of tourism organizations as it is incontestable to replicate the knowledge sharing for the competitors, therefore it is the most competitive as well as a strategic advantage for the tourism firms which distinguish from others on basis of knowledge sharing and knowledge transferring tactics. It is very significant to have a clear concept in understanding the very nature of the knowledge and kinds of knowledge that are applicable in the domain of tourism. In other words, knowledge is known as the sets of skills and experiences of the intellectual capital with the addition of up-to-date information to take decisive actions on basis of the reliable sources (Westeren et al., 2008).

Tacit knowledge cannot be codified, disseminated and digitized but explicit knowledge comparatively can be symbolized and digitized, therefore it is easy to access and transferable. In the perspectives of knowledge management, there is significance of managing tacit and explicit knowledge in the field of tourism for creating and transferring knowledge within the organization or destination to identify the significance of knowledge and medium of knowledge sharing so that the significant measures should be taken for prompt actions in the context of exchanging and sharing knowledge. Hence, knowledge sharing and transferring is an instrumental tool in fostering knowledge sharing culture to enhance the creativity and innovation for signifying the business tourism in broader prospects.

Figure 5.

REFERENCES

Airey, D., & Johnson, S. (1999). The content of tourism degree courses in the UK. *Tourism Management*, *20*(2), 229–235. doi:10.1016/S0261-5177(98)00079-X

Allee, V. (2009). *The future of knowledge*. Routledge. doi:10.4324/9780080507156

Altinay, L., Sigala, M., & Waligo, V. (2016). Social value creation through tourism enterprise. *Tourism Management, 54*, 404–417. doi:10.1016/j.tourman.2015.12.011

Aquino, R. S., Lück, M., & Schänzel, H. A. (2018). A conceptual framework of tourism social entrepreneurship for sustainable community development. *Journal of Hospitality and Tourism Management, 37*(July), 23–32. doi:10.1016/j.jhtm.2018.09.001

Baerenholdt, J. O., & Haldrup, M. (2006). Mobile networks and place making in cultural tourism: Staging Viking ships and rock music in Roskilde. *European Urban and Regional Studies, 13*(3), 209–224. doi:10.1177/0969776406065431

Baggio, R., & Cooper, C. (2010). Knowledge transfer in a tourism destination: The effects of a network structure. *Service Industries Journal, 30*(10), 1757–1771. doi:10.1080/02642060903580649

Bandura, A. (1997). *Self-efficacy: The exercise of control*. Macmillan.

Baumol, W. J. (2002). *The free-market innovation machine: Analyzing the growth miracle of capitalism*. Princeton University Press. doi:10.1515/9781400851638

Beaverstock, J. V, & Budd, L. (2013). *International business travel in a digital world economy*. Academic Press.

Benn, S. H., Bubna-Litic, D. C., & Eckstein, D. (2001). Is the MBA sustainable? Degrees of change. In *Australian and New Zealand Academy of Management Conference*. Massey University.

Beritelli, P. (2011). Cooperation among prominent actors in a tourist destination. *Annals of Tourism Research, 38*(2), 607–629. doi:10.1016/j.annals.2010.11.015

Biletska, I. (2011). *Business tourism in the context of international tourism development*. Academic Press.

Blackler, F. (2002). Knowledge, knowledge work and organizations. *The Strategic Management of Intellectual Capital and Organizational Knowledge*, 47–62.

Blackshaw, P. (2004). *Consumer-generated media (CGM) 101: Word-of-mouth in the age of the web-fortified consumer.* Http://Www. Nielsen-Online. Com/Downloads/Us/Buzz/Nbzm_wp_CGM101. Pdf

Blau, P. M. (1964). *1964 Exchange and power in social life New York*. Wiley.

Brookfield, S. D., & Holst, J. D. (2010). *Radicalizing learning: Adult education for a just world*. John Wiley & Sons.

Buhalis, D. (1998). Strategic use of information technologies in the tourism industry. *Tourism Management, 19*(5), 409–421. doi:10.1016/S0261-5177(98)00038-7

Buhalis, D., & Costa, C. (2006). *Tourism management dynamics: trends, management and tools*. Routledge. doi:10.4324/9780080455907

Buhalis, D., & Law, R. (2008). Progress in information technology and tourism management: 20 years on and 10 years after the Internet—The state of eTourism research. *Tourism Management, 29*(4), 609–623. doi:10.1016/j.tourman.2008.01.005

Cabrera, E. F., & Cabrera, A. (2005). Fostering knowledge sharing through people management practices. *International Journal of Human Resource Management, 16*(5), 720–735. doi:10.1080/09585190500083020

Cainelli, G., Evangelista, R., & Savona, M. (2005). Innovation and economic performance in services: A firm-level analysis. *Cambridge Journal of Economics, 30*(3), 435–458. doi:10.1093/cje/bei067

Coakes, E., Willis, D., & Clarke, S. (2001). *Knowledge management in the sociotechnical world: the graffiti continues.* Springer Science & Business Media.

Cohen, W. M., & Levinthal, D. A. (1990). Absorptive capacity: A new perspective on learning and innovation. *Administrative Science Quarterly, 35*(1), 128–152. doi:10.2307/2393553

Connolly, T., Thorn, B. K., & Heminger, A. (1992). Discretionary databases as social dilemmas. *Social Dilemmas: Theoretical Issues and Research Findings, 7*, 199–207.

Cooper, C. (2006). Knowledge management and tourism. *Annals of Tourism Research, 33*(1), 47–64. doi:10.1016/j.annals.2005.04.005

Cooper, C. (2015). Managing tourism knowledge. *Tourism Recreation Research, 40*(1), 107–119. doi:10.1080/02508281.2015.1006418

Czernek, K. (2017). Tourism features as determinants of knowledge transfer in the process of tourist cooperation. *Current Issues in Tourism, 20*(2), 204–220. doi:10.1080/13683500.2014.944107

Darr, E. D., Argote, L., & Epple, D. (1995). The acquisition, transfer, and depreciation of knowledge in service organizations: Productivity in franchises. *Management Science, 41*(11), 1750–1762. doi:10.1287/mnsc.41.11.1750

Davenport, T. H., & Prusak, L. (1998). *Working knowledge: How organizations manage what they know.* Harvard Business Press.

Denicolai, S., Cioccarelli, G., & Zucchella, A. (2010). Resource-based local development and networked core-competencies for tourism excellence. *Tourism Management, 31*(2), 260–266. doi:10.1016/j.tourman.2009.03.002

Dredge, D., Benckendorff, P., Day, M., Gross, M. J., Walo, M., Weeks, P., & Whitelaw, P. (2011). *Conceptualising the perfect blend in the tourism and hospitality education curriculum space.* Academic Press.

Dutta, S., & Mia, I. (2011). The global information technology report 2010–2011. In *World Economic Forum* (Vol. 24). Academic Press.

Dvir, R., & Pasher, E. (2004). Innovation engines for knowledge cities: An innovation ecology perspective. *Journal of Knowledge Management, 8*(5), 16–27. doi:10.1108/13673270410558756

Fidgeon, P. R. (2010). Tourism education and curriculum design: A time for consolidation and review? *Tourism Management, 31*(6), 699–723. doi:10.1016/j.tourman.2010.05.019

Framke, W. (2002). The destination as a concept: A discussion of the business-related perspective versus the socio-cultural approach in tourism theory. *Scandinavian Journal of Hospitality and Tourism, 2*(2), 92–108. doi:10.1080/15022250216287

Franzidis, A. (2018). An examination of a social tourism business in Granada, Nicaragua. *Tourism Review*.

Glasser, H., Calder, W., & Fadeeva, Z. (2005). *Definition: research in higher education for sustainability*. Halifax, Nova Scotia: Document Prepared for the Halifax Consultation.

Gouldner, A. W. (1960). The norm of reciprocity: A preliminary statement. *American Sociological Review, 25*(2), 161–178. doi:10.2307/2092623

Graham, B., & Vowles, T. M. (2006). Carriers within carriers: A strategic response to low-cost airline competition. *Transport Reviews, 26*(1), 105–126. doi:10.1080/01441640500179377

Grant, R. M. (1996). Toward a knowledge-based theory of the firm. *Strategic Management Journal, 17*(S2), 109–122. doi:10.1002mj.4250171110

Grant, R. M. (2000). Shifts in the world economy: the drivers of knowledge management. *Knowledge Horizons: The Present and the Promise of Knowledge Management*, 27–53.

Grant, R. M., & Baden-Fuller, C. (2004). A knowledge accessing theory of strategic alliances. *Journal of Management Studies, 41*(1), 61–84. doi:10.1111/j.1467-6486.2004.00421.x

Greenspoon, J. S. (2008). *Sustainability in Graduate Business Curriculum: Implications for Business Schools*. ProQuest.

Gretzel, U., & Yoo, K. H. (2008). Use and impact of online travel reviews. *Information and Communication Technologies in Tourism, 2008*, 35–46. doi:10.1007/978-3-211-77280-5_4

Hall, M. C., & Williams, A. (2008). *Tourism and innovation*. Routledge. doi:10.4324/9780203938430

Hemsley, J., & Mason, R. M. (2012). The nature of knowledge in the social media age: Implications for knowledge management models. In *2012 45th Hawaii International Conference on System Sciences* (pp. 3928–3937). IEEE.

Hjalager, A.-M. (1996). Tourism and the environment: The innovation connection. *Journal of Sustainable Tourism, 4*(4), 201–218. doi:10.1080/09669589608667268

Hjalager, A.-M. (2002). Repairing innovation defectiveness in tourism. *Tourism Management, 23*(5), 465–474. doi:10.1016/S0261-5177(02)00013-4

Hjalager, A.-M. (2010). A review of innovation research in tourism. *Tourism Management, 31*(1), 1–12. doi:10.1016/j.tourman.2009.08.012

In pictures: Prince William, Kate bid farewell to Pakistan after action-packed 5-day tour. (2019). Retrieved from https://www.dawn.com/news/1511580

Inui, Y., Wheeler, D., & Lankford, S. (2006). Rethinking tourism education: What should schools teach. *Journal of Hospitality, Leisure, Sport and Tourism Education*, 5(2), 25–35. doi:10.3794/johlste.52.122

James, R. (2002). *Academic standards and the assessment of student learning: Some current issues in Australian Higher Education*. Centre for the Study of Higher Education.

Jensen, C. F., Hansen, J., & Metz, L. (1993). *Turistvirksomhed*. Copenhagen: Akademisk Forlag.

Jeppesen, L. B., & Frederiksen, L. (2006). Why do users contribute to firm-hosted user communities? The case of computer-controlled music instruments. *Organization Science*, *17*(1), 45–63. doi:10.1287/orsc.1050.0156

Ladkin, A. (2006). Conference tourism–MICE market and business tourism. In *Tourism business frontiers* (pp. 78–88). Routledge. doi:10.1016/B978-0-7506-6377-9.50014-6

Lohmann, M. (2004). New demand factors in tourism. In *European Tourism Forum* (Vol. 10). Academic Press.

Longhi, C., & Keeble, D. (2000). *High-Technology Clusters and Evolutionary Trends in the 1990s, in the High-Technology Clusters, Networking and Collective Learning in Europe*. Ashgate.

Mathieson, A., & Wall, G. (1982). *Tourism, economic, physical and social impacts*. Longman.

Mayer, M. (2009). Innovation as a success factor in tourism: Empirical evidence from western Austrian cable-car companies. *Erdkunde*, *63*(2), 123–139. doi:10.3112/erdkunde.2009.02.02

McKercher, B. (2002). The future of tourism education: An Australian scenario? *Tourism and Hospitality Research*, *3*(3), 199–210. doi:10.1177/146735840200300302

Medina-Muñoz, D., & García-Falcón, J. M. (2000). Successful relationships between hotels and agencies. *Annals of Tourism Research*, *27*(3), 737–762. doi:10.1016/S0160-7383(99)00104-8

Mill, R. C., & Morrison, A. M. (1985). *The Tourism System*. Prentice Hall.

Miller, K. D., Fabian, F., & Lin, S. (2009). Strategies for online communities. *Strategic Management Journal*, *30*(3), 305–322. doi:10.1002mj.735

Miralles, P. (2010). Technological innovation, a challenge for the hotel sector. *UPGRADE: The European Journal for the Informatics Professional*, *11*(2), 33–38.

Nonaka, I., & Takeuchi, H. (1995). *The knowledge-creating company: How Japanese companies create the dynamics of innovation*. Oxford University Press.

Nonaka, I., & Takeuchi, H. (2007). The knowledge-creating company. *Harvard Business Review*, *85*(7/8), 162.

Nonaka, I., & Takeuchi, H. (1996). The knowledge-creating company: How Japanese companies create the dynamics of innovation. *Long Range Planning*, *4*(29), 592. doi:10.1016/0024-6301(96)81509-3

Ottenbacher, M. C. (2007). Innovation management in the hospitality industry: Different strategies for achieving success. *Journal of Hospitality & Tourism Research (Washington, D.C.)*, *31*(4), 431–454. doi:10.1177/1096348007302352

Palacios Marqués, D., & José Garrigós Simón, F. (2006). The effect of knowledge management practices on firm performance. *Journal of Knowledge Management*, *10*(3), 143–156. doi:10.1108/13673270610670911

Piller, F. T., & Reichwald, R. (2009). Wertschöpfungsprinzipien von Open Innovation. In *Kommunikation als Erfolgsfaktor im Innovationsmanagement* (pp. 105–120). Springer. doi:10.1007/978-3-8349-8242-1_5

Polanyi, M. t. (1966). *1966 The tacit dimension*. London: Routledge & Kegan Paul.

Porter, B. A., Orams, M. B., & Lück, M. (2018). Sustainable entrepreneurship tourism: An alternative development approach for remote coastal communities where awareness of tourism is low. *Tourism Planning & Development, 15*(2), 149–165. doi:10.1080/21568316.2017.1312507

Quandt, C., Ferraresi, A., Kudlawicz, C., Martins, J., & Machado, A. (2017). Social innovation practices in the regional tourism industry: Case study of a cooperative in Brazil. *Social Enterprise Journal, 13*(1), 78–94. doi:10.1108/SEJ-12-2015-0038

Quintane, E., Mitch Casselman, R., Sebastian Reiche, B., & Nylund, P. A. (2011). Innovation as a knowledge-based outcome. *Journal of Knowledge Management, 15*(6), 928–947. doi:10.1108/13673271111179299

Rogers, E. M. (1995). Diffusion of Innovations: modifications of a model for telecommunications. In *Die diffusion von innovationen in der telekommunikation* (pp. 25–38). Springer. doi:10.1007/978-3-642-79868-9_2

Rogers, T. (2013). Conferences and conventions: A global industry (3rd ed.). Routledge. doi:10.4324/9780203119402

Rogerson, C. M. (2015). Unpacking business tourism mobilities in sub-Saharan Africa. *Current Issues in Tourism, 18*(1), 44–56. doi:10.1080/13683500.2014.898619

Schmookler, J. (1966). *Invention and economic growth.* Academic Press.

Scott, N., & Ding, P. (2008). Management of tourism research knowledge in Australia and China. *Current Issues in Tourism, 11*(6), 514–528. doi:10.1080/13683500802475802

Shaw, G., & Williams, A. (2009). Knowledge transfer and management in tourism organisations: An emerging research agenda. *Tourism Management, 30*(3), 325–335. doi:10.1016/j.tourman.2008.02.023

Shaw, G., Williams, A. M., & Nelson, R. R. (1995). Critical issues in tourism: A geographical perspective. *The Canadian Geographer. Geographe Canadien, 39*(3), 283.

Sheldon, P., Fesenmaier, D. R., & Tribe, J. (2009). *White paper: The Tourism Education Futures Initiative (TEFI): Activating change in tourism education.* Academic Press.

Sheldon, P. J. (2007). 18 Tourism information technology. *International Handbook on the Economics of Tourism*, 399.

Sheldon, P. J., Pollock, A., & Daniele, R. (2017). Social entrepreneurship and tourism: Setting the stage. In *Social Entrepreneurship and Tourism* (pp. 1–18). Springer. doi:10.1007/978-3-319-46518-0_1

Skyrme, D., & Amidon, D. (1997). The knowledge agenda. *Journal of Knowledge Management, 1*(1), 27–37. doi:10.1108/13673279710800709

Smith, S. L. J. (1989). Tourism analysis. *Tourism Analysis.*

Spender, J.-C. (2008). Organizational learning and knowledge management: Whence and whither? *Management Learning, 39*(2), 159–176. doi:10.1177/1350507607087582

Sundbo, J. (1998). *The theory of innovation: enterpreneurs, technology and strategy.* Edward Elgar Publishing.

Swarbrooke, J., & Horner, S. (2012). *Business travel and tourism*. Routledge. doi:10.4324/9780080490601

Teece, D. J. (2007). Explicating dynamic capabilities: The nature and microfoundations of (sustainable) enterprise performance. *Strategic Management Journal, 28*(13), 1319–1350. doi:10.1002mj.640

Tribe, J. (1999). The concept of tourism: Framing a wide tourism world and broad tourism society. *Tourism Recreation Research, 24*(2), 75–81. doi:10.1080/02508281.1999.11014879

Von Hippel, E. (1986). Lead users: A source of novel product concepts. *Management Science, 32*(7), 791–805. doi:10.1287/mnsc.32.7.791

Wang, Y., & Fesenmaier, D. R. (2003). Assessing motivation of contribution in online communities: An empirical investigation of an online travel community. *Electronic Markets, 13*(1), 33–45. doi:10.1080/1019678032000052934

WCED. (1987). World commission on environment and development. *Our Common Future, 17*, 1–91.

Weidenfeld, A., & Hall, C. M. (2014). Tourism in the development of regional and sectoral innovation systems. *The Wiley Blackwell Companion to Tourism*, 578.

Weidenfeld, A., Williams, A. M., & Butler, R. W. (2010). Knowledge transfer and innovation among attractions. *Annals of Tourism Research, 37*(3), 604–626. doi:10.1016/j.annals.2009.12.001

Weiermair, K. (2006). *Product Improvement or Innovation*. Academic Press.

Zahra, S. A., & George, G. (2002). Absorptive capacity: A review, reconceptualization, and extension. *Academy of Management Review, 27*(2), 185–203. doi:10.5465/amr.2002.6587995

KEY TERMS AND DEFINITIONS

Business Tourism: Business tourism is defined as the act of traveling of people in the perspectives of business deals, business meetings, and business conferences and when people take a long journey out of the country and stay for sponsoring their businesses, it is specified in terms of business tourism.

Destination: The destination is defined in terms of infrastructure, transportation system, manifold attractions, and hospitality facilities within the particular locality where the people are willing to take a tour for purposes of entertainment, businesses, and pleasures.

Knowledge: Knowledge is an amalgamation of experiences, contextual information, expert insights, and values, providing a framework for the evaluation and incorporation of valuable information and new experiences.

Knowledge Management: Knowledge management is an act of sorting out all significant information, values, ideas, and experiences which are shared by people to identify which is better for value-creation for the organization and intellectual assets.

Role of Business School: The main objective of business schools is to develop graduates who are capable to develop critical and creative thinking which would be transformed into efficient and innovative business practices in the domain of tourism.

Social Media: In the digital world, social media is an instrumental tool for sharing and transferring the relevant information rapidly among potential customers for the sponsorship of the specified destinations for business tourism.

Theory of Innovation: The very idea of innovation is taken into consideration by applying innovative ideas in correspondence with the new requirements for fostering the up-to-date knowledge and pioneering the unique ideology or theory.

Tourism: Tourism implies recreational tour or business tours around the globe through the provision of tourism organization for operationalizing the specific tours to the desired destination along with all attractions, accommodation and guiding or entertaining stuff.

Chapter 14
Importance of Creativity and Innovation for Gastronomy and Business Tourism

Nurgül Boz
Pamukkale University, Turkey

Hande Mutlu Ozturk
https://orcid.org/0000-0002-4404-0106
Pamukkale University, Turkey

ABSTRACT

Gastronomy examines, researches, and applies the food and beverage culture of a society in the historical process. Gastronomy is also a field of science that includes elements of science and art. The aim of gastronomy education is to train experts with the knowledge and equipment needed by the sector. Creativity is an important parameter in the field of gastronomy. The concept of creativity can be defined in different ways. In a simple way, creativity is used in the meaning of creating, revealing, and discovering the unknown. The Torrance scale is one of the most widely used methods for measuring creativity. In this chapter, the effect of gender, the type of school graduated at high school, and pre-school education on creativity was examined for undergraduate students studying gastronomy at the universities. The importance of gastronomy in business tourism is increasing. The development of creative ideas and thoughts by gastronomy employees can positively affect gastronomy tourism.

INTRODUCTION

With very general definition, tourism is a social, cultural and economic activity that requires people to move to countries or places outside their usual environment for personal or commercial/professional purposes. These people are called visitors or tourists. While defining the tourism market, it can be evaluated in two categories: business tourism market and private tourism market by taking into consideration the travel objectives, goals and financing methods.

DOI: 10.4018/978-1-7998-3142-6.ch014

As in the other sector, increasing competitiveness forces businesses to raise the value chain and forces them to develop new value sources (Pine and Gilmore, 1999). Today, as in all areas, there is differentiation and change in the tourism sector. By developing new methods, companies, cities, regions or countries are trying to revive tourism. In this context, as in other sectors, creativity new ideas or innovation are becoming important for tourism sector.

In recent years, the concept of 'innovation' has started to be widely discussed around the world. Innovation and creativity are today a key factor in the economic competitiveness of tourism destinations in countries, cities or firms. In order to develop new, creative ideas and products, creative individuals should be trained. Creativity gains great importance in the evolution of new products. In the process of developing a product, especially at the outset, a number of innovative solutions need to be developed, such as the specific process, organization, use of material resources.

The concept of "creative city" or "creative region is also discussed in the literature as an economic resource (Porter, 1998). Richards and Wilson (2006) discussed the reasons why creativity is more popular than traditional cultural approaches to development, and emphasizes the relationship between creativity, change and innovation.

This new area of hypothesis offers opportunities for various strategies for creative tourism development (Richards and Wilson, 2007):

- Tourism based on creative media consumption;
- Tourism based on the creative production of consumers;
- Traditional forms of tourism consumed or produced more creatively.

Creative activities may consist of apparently ordinary learning to improve local production rather than gaining more flashy skills. Alternatively, creativity in rural tourism can include community performances based on, for example, hybrid nature - eco-experience or adventure, and suggest new ways of understanding creativity. Tourists can be offered the opportunity to develop their creative potential by actively participating in rural areas. Rural tourism is gaining importance with different vinovative approaches. Any self-respecting country house hotel now offers additional creativity: it learns to cook country-style cooking, fishing, shooting, riding, and using various forms of environmental art (Richards and Wilson, 2007).

In urban areas, on the other hand, new museums and big art galleries, attractive casinos, large and different cinemas, neighborhood art galleries, experimental performance areas, spesific theaters, festival/event programming, small jazz and dance clubs and independently operated cafes are becoming more interesting.

In general, tourism makes a major contribution to the economic development of countries and the improvement of tourism products. Competitiveness in the tourism sector is constantly increasing. In order to achieve success in this competitive tourism market in the world and attract as many tourists as possible, it is very important that local authorities make a difference in tourism. The competencies responsible for the management of the country or the city may require a lot of innovation in different areas.

One of the most important types of tourism is business tourism. Business tourism is one of the issues that need to be studied in the tourism sector. In the business tourism market, it is very important that the country or city can create and present attractive products. It is noteworthy to meet customers' perceptions and high demands, and to offer products that are unique and tailored to their meeting characteristics.

The main purpose of a business visitor for a tourism trip is a visit for business and professional purposes. Compared to regular tourism, the number of participants in business tourism is less, their expectations from travel are different and their freedom to choose regions, cities or countries is limited. Business tourism destinations are more likely to be regions developed mainly for business purposes (countries, cities, industrial zones, etc.). Comparing an average business tourist with an average leisure tourist, business tourists are richer and can be expected to spend more money.

Both business tourists and leisure tours visiting a country or a region spend time and money for eating and drinking. Local food or gastronomy is known to be an important component of the motivation that attracts tourists.

In this chapter, the importance of creativity and creativity in tourism sector are emphasized. The parameters affecting creativity were analyzed. The factors that affect the creativity of gastronomy students are discussed. The main aim of this study to show the to show the creativity levels of gastronomy students and to determine the most effective parameters on creativity.

BACKGROUND

Business Tourism and Creativity

In general, tourism research shows that a competitive tourism sector improves the tourism industry as well as the quality of life of the people living in the tourism region (Dwyer and Kim, 2003, Kim et al., 2006). Bradley et al. (2002) describe business tourism as travel to attend corporate or association meetings, conferences, congresses or congresses or public or commercial exhibitions. They also acknowledge that business tourism emerges as an important sub-segment of the tourism industry, both in terms of travel volume and expenditure generated. Kotler et al. (1993) acknowledged that business tourism is especially important for marketing interviews and covers an important target audience.

Scott and Lodge (1985), in their study, describe national competitiveness as the ability of a country to gain increased revenue from its resources while gaining the ability to create, produce, distribute and serve products in international trade. Hong (2008) defined the competitiveness of the tourism sector. In this definition, a destination is considered as creating, integrating and presenting tourism experiences, including value added goods and services that are considered important by tourists. In the literature, competitiveness in tourism sector has been evaluated as a very important factor for the success of touristic places and sustainability of tourism (Gooroochurn and Sugiyarto, 2005, Mangion et al., 2005, Chen, 2008).

Hjalager (2010) focused on innovation in tourism and addressed several categories of innovation - product, process, management, marketing and corporate. The areas of innovation such as the role of entrepreneurship, technology repulsion, and the presence of regional industrial clusters were also evaluated in the study.

In their study, Aramendia Muneta and López (2013) analyzed the effects of Information Area and Communication Technology (ICT) on the competitiveness, innovation levels, productivity levels and market share of tourism in the tourism sector. The study deals with Accommodation, Gastronomy and Travel Agencies. In the study, it was shown that the use of different ICTs had little effect on the productivity increase as well as competition, but it had an effect on increasing the market share of the firms in general.

Creativity is a subject that scientists have been working on and researching for a long time. What is Creativity? Who are creative people? Can creativity be improved? similar questions are still being asked today. The concept of creativity has been perceived and defined in different ways by the people and places it is used. Although some definitions have similar points, this concept has no common definition (Bulut, 2014). Therefore, researchers in different fields related to the concept of creativity have different definitions.

Creativity is used in the simplest sense in terms of creating innovation, revealing and discovering the unknown. There are different perspectives on the concept of creativity. In these perspectives, sometimes the process, sometimes the product and sometimes some kind of personality structure are emphasized. However, the common view of all these different perspectives is that creativity reveals innovations (Kesici and Önçel, 2015; Öncü, 1992).

Creativity is the ability to imagine something that is not there, to do something in different ways than anyone else, and to develop new ideas about it. In other words, creativity is to see what everyone sees, but to think about it differently. Creativity is to look at events and objects differently than anyone else and to develop different approaches (Atay, 2009).

Creativity is not a rare talent. It should be regarded as a cognitive skill that can be developed and nourished by all people. Sometimes a person may not even realize the limits of his or her own power. For this reason, creativity is defined as a cognitive ability that has emerged as a new, original and skill-based product, or has not yet turned into a product, which includes a unique problem-solving process, and which uses the intelligence elements of the individual as original and productive (Aslan, 2001a).

According to San (1979), creativity is a skill that exists in every individual and can be found in every part of human life, a process that covers a wide area ranging from daily life to scientific studies, a form of attitude and behavior. Hennessey and Amabile (1987) emphasized that it is important to define the creativity in terms of the exact answer or the product presented. If a product or answer is unique, if it contains an open-ended problem / task that can provide many solutions, if it is true or valuable, it is seen as creative (Filiz, 2013).

Torrance, on the other hand, defines creativity as a set of skills, telents, motivation and the all connections established to deal with a problem (Torrance, 2004: 6). Creativity has emerged as a new, original and skill-based feature. It includes a unique problem-solving process that has not yet turned into a product. It is a cognitive ability that uses the intelligence elements of the person in an original and productive manner (Aslan, 2001b).

Creativity is not a spontaneous phenomenon. In creativity, different ideas and different solutions must be revealed. In other words, there should be intuitive and imaginary thinking in addition to fluent and logical thinking in every stage of creativity. When considered in this way, different and original ideas emerge and if these ideas reach a purpose, this can be perceived as creativity (Demirci, 2007).

With more comprehensive identification, creativity is breaking down existing patterns, being open to the lives of others, going out of the ordinary, discovering new ways, looking at life from different perspectives, acting on the unknown, wondering, breaking the imposed line of thought and revealing a new line of thought, to provide different alternative solutions for the problem, to leave the path of others, to find something new that leads to other things, to establish a new relationship or to establish a relationship between existing thoughts, to create a new idea, to invent a new technique or method unknown or to make additions (Ozerbas, 2011).

Creativity manifests itself in areas such as individual traits, intelligence, personality and physical development. The idea that creativity comes from birth and that the non-creative person cannot be creative later is abandoned and the idea that everyone can be creative with good education is now dominant (Özer, 1991). From the past to the present day, the concept of creativity is the most important mainstay of plastic arts. The recognition of artists as gifted, genius and non-human individuals has been one of the factors that enabled the concept of creativity to be identified with fine arts until the 19th century (Baltacı, 2013).

Art, which is a highly debated phenomenon of human life, has been evaluated same with creativity for a long time and creativity has been seen as a trans-physical, trans-natural phenomenon for many centuries (Sevimli, 2015). Creativity is not just a talent that artists have. Appropriate places and environmental conditions need to be provided for the development and progress of creativity. A creative environment should not be restrictive, authoritarian, or overly structured. The trainer, who plays an important role in revealing creativity, should be self-confident, knowledgeable, natural, sincere and full of enthusiasm. The first condition for children to be creative and productive is that they are free and can use their tools without hesitation and fear (Aral, 1999).

Creative education should aim to create people with the ability to create new and original products, not repetitions of what other generations do. It can be said that creativity is necessary and important in all areas of life. Therefore, the creativity potential of each individual should be revealed and creative thinking skills should be gained to the person. Therefore, it is important for individuals to know themselves and know how to develop or improve their creativity (Temizkalp, 2010). In the 21st century, demands in the tourism and food sector are changing rapidly. In order to maximize satisfaction for changing and differentiating demands, it has become compulsory to improve creativity performance. Worldwide, the gastronomy sector has become an economic force. Therefore, there is a need for creative individuals who understand, interpret, apply, and have high quality workforce in this field.

In the definitions of creativity, in a creative person; in the first place, there should be curiosity, patience, ability to make discoveries, adventurous thinking, thinking with images, a personality that cannot escape from experiments and researches and can reach conclusive judgments. In a creative person, it is observed that the creation process occurs individually. The interest in artistic creativity can also be perceived and interpreted by various social groups that vary depending on many variables, such as income and education levels. Income levels and occupational differences between people can shape and format people's lives in many ways. The reflections of this situation can be observed in the educational environment. The success of a child in a social group may be dependent on many variables such as income level and occupation of his / her parents (Konak, 2008).

MAIN FOCUS OF THE CHAPTER

The market share of business tourism in international tourism is constantly growing and business tourists are the most desirable tourists in destinations in the world, and this sector is expected to grow even more in the future. According to WTO (2007), business tourism provides high quality and profit and can be considered as an important part of economic development strategy. On the other hand, the business tourism sector is resistant to the types of events that affect leisure tourism and economic crises. Therefore, there is great competition in the market for the development of business tourism and this competition is expected to increase in the future.

To be competitive in the tourism sector, it may be needed to use many tools. One of them is the development of creative ideas in gastronomy and culinary arts. It may be useful to offer new, creative attractive and different foods oor flavors to the business people to visit or organize meeting in a region.

Today, gastronomy has become more important and as a result many universities have opened programs and courses in gastronomy (Santich, 2004). Creative ideas in the field of gastronomy can create a popular perception of gastronomy for a region or a country. Gastronomy is not only food, but also art of pleasure and food, and like everything else, tourism or gasronomy is about perception.

Creative ideas in the field of gastronomy can create a popular perception of gastronomy for that region. Because gastronomy is not only food, but also art of pleasure and food, and like everything else, tourism or gasronomy is about perception. If a good perception can be created, this will be the reason for choosing that place. The same is true for business tourism. Business people will also look for gastronomy as well as destinations while choosing a location for the meeting.

Creative ideas in the field of gastronomy can create a positive perception for the promotion of gastronomic products of a region. At the same time, creative innovations in the field of gastronomy can be influential for people choosing a destination. Creative ideas can create a popular perception of gastronomy. Because gastronomy is not only food, but also art of pleasure and food, and like everything else, tourism or gasronomy is about perception. If a good perception of gastronomy in a region can be created, this would be an additional reason for choosing the place. The same is true for business tourism. Business people can also follow the food trends of a region as well as a destination when they prefer a meeting.

CREATIVE THINKING PROCESS

Creativity is a process formed and emerged as a result of stages. Passing each step properly can ensure that the product to be produced is of the desired quality. In this direction, the stages of the creative thinking process are listed below.

Preparation Phase

This stage is the first step of the creative process ladder. There are many factors and data to solve the problem. It is a stage where both questions and answers are intertwined. At this stage, the elements of the problem are tried to be determined, and the information to be obtained by the people to find a solution is presented. The proliferation of information makes it easier to produce creative ideas. Organizing and evaluating this information constitutes the preparatory stage of creativity. The preparation stage is the longest stage of the creativity process (Baki, 2004; Çidem, 2016).

Incubation Phase

During the incubation phase, the problem is reversed. The problem is left to the examination of the mind. This period can last for minutes or weeks or years, as in the preparation phase. Meanwhile, skills such as pensive thinking, deep thinking, subconscious processes, visualization and sensory perception work. During the incubation phase, the individual does not think consciously about the problem. In the meantime, while the individual dives into other activities, the mind of the individual continues to think about the problem. In this period, the individual began to combine different ideas and opinion, identify the problem and produce solutions. This stage is considered as the stage in which the individual does nothing to solve the problem (Demirci, 2007).

Enlightenment Phase

At this stage, ideas are born to provide a basis for creativity. These ideas can be parts of the product or result, or they can be the result itself. This stage is a flare stage in which the parts that solve the problem are gathered together and provide integrity. This stage is momentary and develops in the wealth of insights. It may take several minutes or several hours. At this stage, the upper quadrant of the right sphere of the brain is active. The discoveries suddenly shine with an exclamation "Hah!". At this moment, this formation immediately registers to the brain, the lower part of the brain left and right part of the upper part of the fast sphere is experienced and provides the result of the validation stage passes (Üstündağ, 2009).

Realization-Verification phase:

The realization-verification phase consists of a series of activities to understand and demonstrate whether the outcome of the enlightenment phase will meet the requirements and whether it will meet the criteria determined during the preparation phase. At this stage, the left hemisphere of the brain is active. Sometimes the thoughts that occur during the enlightenment phase, the solutions that are realized, may not constitute the solution of the problem.

In this case, the person returns to the incubation stage and returns to the process of developing new solutions (Bulut, 2014).

CREATIVE PERSONALITY FEATURES

Bernard Shaw says that the intelligent man tries to adapt to the world. Those who are not smart try to adapt the world to himself. As a result, all development and progress is due to the unwise. Bernard Shaw emphasizes the characteristics of being incompatible and original in creative people. On the other hand, one of the definitions of intelligence is the ability to adapt. These two different views emphasize the differences in personality traits of creative people (Sevimli, 2015).

Csikszentmihalyi (1996) describes the characteristics of creative people that distinguish them from other people with the word 'complexity'. According to Csikszentmihalyi (1996), the characteristics common to creative individuals are listed below.

- Creative people have a large amount of physical energy, but are often quiet and resting,
- Creative people tend to be sociable and naive at the same time,
- Creative people combine fun and discipline, responsibility and irresponsibility,
- Creative people alternate between imagination and fantasy and offer a deep sense of reality,
- Creative people tend to be both inward and outward,
- Creative people can be humble and arrogant at the same time,
- Creative people can be both rebellious and conservative,
- Most creative people are very passionate about their work, but they can still be very objective about it,
- Creative people escape to some degree from strict gender roles,
- The openness and sensitivity of creative people often make them suffer, but they can also make them enjoy (Atilla, 2017).

When the creative individual and the non-creative individual are compared, the non-creative individual accepts the truth as it is, and the creative individual establishes connections between the truths and tries to arrange the truths in new ways. However, while the non-creative individual has fixed opinions, the creative individual is in constant struggle with these opinions (Aksüzek, 2008; Çidem, 2016).

IMPORTANCE OF CREATIVITY

From the existence of mankind to the point reached in the 21st century, human beings have tried to solve the problems they face in order to sustain their lives in the best conditions and designed the nature to raise their living standards. Whether technological products or works of art, human beings have always produced and felt the need to express themselves. The basis of this production process is the creativity and creative thinking skills that exist in every person (Keleşoğlu, 2017). Is creativity innate or can it be won later? It is possible to say yes to both parts of this question. Some people have a combination of innate creativity and superior intelligence. When these people focus on their creative potential, motivation and a goal, they make contributions to make the world better for everyone. Ibn Khaldun, Ibn Sina, Farabi, Edison, Einstain are some examples of the combination of creativity and superior brain power. It is certain that everyone's creative potential can be developed within genetic limits. The programs used to develop superior brain power in history and today show that when people are given the opportunity to use and process their innate creativity potential, wonderful results emerge (Özden, 2003; Güneştekin, 2011).

Creativity being in the daily life; moving people away from mediocrity, uniformity; It is said that its importance is undeniable due to the fact that it sheds light on finding the most suitable solutions to the problems encountered. For this reason, it is thought that understanding and using the importance of creativity in an important sector such as education will contribute to the future (Ayan, 2017: 3). Art, which is a highly debated phenomenon of human life, has been kept for a long time with creativity and creativity has been seen as a trans-physical phenomenon for many centuries. Ernst Fischer says that art was originally magic, a talismanic tool for dominating an unknown world. This sorcery duty of art has gradually become a duty to shed light on social relations, to enlighten the people in the concentrated societies, and to help people recognize and change social realities (Sevimli, 2015).

Creativity is involved in all kinds of work and engagement with all emotional and mental activity. Creative ability, constitutes the basis of all aspects of human life and development (Durukan, 2006). The aim of the education is not to be a transfer of knowledge, but to create different, extraordinary, new and original ideas by using the knowledge acquired and to adopt this creative thinking structure to people. In order to achieve these goals, the importance of creativity is increasing day by day, and this situation creates an environment for creativity to take place in education programs more (Ayan, 2017).

DEFINITION AND EXPLANATION OF THE IMPORTANCE OF CREATIVITY IN THE FIELD OF GASTRONOMY AND culinary ARTS

In the 21st century, international competition is increasing, and societies need creative individuals to make a difference in this competition. In order to educate creative individuals, education process is one of the important factors in supporting creativity. Educational processes are questioned in the researches. In order to educate creative individuals, studies are done on what needs to be done.

Contemporary education aims to gain skills with knowledge transfer and develop interests and abilities. Education aims to bring constructive, creative and productive individuals to society by revealing the creativity ability of the individual and to reach societies that produce new products both in science and technique and in intellectual, artistic and cultural fields (Aral, 1999). Art has been kept for a long time with creativity (Sevimli, 2015). One of the areas that requires a high level of creativity is gastronomy. Most dictionaries define gastronomy as the art and science of delicate eating. This focus on art and science, in practical terms, can be explained as skill and knowledge, which is closely linked to the origins of terms (Santich, 2004).

On the other hand, the concept of kitchen includes both art and science. Learning the science part of the kitchen can be achieved with good education. The artistic part can develop with the creativity of the cook. The secret of growing excellent cooks is to ensure that individuals with a strong creative interest in the kitchen are attracted to the culinary profession. It is very important to ensure that individuals have a good cookery education and that the scientific aspect of the kitchen is understood. Culinary education starts with the theoretical knowledge given to the chefs at school. When the cooks start working in the kitchen, they begin to add practical knowledge to this theoretical knowledge (Çakır, 2010).

When creative and non-creative individuals are compared, the non-creative individual accepts the truth as it is, and the creative individual establishes connections between the truths and tries to organize them in new ways. However, the non-creative individual has fixed views, the creative individual is in constant struggle with these views (Santich, 2004).

The concept of gastronomy is a cross-section of art. A craftsman, taster or restaurant owner who is not an artist cannot appreciate the subtleties of this concept and cannot impose the necessary meaning on the concept of gastronomy. For this reason, the care taken while preparing the food still raises aesthetic concerns in presentation and visuality (Dilsiz, 2010).

Since the applications in the kitchen section require hand labor and aesthetics, it includes artistic activities (Çifci, 2014). In the kitchen, where many products and materials are studied, success can only be achieved with creativity. In the kitchen, the chef is like a conductor. Good cooks seem to be a painter or sculptor in terms of their personality (Horng and Hu, 2008; Akt: Kesici and Önçel, 2015: 24). According to Hegarty and O'Mahony (2001), a work of art is expected to be beautiful and to meet established standards of performance and composition. The same can be said to produce food in culinary arts based on the aesthetic demands of the raw materials used by the chef to create flavor, color harmony, warmth and a dish (Akt: Atilla, 2017).

People are constantly following their food and travel blogs and schedules and trying to make their favorite food at home or going to eat a favorite food on site. Clearly; The journey for good, beautiful and original food is called gastronomic tourism. This market is enormous. Namely, according to TÜRSAB 2015 gastronomy report, the tourists coming to Turkey in 2014 spent 34.3 billion dollars. $ 6 billion 523 million of this expenditure was spent on eating and drinking. Its share in total expenditures is 19%. The money spent on eating and drinking is 1/5 of the total expenditure. This means a restaurant bill of $ 157.5 per tourist. their spend to eat and drink of the Turks abroad is more than 36.5 dollars from tourists coming to Turkey on holiday. In 2014, per capita expenditure on food and drink is 194 dollars. This corresponds to a share of 28-30 percent in total expenditures. (Sunnetcioglu and Ozkok, 2017).

In many areas, various tests and scales are being developed to measure creativity. One of these studies tested 10 areas of creativity. As a result of this test, creativity areas were examined separately as 3 main areas. These areas are named as lively, performance-based and scientific. The creative talent questionnaire was also scientifically classified in the kitchen. A person who does trial in the kitchen is no different than a scientist working in a laboratory. It includes similar processes as the mode of application. The results can be considered as inventions (Carson, Peterson and Higgins: 2005; Act: Çifci, 2014: 47).

Tourism business, like other enterprises, should be able to keep up with the recent changes. Change, innovation and creativity affect every field from past to present. Therefore, tourism businesses need to renew their services to meet the expectations of changing customers. However, institutionalized bussiness can keep up with these innovations with their culture and creative employees (Tekin and Çidem, 20017). In addition, culinary professions become areas where food is transformed into art, not just the areas where food is used. In this context, new ideas and creativity gain importance.

In gastronomy there is a certain hierarchy, rules-based configurations and traditional approaches to production and organization. Despite this structured system, with the recent developments, creative ideas have started to be developed in restaurants around the world. Designed as an architectural structure, presented visually as an appetizer and composed as a symphony that arouses emotions, aesthetically beautiful dishes are being made and served. It takes hours for meticulously prepared, cooked and presented dishes. It also offers new and creative experiences (Pedersen, 2012).

Hall (2012) indicated that "food consumption is integral to tourism and its economic impact can be important not only for immediate businesses that directly provide food for tourists (such as hotels, restaurants and attractions), it can also have significant economic impact throughout the food supply chain", especially if meals are provided locally.

As the culinary industry plays a dynamic and fundamental role in both tourism and retail, culinary creativity can serve as part of a solid environment to help conceptualize creativity and innovation in the tourism and retail sectors. The kitchen industry depends on eating as one of our daily routines. Eating is a personal and social act.

As the culinary industry plays an important role in the tourism and retail sectors, culinary creativity can help conceptualize creativity and innovation in the tourism and retail sectors. The culinary industry is among the daily routines of eating and is both a personal and social act (Brown et al. 2010).

Chossat and Gergaud (2003) stated that culinary creativity requires radical changes in the traditional cooking style. The goal of creativity in the culinary industry is to increase the sales and profits, as well as to ensure the quality and recognition of the organization.

MATERIAL AND METHODS

In this study, it was aimed to measure the parameters affecting the creativity levels of gastronomy students. During the study, demographic information was collected from the students with a form and Torrance Creative Thinking Test was applied to the students. During the implementation, the participants were given general information about what to do, the importance of research and confidentiality. Then, the application of the scales was started. The application lasted approximately 50-60 minutes. Then, three forms filled in by each participant were brought together and collected for evaluation.

After the application, Torrance Creative Thinking Test-Verbal Form A and Torrance Creative Thinking Test-Shape Form A, which were used to determine the creativity levels of the participants, were evaluated. The scores of the students were evaluated by using the Torrance Creative Thinking Tests Turkish Version Scoring Guide.

After the application, Torrance Creative Thinking Test-Verbal Form A and Torrance Creative Thinking Test-Shape Form A, which were used to determine the creativity levels of the participants, were evaluated. The scores obtained from the tests were evaluated by using the Torrance Creative Thinking Tests Turkish Version Scoring Guide and were recorded on a fluent scoring paper for each student. After completing the scores of Torrance Creative Thinking Test-Verbal Form A and Torrance Creative Thinking Test Figure Form A, all the collected data were transferred to computer and evaluated with the statistical program. Fluency, originality and flexibility scores of the participating students were calculated and evaluated.

Torrance Test for Creative Thinking (TTCT)

Creative Thought Tests, developed by Ellis Paul Torrance, was first published in 1966. The scale is a tool that measures the performance of creative thinking and has been developed as four forms: Verbal Form A, Verbal Form B, Formal Form A and Formal Form B. Verbal Form A and B, Figure A and B are parallel to each other. It is a scale that has been widely used since 1966 worldwide. (Lion and Puccio, 2006). The goal of Torrance was to develop a reliable and valid test for creative thinking skills that could be applied from kindergarten to childhood. The scores obtained by Torrance test can be considered as an indicator of creative thinking potential, but this is not enough to reveal creativity.

Participants

In the study, the scores of 154 students who were included in the study were evaluated by the Torrance creative thinking test for gender, the type of school they graduated from and the status of pre-school education. Ratings; Fluency, flexibility and originality were compared. Fluency refers to the ability to assess whether a response is related to a hypothetical situation. Originality sub-size; expresses the ability to think of something unusual, new or unique. Flexibility refers to the ability to use information and objects in different ways.

SOLUTIONS AND RECOMMENDATIONS

The aim of the study is to examine the creativity levels of gastronomy students. Findings are given below. A total of 154 students participated in the study. The answers of the students to the Personal Information Form were recorded by gender, whether they had pre-school education or not and the type of school they graduated from. The distribution of students according to gender variable is shared in Table 1. 57.1% of the students in the sample were female and 42.9% were male. The number of female students is 88 and the number of male students is 66. The type of school in which the students in the sample graduated; 16.2% are vocational high schools and 83.8% are normal high schools - Anatolian high schools. The number of vocational high school students is 25 and the number of normal high school - Anatolian high school students is 129 (Table 2). The students in the sample; 37.7% had pre-school education, 62.3% had no pre-school education. The number of those who attend preschool education is 58 and the number of those who do not attend preschool education is 96 (Table 3).

Table 1. Distribution of Sample Group by Gender Variable

Gender	F	%
Female	88	57,1
Male	66	42,9
Total	154	100,0

Table 2. Distribution of Sample Group According to School Type Variable

Type of School Graduated	F	%
Vocational High School	25	16,2
Normal High School - Anatolian High School	129	83,8
Total	154	100,0

Table 3. Distribution of Sample Group by Preschool Education

Preschool Education Status	F	%
Yes	58	37,7
No	96	62,3
Total	154	100,0

According to the findings,

When examined in terms of fluency subscale, male students mean score of Torrance creative thinking test is higher than female students mean score, but it does not make a statistically significant difference on gender fluency score (Table 4). According to the type of school graduated, students' scores obtained from the Fluency dimensions of Torrance Creativity Test were compared with T-test according to their average (Table 5). As a result of the analysis, a significant difference was observed in the total scores of the Torrance Creativity Test (, 000) (** $p < .01$). This shows that students' learning fields are related to their creativity. The verbal fluency scores of the vocational high school students who preferred the gastronomy department were higher than those of the Anatolian and normal high school students. It was concluded that among the 3 different factors examined (gender, type of school graduated and attendance to pre-school education), the highest score was Fluency. On the other hand, significant differences were found at $p < .05$ level according to whether they have attended to pre-school education or not. Fluency scores of the students who attended preschool education were higher than those who did not attend (Table 6).

This result is supported by many studies. In the study carried out in order to compare the formal creativity of kindergarten (6 years) children in terms of gender variable, it was concluded that there was no significant difference between the averages of creativity, fluency, flexibility, originality between girls and boys (Öncü, 2000).

In the study, it was found that the highest average total fluency and the lowest average total flexibility. Aslan (2001b) collected data from study groups randomly obtained from different age and education levels in his study to construct the Turkish version of the test. In the study, it was found that the verbal mean scores of students of different educational level decreased from verbal fluency to verbal flexibility. This finding supports our research. Verbal flexibility shows one's ability to look at things from different angles and produce different ideas. If the verbal flexibility score average is lower than other verbal scores, it means that this feature is less than verbal fluency and verbal originality abilities.

When examined in terms of flexibility sub-dimension; there is no statistically significant difference between the scores of female students and male students in this sub-dimension, but the average score of girls is higher than the average score of boys (Table 4). Also, there was no statistically significant difference between those who attanded pre-school education and those who did not. However, the average of score for the student attanded the pre-school education was higher. If the situation is examined in terms of the school they graduated from; No significant difference was found when the mean scores of the students' verbal originality dimensions and the T-test scores were compared (Table 5). Among the sub-dimensions, the lowest level was found to be the flexibility level. This shows that the students who participated in the research could not use their ability to produce a large number of ideas in terms of dealing with different aspects of the events. Current studies support the findings.

In the study conducted by Gülözge (2006) in order to determine the creativity levels of prospective classroom teachers according to their own perceptions, no significant difference was found in the level of creativity of prospective teachers in terms of the type of high school they graduated. The findings support our study.

In another study, the effect of problem-based learning approach on creative thinking levels in science teaching was evaluated. In the study, it was determined that the creative thinking pre-test scores of the prospective teachers in the experimental group did not differ significantly according to the types of secondary schools they graduated (Yaman & Yalçın, 2005).

When examined in terms of originality sub-dimension; It is seen that the average score of female students is higher than the average score of male students. However, according to T test results, there was no significant difference on the originality score of the gender (Table 4). The type of school graduated and the attendance to preschool education do not create a significant difference (Table 5). However, if compared in terms of their average score, the creativity of students who have graduated from vocational high school is higher than the students who have graduated from normal high school and Anatolian high school. The mean scores of the students who have continued pre-school education in this sub-dimension were found to be higher than the average score of the students who did not attend preschool education (see Table 6).

Özyurt (2011) examined the relationship between creativity levels of students attending private school and SBS achievement. In the study, it was found that there was no significant difference in the fluency, originality and flexibility sub-dimensions of students' creativity according to gender variable.

In a study, the effects art education on the creativity for adolescents were examined. In this study, the effects of whether they have taken art education before or not, the gender, the class they and the branches of art education attend were examined. In the study, it was stated that gender has no effect on creativity (Aral, 1999).

Table 4. Creative Thinking Scores by Gender

Gender	N	Fluency	Flexibility	Originality
Female	88	50,1818	27,9091	45,5568
Male	66	54,8485	25,1970	38,6061

Table 5. Creative Thinking Scores by Type of High School Graduated

School Type	N	Fluency	Flexibility	Originality
Vocational High School	25	72,7200	25,8000	44,3200
Normal High School - Anatolian High School	129	48,2016	26,9302	42,2403

Table 6. Creative Thinking Scores by Preschool Education

Gender	N	Fluency	Flexibility	Originality
Attended to Pre-school education	58	63,2586	28,1034	45,4138
Not attended to pre-school education	96	45,4896	25.9271	40,8646

FUTURE RESEARCH DIRECTIONS

In some cases, tourists want to taste local food and taste local delicacies, but in some cases they expect different presentations or tastes. In order to make different tastes or presentations, creativity and creative ideas are needed. For this, it is important to develop the creativity of the people working in the field of gastronomy. With regard to food and beverage, the expectations of business tourists may differ from the expectations of leisure tourists. For this reason, taking into consideration the social, cultural and economic expectations of tourists, it is important to prepare and serve food and beverages. In this chapter, some of the parameters that affect the creativity of students studying at the university in the field of gastonomy are examined.

Today, gastronomy is very popular in all societies. It has started to be one of the determining factors in destination preferences. Developing local foods and flavors and presenting them with innovative methods may increase the interest the the field of gastronomy. Gastronomy can be regarded as an art. It appeals to both the eye and the palate. Instead of serving local taste, flavor or food as it is, presenting it with new taste, flavor or appearance can make that dish or region popular.

Producing new and creative products can only be done by people with creative features. It is very important to determine the parameters that affect the creativity of students studying in the gastronomy and culinary arts departments of universities. Therefore, it is important to know the parameters that affect creativity.

In this chapter, the parameters affecting the creativity of the students studying in the gastronomy and culinary arts departments of the universities were examined for gender, the type of school they graduated from high school and whether they attend preschool education or not.

However, the study should be further expanded. Many parameters that affect creativity should be examined. For example, age, occupation of parents, income level of parents, educational status of parents, number of siblings, growing up in rural or urban areas, growing up in large or small cities, number of social and cultural activities they participated during childhood and adolescence etc. At the same time, the importance of gastronomy for the business tourism can be studied in another studies.

CONCLUSION

Based on the results of the research, the following recommendations can be made: Further research can be done to examine the creativity of students in different age groups and experimental studies. It can be conducted to determine the factors that affect students' creative skills. In the studies to be conducted, various researches can be carried out by using variables such as cognitive development processes, different high level thinking skills and learning styles.

In order to educate creative individuals in the field of gastronomy, creativity aspect should be developed starting from childhood. Creativity develops depending on the people interacted throughout life. In this case, parents, familiy and teachers, who are the closest, can be seen as the most influential in creativity. In order to develop creativity, one should be very careful and avoid any restrictive behaviors. Opportunities should be provided for students to think and express their thoughts. Their supportive behavior will allow them to leave lasting traces in a positive way throughout their life. The creativity of gastronomy students may be improved by the education they receive at the university. However, the process and creativity of creativity until the university is very effective. The creativity of gastronomy students may be improved by the education they receive at the university. However, it should not be forgotten that the processes and trainings until university education are highly effective on creativity.

REFERENCES

Aksüzek, S. (2008). İşletmelerde Rekabet Avantajı Olarak Yaratıcı Düşünceden Yararlanılması Üzerine Bir Araştırma (Yayımlanmamış yüksek lisans tezi). Gazi Üniversitesi Sosyal Bilimler Enstitüsü, Ankara.

Aral, N. (1999). Sanat Eğitimi - Yaratıcılık Etkileşimi. *Hacettepe Üniversitesi Eğitim Fakültesi Dergisi Ankara, 15*, 11–17.

Aramendia Muneta, M. E., & Ollo López, A. (2013). ICT Impact on tourism industry. *International Journal of Management Cases, 15*(2), 87–98.

Aslan, E. (2001a). *Kavram Boyutunda Yaratıcılık*. Türk Psikolojik Danışma ve Rehberlik Dergisi 16.

Aslan, E. (2001b). Torrance Yaratıcı Düşünce Testi'nin Türkçe Versiyonu. *M.Ü. Atatürk Eğitim Fakültesi Eğitim Bilimleri Dergisi, 14*, 19–40.

Aslan, E., & Puccio, G. J. (2006). Developing and Testing A Turkish Version of Torrance's Tests of Creative Thinking: A Study Of Adults. *The Journal of Creative Behavior, 40*(3), 163–178. doi:10.1002/j.2162-6057.2006.tb01271.x

Atay, Z. (2009). *Okul öncesi eğitim kurumlarına devam eden 5-6 yaş öğrencilerinin yaratıcılık Düzeylerinin yaş, cinsiyet ve ebeveyn eğitim Durumlarına göre incelenmesi: Ereğli örneği. (Yayımlanmamış yüksek lisans tezi).* Konya: Selçuk Üniversitesi Sosyal Bilimler Enstitüsü.

Atilla, O. (2017). Müziğin Yaratıcı Mutfak Uygulamaları Üzerindeki Etkilerinin Belirlenmesi (Yayımlanmamış yüksek lisans tezi). Anadolu Üniversitesi Sosyal Bilimler Enstitüsü, Eskişehir.

Ayan, B. E. (2017). *Okul Öncesi ve İlkokul Öğretmenlerinin Yaratıcılığa İlişkin Görüşleri ve Öğrencilerinin Yaratıcılık Performansları Arasındaki İlişki* (Yayımlanmamış yüksek lisans tezi). Kocaeli Üniversitesi Sosyal Bilimler Enstitüsü, Kocaeli.

Baki, S. (2004). *Küresel Ekonomik Sistemde Rekabet Gücünün Geliştirilmesine Yönelik Yaratıcılık Stratejilerinin İncelenmesi* (Yayımlanmamış yüksek lisans tezi). Kocaeli Üniversitesi Sosyal Bilimler Enstitüsü, Kocaeli.

Baltacı. (2013). Çocuğun Yaratıcılık Performansında Görsel Değerlerin Yeri (Yayımlanmamış doktora tezi). Ondokuz Mayıs Üniversitesi Eğitim Bilimleri Enstitüsü, Samsun.

Bradley, A., Hall, T., & Harrison, M. (2002). Selling cities: Promoting new images for meetings tourism. *Cities (London, England), 19*(1), 61–70. doi:10.1016/S0264-2751(01)00046-4

Brown, L., Edwards, J., & Hartwell, H. (2010). A taste of the unfamiliar. Understanding the meanings attached to food by international postgraduate students in England. *Appetite, 54*(1), 202–207. doi:10.1016/j.appet.2009.11.001 PMID:19896514

Bulut, Y. (2014). Meslek Yüksekokulu Öğrencilerinin Yaratıcılık Açısından Kendilerini Değerlendirmeleri İle Yaratıcılık Düzeyleri Arasındaki İlişkinin İncelenmesi (Yayımlanmamış yüksek lisans tezi). Gazi Üniversitesi Eğitim Bilimleri Enstitüsü, Ankara.

Bulut, Y. (2014). Meslek Yüksekokulu Öğrencilerinin Yaratıcılık Açısından Kendilerini Değerlendirmeleri İle Yaratıcılık Düzeyleri Arasındaki İlişkinin İncelenmesi (Yayımlanmamış yüksek lisans tezi). Gazi Üniversitesi Eğitim Bilimleri Enstitüsü, Ankara.

Çakır, M. (2010). Otel İşletmelerinin Mutfak Bölümünde İstihdam Edilen Personelin Eğitim Sürecinin Değerlendirilmesi: İstanbul'daki 5 Yıldızlı Zincir Otellere Yönelik Bir Alan Çalışması (Yayımlanmamış yüksek lisans tezi). İstanbul Üniversitesi Sosyal Bilimler Enstitüsü, İstanbul.

Carson, S. H., Peterson, J. B., & Higgins, D. M. (2005). Reliability, Validity, and Factor Structure of the Creative Achievement Questionnaire. *Creativity Research Journal, 17*(1), 37–50. doi:10.120715326934crj1701_4

Chen, C. F. (2006). Applying the analytical hierarchy process (AHP) approach to convention site selection. *Journal of Travel Research, 45*(2), 167–174. doi:10.1177/0047287506291593

Chossat, V., & Gergaud, O. (2003). Expert opinion and gastronomy: The recipe for success. *Journal of Cultural Economics, 27*(2), 127–141. doi:10.1023/A:1023432502059

Çidem, G. (2016). Mutfak Çalışanlarının Örgüt Kültürü Algıları ve Yaratıcılık Süreçleri Arasındaki İlişkiler: Kemer Bölgesi Örneği (Yayımlanmamış yüksek lisans tezi). Süleyman Demirel Üniversitesi Sosyal Bilimler Enstitüsü, Isparta.

Çifci, M. A. (2014). İş Ortamının Yaratıcılık Performansına Etkisi Mutfak Çalışanları Üzerine Bir Uygulama (Yayımlanmamış yüksek lisans tezi). Akdeniz Üniversitesi Sosyal Bilimler Enstitüsü, Antalya.

Csikszentmihalyi, M. (1996). *Creativity: Flow and the Psychology of Discovery and Invention.* Harper-CollinsPublishers. https://books.google.com.tr/books?id=K0buAAAAMAAJadresindenerişildi

Demirci, C. (2007). Fen Bilgisi Öğretiminde Yaratıcılığın Erişi Ve Tutuma Etkisi. *Hacettepe Üniversitesi Eğitim Fakültesi Dergisi*, (32), 65-75.

Dilsiz, B. (2010). Türkiye'de Gastronomi ve Turizm (İstanbul Örneği) (Yayımlanmamış yüksek lisans tezi). İstanbul Üniversitesi Sosyal Bilimler Enstitüsü, İstanbul.

Durukan, H. (2006). Okul Yöneticisinin Vizyoner Liderlik Rolü. *Ahi Evran Üniversitesi Kırşehir Eğitim Fakültesi Dergisi*, *7*(2), 277–286.

Dwyer, L., & Kim, C. (2003). Destination competitiveness: Determinants and indicators. *Current Issues in Tourism*, *6*(5), 369–414. doi:10.1080/13683500308667962

Filiz, F. (2013). Kimya Dersleri İçin Bilimsel Yaratıcılık Ölçeğinin Geliştirilmesi Ve Genel Yaratıcılık İle Bilimsel Yaratıcılık Arasındaki İlişkinin Belirlenmesi (Yayımlanmamış yüksek lisans tezi). Balıkesir Üniversitesi Fen Bilimleri Enstitüsü, Balıkesir.

Gooroochurn, N., & Sugiyarto, G. (2005). Competitiveness indicators in the travel and tourism industry. *Tourism Economics*, *11*(1), 25–43. doi:10.5367/0000000053297130

Gülözge, G. (2006). *Sınıf Öğretmeni Adaylarının Yaratıcılık Düzeylerinin Çeşitli Değişkenler Açısından İncelenmesi* (Yayımlanmamış yüksek lisans tezi). Pamukkale Üniversitesi Sosyal Bilimler Enstitüsü.

Güneştekin, F. (2011). İlköğretim 1-5. Sınıf Öğrencilerinin Yaratıcılığının Bazı Demografik Değişkenler Açısından İncelenmesi (Yayımlanmamış yüksek lisans tezi). Akdeniz Üniversitesi Sosyal Bilimler Enstitüsü, Antalya.

Hall, C. M. (2012, February). Boosting food and tourism-related regional economic development. In Food and the Tourism Experience: The OECD-Korea Workshop; OECD Studies on Tourism (pp. 49-62). OECD.

Hegarty, J. A., & Barry O'Mahony, G. (2001). Gastronomy: A phenomenon of cultural expressionism and an aesthetic for living. *International Journal of Hospitality Management*, *20*(1), 3–13. doi:10.1016/S0278-4319(00)00028-1

Hennessey, B. A. & Amabile, T. M. (1987). *Creativity and Learning: What Research Says to the Teacher.* ERIC Document Reproduction Service No. ED 312-835. Advance online publication. https://www.eric.ed.gov/PDFS/ED312835.pdf

Hjalager, A. M. (2010). A review of innovation research in tourism. *Tourism Management*, *31*(1), 1–12. doi:10.1016/j.tourman.2009.08.012

Hong, S. W. C. (2008). *Competitiveness in the tourism sector: a comprehensive approach from Economic and Management points.* Springer Science & Business Media.

Horng, J.-S. & Hu, M.-L. (2008). The Mystery in the Kitchen: Culinary Creativity (C. 1-2, C. 20). *Creativity Research Journal*. doi:10.1080/10400410802060166

Keleşoğlu, S. (2017). Öğretmen Eğitiminde Yaratıcı Düşünme ve İnovasyon Eğitim Programının Tasarımı, Denenmesi ve Değerlendirilmesi (Yayımlanmamış doktora tezi). Gazi Üniversitesi Eğitim Bilimleri Enstitüsü, Ankara.

Kesíci, M. & Önçel, S. (2015). Aşçıların Mesleki Özerklikleri İle Yaratıcılık Süreci İlişkisi. *İşletme Fakültesi Dergisi, 16*(1), 23-45.

Kim, H. J., Chen, M. H., & Jang, S. C. S. (2006). Tourism expansion and economic development: The case of Taiwan. *Tourism Management, 27*(5), 925–933. doi:10.1016/j.tourman.2005.05.011

Konak, A. (2008). İlköğretim 6. Sınıf Öğrencilerinin Sanatsal Yaratıcılık Düzeyleri (Yayımlanmamış yüksek lisans tezi). Süleyman Demirel Üniversitesi Sosyal Bilimler Enstitüsü, Isparta.

Mangion, M. L., Durbarry, R., & Sinclair, M. T. (2005). Tourism competitiveness: Price and quality. *Tourism Economics, 11*(1), 45–68. doi:10.5367/0000000053297202

Öncü, T. (1992). Yaratıcılığın Betimlenmesi Ve Yaratıcılık Üzerine Çevresel Etkiler. *Ankara Üniversitesi Dil ve Tarih-Coğrafya Fakültesi Felsefe Bölümü Dergisi, 14*(1), 255–264.

Öncü, T. (2000). Anasınıfı (6 Yaş) Düzeyindeki Çocukların Şekilsel Yaratıcılıklarının Cinsiyet Değişkeni Açısından Karşılaştırılması. *Ankara Üniversitesi Dil ve Tarih-Coğrafya Fakültesi Dergisi, 40*(1-2), 25–34.

Özden, Y. (2003). *Öğrenme ve Öğretme*. Ankara: Pegem Yayıncılık.

Özer, Z. (1991). Yaratıcılığa Giden Yolda Beyin Fırtınası. *Bilim ve Teknik*, (348), 50.

Özerbaş, M. A. (2011). Yaratıcı düşünme öğrenme ortamının akademik başarı ve bilgilerin kalıcılığa etkisi. *Gazi Üniversitesi Gazi Eğitim Fakültesi Dergisi, 31*(3), 675–705.

Pedersen, L. B. (2012). *Creativity in Gastronomy: exploring the connection between art and craft* Unpublished Thesis). Copenhagen.

Pine, B. J., & Gilmore, J. H. (2011). *The experience economy*. Harvard Business Press.

Porter, M. E. (1998). Clusters and the new economics of competition. *Harvard Business Review, 76*(6), 77-90.

Rein, I., Kotler, P., & Haider, D. (1993). *Marketing places: attracting investment, industry, and tourism to cities, states, and nations*. Free Press.

Richards, G., & Wilson, J. (2006). Developing creativity in tourist experiences: A solution to the serial reproduction of culture? *Tourism Management, 27*(6), 1209–1223. doi:10.1016/j.tourman.2005.06.002

Richards, G., & Wilson, J. (Eds.). (2007). *Tourism, creativity and development* (Vol. 10). Routledge. doi:10.4324/9780203933695

San, İ. (1979). Yaratıcılık, Iki Düşünce Biçimi Ve Çocuğun Yaratıcılık Eğitimi. *Ankara Ünversitesi Eğitim Bilimleri Fakültesi Dergisi, 12*(1-4), 177–190.

Santich, B. (2004). The Study of Gastronomy and Its Relevance to Hospitality Education and Training. *International Journal of Hospitality Management, 23*(1), 15–24. doi:10.1016/S0278-4319(03)00069-0

Scott, B. R., & Lodge, G. C. (1985). US competitiveness in the world economy. *The International Executive, 27*(1), 26–26. doi:10.1002/tie.5060270112

Sevimli, A. D. (2015). Yaratıcılık ve Sanat Bağlamında Sinema Bir Örnek: Akira Kurosawa (Yayımlanmamış doktora tezi). İstanbul Üniversitesi Sosyal Bilimler Enstitüsü, İstanbul.

Sünnetçioğlu, S. & Özkök, F. (2017). Güzel, Özgün ve İyi Yemeğin Peşindeki Yolculuk: Gastronomi. *Journal of Awareness, 2*, 585–596.

Tekin, Ö. A., & Çidem, G. (2017). The Relationship between Creativity Processes and Organizational Culture in Five-Star Hotel Cuisines. *Journal of Tourism and Gastronomy Studies, 324*, 342.

Temizkalp, G. (2010). *Öğretmen Adaylarının Yaratıcılık Düzeyleri* (Yayımlanmamış yüksek lisans tezi). Mehmet Akif Ersoy Üniversitesi Sosyal Bilimler Enstitüsü, Burdur.

Torrance, E. P. (2004). Great Expectations: Creative Achievements of the Sociometric Stars in a 30-Year Study. *The Journal of Secondary Gifted Education, 16*(1), 5-13. https://files.eric.ed.gov/fulltext/EJ682714.pdf

Üstündağ, T. (2009). Yaratıcılığa yolculuk (6. Baskı.). Ankara: Pegem Akademi.

WTO. (2007). *A practical guide to tourism destination management*. World Tourism Organization.

Yaman, S. & Yalçın, N. (2005). Fen Bilgisi Öğretiminde Probleme Dayalı Öğrenme Yaklaşımının Yaratıcı Düşünme Becerisine Etkisi. *İlköğretim-Online, 4*(1), 42-52.

ADDITIONAL READING

Ateljevic, J., & Li, L. (2017). Tourism entrepreneurship–concepts and issues. In *Tourism and entrepreneurship* (pp. 30–53). Routledge. doi:10.4324/9780080942728

Buhalis, D., Costa, C., & Ford, F. (2006). *Tourism business frontiers*. Routledge. doi:10.4324/9780080455914

Della Corte, V., D'Andrea, C., Savastano, I., & Zamparelli, P. (2017). Smart cities and destination management: Impacts and opportunities for tourism competitiveness. *European Journal of Tourism Research, 17*, 7–27.

Emmendoerfer, M. L., Emmendoerfer, L., & Ashton, M. S. G. (2016). *Analysis of the Heritage Requalification Process to the Recognition of a UNESCO Creative City of Gastronomy. In 4 th Internatıonal Conference the Unesco Unıwın Network* (pp. 163–173).

Hall, M. C., & Williams, A. (2008). *Tourism and innovation*. Routledge. doi:10.4324/9780203938430

Lohmann, G., & Netto, A. P. (2016). *Tourism theory: Concepts, models and systems*. Cabi.

Raj, R., Griffin, K., & Morpeth, N. D. (Eds.). (2013). *Cultural tourism*. CABI. doi:10.1079/9781845939236.0000

Sotiriadis, M., & Gursoy, D. (Eds.). (2016). *The handbook of managing and marketing tourism experiences*. Emerald Group Publishing Limited. doi:10.1108/9781786352897

Swarbrooke, J., & Horner, S. (2012). *Business travel and tourism*. Routledge. doi:10.4324/9780080490601

Virfinija, J. (2016). *Interaction between cultural/creative tourism and tourism/cultural heritage industries* (pp. 137–156). Rijeka, Croatia: Tourism–From Empirical Research Towards Practical Application, InTech.

KEY TERMS AND DEFINITIONS

Business Tourism: Business tourism can be defined as activities include meetings, and attending conferences and exhibitions.

Creativity: Creativity is creating innovation, revealing and discovering the unknown.

Gastronomy: The practice or art of choosing, cooking, and eating good food. The art or science of good eating.

Innovation: The action or process of innovating a new method, idea, product, etc.

Torrance Test: A test used for measuring or defining of creativity.

Chapter 15
The Exploration of New Avenues Toward Better Business Management:
A Business Tourism Perspective

Saba Ahmed
Capital University of Science and Technology, Islamabad, Pakistan

ABSTRACT

Business tourism is type of tourism that involves work rather than enjoyment. It is becoming the most income-generating part of tourism. Business tourism has different types that are meetings, incentives, convention, and exhibition (MICE); congress tourism; event tourism; sport tourism; and business agencies. Meetings, incentives, convention, and exhibition is a detailed and comprehensive type of business tourism. The incentive traveling is that of business tourism that is offered by the company to their employees. Exhibitions are those types of traveling that are organized to introduced new products, services, and information to the business tourists. Convention consists of the congresses and conferences. Congresses are the larger-scale business meetings, and conferences are the smaller-scale business meetings while complexes are the combination of small exhibitions and conferences. There is a need to understand that leisure tourism is different from business tourism.

INTRODUCTION

Tourism is a cultural and social activity in which people move from country to country for entertainment, family tour, world exploration and business activities with the minimum period of twenty-four-hour and maximum periods of six months for leisure and pleasure. The world tour comes from the Latin word tornus that means a tool for making a circle. Tourism is one of the reckless growing industries and big foreign exchange that causes an increase in the employment level (Tourism Notes, 2017). One of the main types of tourism is Business tourism, which involves tours regarding some business transactions, corporate events, official meetings, conferences and trade fairs. Grove and King (1996) revealed that the

DOI: 10.4018/978-1-7998-3142-6.ch015

main goals of the business, tourism are to avail business challenges, opportunities, lifestyle and financial gains. Companies that are targeting diverse customers and clients are keen to organize business tours. In order to pursue organizational goals, the companies send their ambassadors individually or collectively on business tours.

Considering the little detail of the business tourism types, one is the Corporate events that are concerned with providing entertainment and hospitality to business customers and clients. These can be in the form of lunch, dinner or a grand party in lavish hotels. On the other hand, few other researchers like Pearce (1990) and Frater (1982) claimed that business tourism is concerned with rewards and social networking. The business tourism types consist of meetings and conferences (Dwyer & Mistilis,2000) as well. Conferences and Meetings are types of meeting of a large group of people to discuss a particular agenda. Furthermore, trade fairs are dealing with showcasing the products and services of companies and businesses

Many occasions cause a group of people individually or together to arrange meetings and conferences or to participate in these types of events. Industries and companies that arrange such meetings or events are different from those countries that host business travels. Business tourism is fruitful for the developing and conversion economies. Those countries that arrange business tourism or give special attention to business tourism get a cost-effective advantage and expand their economies. Tourism is expanding largely around the world and causes an increase in occupation globally. Especially for underdeveloped countries business tourism (Forum, 2012) is becoming the major source of employment. Business travel accounts for Nine percent of all international tours as business tourism can be regarded as leisure and work together. Leisure tourists are more cost-sensitive than the business tourists as the business tourists partly enjoyed their work but leisure tourists come for enjoyment leaving their all extra activities. Similarly, the purchasing ability of leisure tourists affects more than business tourists. There is somehow problems occur for business tourists to manage their schedule with business meetings and leisure time. Five out of ten tourists take out their time specifically for tourism as they think to manage both activities. Business tourism is expected to be the most popular type of tourism in the modern era. Industries of developing economies are interested to expand their revenues through business tourism.

How Business Tourism is Different From Leisure Tourism

There are some factors that distinguish business tourism from leisure tourism (Cieślikowski, 2015). The participants of the business tourism are mostly professional and business people while the participants of the leisure tourism are nonprofessionals and tourists. The motivation of the business tourists is to do business, to promote networking and to gain knowledge and information related to the product. While the motivation for leisure tourists is to have fun and self-actualization. The purpose of the visit of business tourists is work-related while the leisure tourists want only leisure. The customer's needs in business tourism are a high level of service and raising demand for their product (Swarbrook & Horner,2012). While the leisure tourists bargains on their customer need. The buyers of business tourism are meeting planners, AMCs, and PEOs while the leisure tourism buyers are travel agents and tour wholesalers. The motivations of buyers in business tourism are cost-effective shows and successful meetings while in the leisure tourism the motivations of buyers are directives and fees.

The decision criteria in business tourism are facilities and increase in business links while the decision criteria in the leisure tourism are destination demand. The buyer service needs in business tourism are accommodation and registration for the event and exhibition services. While the buyer services need in leisure tourism are reservations and itineraries. The distribution channels in business tourism are corporations and associations of AMCs, HQs, and PEOs. While the distribution channels in leisure tourism are GDS, travel agents and wholesalers. The supply chains in business tourism are the venues and event suppliers while the supply chains in leisure tourism are hospitality companies. The market approach in business tourism is sales leads while the market approach in leisure tourism is advertising, marketing, and packaging. The business of the host country is secured in the business tourism through tactical and oriented sales while the business is secured in leisure tourism through strategic and oriented marketing. According to Mckercger et al (2002), the size of groups in business tourism is about three hundred to ten thousand delegates and the size of groups in leisure tourism consists of individuals and small groups. The economic impact of business tourism is trade and investment while the economic impact of leisure tourism is direct spendings. Leisure tourism is experienced while business tourism is not experienced as much as compared with leisure tourism.

Trends and Opportunities of Business Tourism

Business tourism involves various trends (Santibáñez, 2017) like event management companies that manage and create and produce events. Convention centers involve the primary meeting facilities, co-operative transactions, event catering. Hotels that provide the rooms, banquets and catering facilities. Exhibition management companies that accompanied customer, producer, and trade (Morrison,2013). Local and State governments maintenances the MICE market through investment in facilities supports financially to CVB, give protocol, provides security facility and give permission to the special event. Localhost communities provide a formal invitation to those who arrange the meetings, works with the CVB during the offer and design social programs during the event.

Professional conferences organizers (PCO) do contracts with the meeting planner, handle on-site logistics, accommodation, hotels, registration, sponsorships, local conveyance, social program arrangements, entertainment and meeting preparation, etc. Destination management companies (DMC) comprises of the tourism services organization that includes ground transportation, logistics, organizing pre and post tours, and incentive trips. Appealing business tourists through hotel entrance twenty-four hours with proper security facility, increase business center care, pick and drop from the airport and attend the conventions and meetings. Infrastructure for the business tourism conventions.Packages for the business tourism conventions. Accompanying families as the rising number of business tourists bring their families with for enjoyment or trip. Adventure travel as managers becoming more conscious about the adventure venue of the business trip which comprises of team building. New technologies like internet promotion, electronic ticketing, and computerized reservation also part of business tourism trends.

Moreover, Business tourism has various opportunities in business tourism to struggle with hotels to develop different activities for the tourists such as show business, golf option, restaurants, weekend tours, special cultural and amusing events and departure flight information (Etiosa, 2012). These opportunities exist to struggle with hotels to raise their standards with their business maintenance services. Produce one to three-hour trip packages for people. Propagate airport tourism means to remain up to date regarding the different schedules and activities of the business tourists. These activities are comprised of internet access, workout facilities, day restrooms that hire for an hourly basis prolonged, shopping options and short tours. Promotion in high-quality services and cost-effective benefits can increase the opportunities of the meetings, exhibitions, and conventions circulation.

Accredited professional conference organizer (PCO) and the accredited exhibition organizer (EO) both have the opportunity to work with the motive of facilitating the business tourists more and more which will ultimately raise their position and value. An opportunity of the accompanying program packages is a cooperation between the hotels and convention centers to produce new bidding for the children and partners during the meetings. It has also the possibility of adventure tourism that provides leisure facilities to the traveler. An opportunity of the specialized site assessment, tour workers and guiders. An opportunity of the consultancy with the focused firms regarding the offer of the document preparation of the family, official proposals, and economic impact assessments and studies of benefits to the firm.

To seek the answer to the above debate the current study investigates the importance of business tourism other than leisure tourism. Secondly, this study incorporates the detail discussion of types of business tourism and its impact on business management. Thirdly, the study explores the best approach for business tourism and its impact on business management. Moreover, the study is going to contribute to the extant literature by focusing on business tourism types in detail and addressing the pros and cons of each type concerning business management. As few studies focused on maximum types of business tourism in detail and investigation of the best approach of business tourism for business management.

THEORETICAL BACKGROUND OF BUSINESS TOURISM

Meetings Incentives Convention and Exhibition (MICE)

Meetings Incentives Convention and Exhibition (MICE) is considered as the largest type of business tourism. It is a component of business tourism. The MICE events are the events that comprise the participants who share a common place to have different business dealings. These gatherings need proper setup of venue and this venue should have all the basic facilities that fulfill the needs of the participants. How the term MICE work we will study it with different formats. The MICE industry offers opportunities in the creation of job, economic and social human resource development (Lau, 2013). This also contributes to obtaining sustainable development that minimizes the ratio of poverty and produces employment. The MICE gives the training of three skills that are serving the industry, marketing, and management of the environment. These skills are very necessary to remain in a competitive market. The business tourism is the labor-intensive bring a higher number of people to work to improve standards of services. The MICE event planners should take important care of the management of infrastructures that involves the transport, Information technology facilities to arrange the meetings (Wanhill, 2012). The more multiple economic effects make the tourism industry more productive. The government plays an important role in giving pieces of training and raising the standards as the government has to spend a large amount of GDP on this industry which may save for the other major purposes. The government does not give always the priority to the MICE events as it has to give value to other sections also, for example, defense and health.

The value of MICE as business tourism is Higher expenditures. The expenditures of the business tourists are more than the expenses of the normal visitor as they like to stay in a standardized hotel and take high-quality food because their purpose of touring is meeting with business clients. As per Travel demand, there are different travel season for destinations as the MICE meetings organized in these destinations and the business tourists go there to attend the business meeting. In peak season MICE does not arrange the meetings as the leisure tourists aim to go to these destinations. The venue country gives opportunities to the MICE in the off-season so that they can enjoy their work. Prior arrangements are the arrangements of MICE that have been done two to three years before as these events are considered as the mega-events. Therefore the selection and bookings of the venue are organized before the time of the event by the event planners. As we know the participants of the MICE events are consist of large numbers and the international conventions required the five hundred rooms for the business clients and big conference room for daily meetings. Similarly, an exhibition also has a large number of members and need many halls for exhibitions.

The participants of the MICE need a high level of services and they expect to have high protocols from the event planners. Many business tourists look for MICE events to have better job opportunities. Compartment set up companies and ornamentation experts are compulsory for the exhibition events while Incentive travelers demand more for the entertainment rather than else. A country that arranges the hosting of the MICE event gives special time and qualified status to its participants. Tourism attraction involves the destination does not attract the business travelers as much as they travel for the business meetings. Hence the countries sometimes do not beautiful for tourism but they are chosen with the aim of easy access and urgent availability. Terminuses are not fixed as they can be changed from business clients. Sometimes venues change due to lack of required facilities in the venue. The event planners of the MICE event are those who are experts to handle the business events and make distinctive arrangements for the business participants. They take care of each and every need of these tourists.

Meetings Travels

When a group of people travel together intending to share knowledge and information related to a specific product, discuss and solve the problems of each other; this gathering is called meeting. Food and beverages are part of the meetings. The time of the meeting fluctuates with the type of meeting. Conferences and business meetings can also be called event tourism. There is the largest industry of association for meetings named meeting industry association. This is the industry in the UK and Ireland. They provide different places for venues and traders in the meetings. Any industry or company wants to arrange or have a conference or meetings, inform this association so that they find out companies that can make arrangements for them. The international congress and convention association also acts as an organization working as the event organizer for the country's business tourism. Event tourism refers to creating a special event that includes business and marketing (Seebaluck, 2015). The main purpose of the event tourism is to increase tourism facilities with business dealings. The first objective is to enhance tourism demand and the second objective is to gather people of business. Events are the momentum factor that generates tourism. Not only generate special tourism but also cause of marketing internationally. These planned events are the propulsion factor that causes decision making tough as an increase in the competitiveness for the destination. Planned events are those events that merge the field of research and study. These are a total of eight fields; Cultural festival, Political and formal, Arts and entertainment, business and trade, Educational and technical, Sport Competition, Frivolous and Private.

There are different types of meetings like Board meetings as these are the meeting conducted by the board of directors of corporation or organization usually held annually. Management meetings arranged among the managers of firms or corporations with the aim of important decision making. Shareholder meetings are conducted among investors who want a share in a company. Seminars arranged to train fresh employees. Meeting with partners, suppliers, and the client is arranged to discuss the business deals and to sort out different problems. Product launches meetings are arranged among the clients, reporters, and the public to introduce a new product in the market. Strategic planning meetings are arranged with the workers of the firm to discuss the future of the organization. Retreats involve the whole firm staff spends two to three days together to discuss the strategic plans, training sessions, and team-building conferences.

Incentive Travel

The incentive travel is that of business tourism which is offered by the company to their employees to hang out and enjoy. It is the complete source of entertainment and gift from the company. This makes the employee more loyal to the company. It also causes happiness for the employees sometimes it counted as a reward for their achievement in different business dealings. Employees become more passionate and enthusiastic about their work. Incentive travel also motivates employees to do their work hard. It encourages their abilities and creates an environment of honesty and loyalty towards the company. There is a need for informal locations to create an atmosphere of localization in incentive traveling. Basically, incentive travel does not have any professional linked activity. Factors that affect the incentive travel decision are the Incentive budget economy, which strengthens the economy of the hosting country. Buying power strengthens the exchange rate of the hosting country that was destined for the trip. Political climate alerts travelers regarding the political climate of the hosting country. Safety and security are about different mishaps or security issues. Value facilities and services are compared with other selections of targeted hosting. The uniqueness of experience accounts that the participants have experienced this trip before or not.

The benefits of Incentive travel to the employee include to facilitate business communication and network occasions for senior management, have better company loyalty, motivate the employees, expand the company socially, create keenness for upcoming company time, strengthen the relationship between the company and its employees and to give equal rights to every employee.

Conventions Travels

Alike to meetings, conventions and conferences are also a congregation of managers with mutual goals and objectives arranged to exchange ideas, opinions and information of joint interest. Organizations and companies arranged the conferences and conventions yearly to discuss the organization's future goals and motives of the present situation. The people who want to participate pay the registration fee and then become part of the convention. The convention travelers also look for the facilities of the leisure. They expect the hosting organization to provide the convention facilities like conference hall, sound system, projector, wireless internet facility, food and beverages, etc. Convention consists of the congresses and conferences. Congresses are the larger scale business meetings and conferences are the smaller scale business meetings while convexes are the combination of small exhibitions and conferences. Conventions also include one to one events that are also called premium networking. The difference between the meeting and convention are Convention is a mega event that lasts for many days and includes also social activities. Association organized the convention through professional and trade connotations, Intended association and societies, assistances, and political parties. Convention held once in a year and generate income for the association. Expenditure on the selection of the hosting of convention.

Exhibitions and Trade

Exhibitions trade are those types of traveling that are organized to introduced new products, services, and information to the business tourists. It also involves potential customers or buyers. The exhibition companies usually hire any specific place that attracts visitors to the new product or service. It arranges or setups meetings with customers or visitors. Business tourists arrange the option of tickets or an invitation for the participation of the visitors. There are often trade carnivals and trade funfairs of different products. Companies and industries participate in these types of functions to remain up to date regarding their product sales and different type of innovations in these products. It also gives the chance to enhance the relationships with colleagues. Through this type of exhibition also up to date regarding the market competition. Exhibitions and trade fairs are of two kinds (Woods, 2010). The first one includes those fairs and function s that hosts and the other one those who want to buy that type of product (Rydalmere, 2019) Organizing such type of functions also raises the importance of business tourism. Trade fairs and functions include events organizing of entertainment technology, lifestyles, oil and gas, safety and security, agribusiness, design, construction materials, irrigation, manufacturing, renewable energy, and autos products, etc. Exhibition and trade fairs provide services of arranging business events, sold such events, marketed and delivered the conferences or meetings around the world. The purpose of the exhibitions is to gain sales leads, to introduce a new product, to create a network for significant contacts, to obtain the latest knowledge and information about the market competition, and raise the value or image of the company on the international level. The exhibition on exposition events intends to gather the number of suppliers to display their work and promote their products and services to the participants in one platform.

Congress Tourism

Congress tourism is a subset of business tourism. This type of tourism is the provision of facilities and services to many ambassadors who attend meetings and congresses, exhibitions, business events, incentive travels, and corporate hospitality. In the life of ordinary people touring brings remarkable change. This professional field concentrated on the destination management and advertising of customer's behaviors (Gračan, 2011). As the event tourism includes planned work and probable effects on the hosted zone. The associations which host such type of business, tourism gives special attention to the high costs that attained from the ambassadors.

It is very important to decide the venue of the congress meetings. The selection of destinations for business meetings is very necessary. Therefore, the host country has to fulfill the requirements of business tourists. Hence the demand for business, tourism and congress destination is increasing. There are some requirements of the host country in the congress tourism that are high capacity costs of applicants [big financiers], Destination congress and leisure image advancement, product enforcement. Congress tourism organization make a demand of such meetings that includes actors also. Professional associations and other initiatives search for the best congress experiences. They also look for advertising strategies that raise the number of customers in the tourism industry (Sung, 2004). There is also demand for some services like food and beverage organizations, entertaining assistance, sports and transport companies, and civic objects. To satisfy customers, requirements tourism organization has to manage different elements for congress destination. A tourism system includes all planning, arranging and promoting the congress experiences. The factors of social and economic affect congress tourism. Hence the pre-congress and post-congress tourism period have another such factor as scientific and conveyance. There is a difference between the material and immaterial parts of congress tourism.

The material part of congress tourism includes necessities like Offices, Conference halls, Exhibitions, and multifunction halls. Technological and translational pieces of equipment such as a microphone, projectors and sound system to fulfill audio and video requirements.High standard and quality hotels and high standard personal.Formal and protocol events and other pleasure trip activities.Large agreement centers for the great numbers of contributors. Food and beverages regarding different international countries to satisfy customer's needs.

The role of the local people in the business meeting is also very important e.g translator. The English language has international value and international language but in some countries, the need for the translator of specific language arises. Therefore the translator is necessary for avoiding language obstacles. The destination management, monitoring arrangements, and getting feedback fall in the immaterial part of congress tourism. There are two types of destinations for congress tourism. The first one is the affirmed tourist destination brand and the second one is the non-affirmed tourist destination brand. Both types need to raise their destination brands as well as work on the congress tourism demand. In the first segment, there should be work on the promotion of business activities that increase demand (Thomas, 2011). The second segment should include the affirmed brands that unify public and private structures in the destination of tourists. The tourist's destination should include congress demand. The non-affirmed destination needs attention towards the attractive perspective to host congress tourism and strengthen the structural and regional characteristics. Affirmed tourist destination involves the confirmed offer of tourist's destination and people have an idea about the destination before going there. Some factors affect destination management like economical geographical and socio-political.

Geographically convenience involves good transport and simplicity to arrive. The social and political aspect that includes the affordability and atmosphere where tourists can live the experience. Economical aspect involves the total cost for roaming to destination and staying there as well. The membership study and their need is also a very important factor. Delegates that go for business tourism use economical components like conveyance, accommodation, registration fees, entertainment fees etc. The reputation congress hosts for customer behavior in congress tourism has been changed socially nowadays. Three factors affect contributor's performance like the Functional contributors i.e Professional motivation for cultural exchange and the increasing knowledge attract these functional types of contributors in congress. The Relational contributors are Professional and interpersonal relationship attracts these types of contributors in congress. They also contribute to tourist's trips and leisure hours. The Symbolic contributor contribute to attain visibility, avoid notoriety and image in congress. They want a higher level of revenues, a good organization for work and good presentations.

The congress tourists customers are interested in the travelings and they have informed profile data, middle high cost, and education. Unlike leisure tourists, congress tourists are more concerned about special hosting services. They prefer a high level of protocol. They don't go with the aim of leisure that's why they don't care so much about money than leisure tourists. It is important to decide the congress consideration in the first segment that is the pre-congress phase and it is helpful in the future. Congress destination has four types of marketing which include Approve destination brands and explain the strategy of positioning, Demand market should be enchanted, Specification of the objectives attraction of the objects for tourists, Procedure and practice should be acted.

Using a structured approach it is necessary to create a list of the congress requirements for the better image and reputation for the tourists. First of all, it is important to create a connection between the delegates and the destination. As the location becomes important for the tourists and also important for the increment in several tourists. Local persons play an important role between the organizer and delegates. As they are the third part that guides the tourists about geographical regions. Exhibitors and sponsors should guide and support to the destination. Therefore it is helpful to take feedback from the tourists about the distance coverage and travel to the destination (Ficarelli, 2013). This feedback is helpful for other delegates to review the successful journies. This tourist experiences in feedback encourage other people to visit their favorite places. The host country should focus on the primary services of the customers as the pick and drop services from the airport to the residence area. These customers can be the employee, tenants and the target markets. There should be a positive relationship between these consumers and functional, marketing and advertising communication media. These factors depend on the transports, infrastructure, sanitization facilities that involve car parking and open spaces. When we combine these factors reality of the place, positioning and personality came in the result. Globally the congress tourism is expanding and making its value. The tourists' companies are working and pay attention to this segment of tourism. Hence, we can say that business tourism comprises the arrangement of meetings and conferences. This activity comprises of delegates and businessmen. Congress tourism requires more protocol and necessities and mostly it is concerned with the location and destination of the meetings.

Conferences and Meetings

Conferences and meetings related to international relations are also part of the business tourism as the incentive in these meetings are the same as of business tourists. The ambassadors and representatives of one country arrange an international meeting with others to solve the country's issues. Most of the time these meetings include the mission of peace-creating between the two nations. There also many issues regarding border countries. In this tour, representatives talk about the peace and welfare of both countries. The other main and important incentives of this tour are to raise the trade and investment opportunities. As the major focus is on the opening of the gate of trade between the two countries (Commonwealth, 2015). These types of conferences and meetings also need special accommodations, transport, security, open spaces, and a healthy clean environment. The host country has to be careful regarding these types of elements that affect the relationship of the countries. Theses conferences mostly talk about Migration as most of the countries have issues regarding migration as people move from one country to another in search of better job opportunities. But they could not attain fully the rights of that country where they prefer to go.

Health is common and very important for every person that's why with the increase in different types of diseases underdeveloped country's scientists look for other developed country scientists to minimizes the dangers of such diseases. Peace and security are common issues for all countries. Each country wants to have peace. Through these meetings, countries sort out their security problems as well to strengthen their relationships. Human rights conferences and meetings arrange to discuss human rights all over the world. Climate change increase in the effects of human activities on the environment results in climate change. These results are global warming, greenhouse effects and ozone depletion etc. Therefore conferences are held on to discuss the solution to these problems. Trade and investment have the purpose of this type of conference are to enhance trade and investment opportunities. International conferences also discuss educational standards equity and raise the level of education globally.

Event Tourism

Event tourism consists of two elements the first one is the professional practice and the second is the academic study. Public institutions and the private organization arranged special and planned celebrations and organized places; this is known as an event. It was about a few years ago that event tourism spread as the spectacle of the established detectable term in the research community and the tourism industry (Skoultsos, 2009). Therefore the successive growth of the even tourism industry can be described as remarkable. Event tourism is associated with specific resources and places where these celebrations are going to hold to attract visitors. There are seven elements of events; the first one is the solid ability to invite tourists, the second one is the ability to promote indigenous development, third one is having confined duration of time, fourth one held repeatedly, the fifth one is the increase the attentiveness of citizens and the image of destination, sixth one is proposing social experience, and the last one is being deliberated in the extraordinary and rare. Different types of events generate healthy revenue that attains from the opportunities of leisureliness, social fascination, and entertainment for the public. The utmost desire of the event managers is to raise tourism to achieve economic development for the country. This event tourism gives monetary benefits and cost-effective as some tourists devote money to the local businesses that are fruitful for the economy.

Events attract tourists, sponsors, and media to a particular place and let them spend on different activities that generate income in the country, it is leveraged for the seasonal demand. Event portfolios are designed for maximum purposes. Events form an optimistic image of the destination and the host country causes marketing as industrializing cities more gorgeous and civilized. Events sentient towns, resorts, parks, hotels, venues, and restaurants by making them more charming and visitable places to take out benefits from them and plays the role of substance for the anticipated development and generation of the long term permanent inheritance.

Events are the significant motivator of tourism and also develop the plan for the destination area. Destination attractiveness and the marketing intention are the key elements of an event. The fundamental goal of an event is to increase global effectiveness to attract the expenditures of the visitors. The event plays a role of bridge raising the numbers of the tourists and the gaps of the seasonally demand because tenants and native tourists provide the replacements of the demand inspired by the events. There is the purpose of the planned events to create a territory of individual and the community initiatives that become greatly territory of the specialists and entrepreneurs.

Event tourism has the typology of the planned events that have the four classifications. The first kind is the business events or MICE portion that need a convention and the exhibition centers and also involves several smaller private celebrations and gatherings held in hotels, restaurants, and resorts. Sports and games also required a special type of facilities like stadiums, grounds, and athletic parks. Festivals and other small events are less sensitive to these types of needs while they can go in different types of venues streets, parks, cinemas, public and private places (McKay, 2012). Corporate market events, government political events, private parties, and fairs and exhibitions are examples of the first kind of typology of event tourism. The second kind comprises of festivals and cultures events. These types of events include the festival places, art centers, museums, galleries, and shrines, etc.

Festival places involve the commemoration places where the memories of old amazing things or stories are celebrated. Art centers involve the parades and carnivals that celebrated to give honor to special persons. Museum and galleries involve the art exhibits and the installation to celebrate the art festivals. Shrines involve religious rights and pilgrimage. The third kind consists of entertainment that is used to rejoice business tourists and to provide them such an atmosphere that relaxes them for some time. Many things come in entertainment for visitors as some of them are concerts halls, cinemas, streets, open-air venues, and parks, etc. Concerts halls include different types of music, dramas, movies, and shows. Outdoor venues include award ceremonies and other business-related official events (Getz, 2007). Streets include the street shows. The fourth kind consists of sports events. These sports or Olympics games are considered very special events that came with a huge number of visitors from all over the world. For this purpose, the event managers planned the proper setups of stadia and arenas, different purpose al fresco facilities. There are professional leagues and participatory in the setup od stadia and arena. Recreational outings, fun, one-off tours, and other tournaments are the subpart of the special purpose outdoor facilities.

Event tourism has three important aspects that are event studies, event management, and itself event tourism. These are called the events in the closely related fields that consists of the study of events with other practical fields involving leisure, hospitality, sports, arts, cinemas, and cultural studies (Iacuone & Zarrilli, 2018). Event studies is an interdisciplinary field that studies the planned events, meetings attached to the event and their experiences. Event studies incorporate event tourism and event management. Event management is an applied and proficient field that dedicated to appreciative and improving the management of planned events. Event tourism is also an applied field that is dedicated to the understanding and improvement of tourism through the management of different types of events.

Event tourism has some typical function that is also called the career paths that has a specific task like Event facilitator or manager facilitators work for the event to help in deciding the destination and to understand their tourism perspectives like funding, advising, and marketing. Event facilitators have an association with convention and exhibition centers and other venues, have cooperation with Olympics, sports, theatres, arts and other industries that organize events. Event tourism Producers producer generates and creates events specially to enhance tourism opportunities and value, producer work with the stakeholder management with several event allies.

Event tourism planner generates different types of strategies for the destination and assimilates event with product improvement and image classifying. Event tourism policy researcher works with the policymaker invasion to facilitate and raise the value of event tourism and researches of the probability studies, demand estimating, impact assessments and performance evaluations. Event bidding comprises of bid or offer related to the events, makes the relationship lead to generating events for the destination and bidding consists of developing risk assessments and realistic studies for each possible bid or offer. Event services provide crucial and special services to events, for example, transportable and logistics; lodgings and venues reservation; worker contacts.

The portfolio of the event consists of a type, season, target market, and value (Zima, 2012). According to the portfolio of events, tourism events are categorized into four types explain as follows. Occasional mega-events are sports and Olympics; pilgrimage, arts, and cultural festivals. It has high tourist demand, high value, high-income generation, high chances of business expansions, economic benefits, growth, and development of the country. Periodic hallmark events are those that occur with some gap of the period and these also have high tourist demand, high value, high-income generation, high chances of business expansions, etc. Regional events are periodic that occurs one time and these events have fewer tourists demand, less value, less income generation, less economic benefit, etc. Local events are also periodic that occurs one time and these events have fewer tourists demand, less value, less income generation, less economic benefit, etc.

Sport Tourism

Sports tourism is one of the fastest-growing types of the tourism industry. Sports tourism is a globally expanding and well-established industry. Sports tourism is considered as 'big business' in the modern age. Sports tourism has been considered a significant and most popular element in the world of tourism and becomes a vastly competitive niche business (corporation, 2013). The interest, attraction and participation of people towards sports enhance leisure tourism as well as increase the competition and also economic benefits. The international multi-sports events are the Olympics, Paralympics games, commonwealth games, international students games, North American indigenous games, world cups of football matches, world cups of cricket matches, etc.

Sports tourism has the following benefits from a business point of view as Sports tourism has increased the opportunities for the organization for the sports facilities, expansion of officials and instructors as well as the capacity of athletes to achieve their prospective. Social and community legacies include increase the fitness level and health; community level and the cultural festivities; and also play an important role in the youth's progress. Economic development is the revenue that is generated from sports tourism contains the larger part of the GDP of any country. Accommodation, transport, hoteling, food and beverages, etc spendings are included in this revenue. There are two types of revenues, the first one is direct event revenues and the second one is indirect event revenues. The direct event revenues comprise of admission fees, sponsorships, on-location sales, etc. The indirect event revenues also called induced economic activities. It comprises of visitors and businesses re-expenditure revenue from an event in the community. Sports tourism causes the families of the participants to come back for future events or future vacation plans.

Sports tourism not only brings leisure for the visitors but it also causes income generation for a country. The increase in the number of visitors and fans of different games and sports increases the value of tourism as they cause high economic development (Neirotti, 2007). The per head number of ticket prices increases especially for these events that ultimately result in economic development. There is also chances of economic and social development to hosting the sports. Having hosting raise the bidding for tourism to that country because sports are mostly liked by everyone. Each ordinary and special person like the games and want to participate, to watch matches or games lively. These desires take the people to the hosting country where the people not only enjoyed the games but also leisure. Hence hosting such sports and games not only bring the visitors of leisure but also causes the business for the other managers (Virgil Niculaa, 2014). Hosting the international sports tournaments are not only producing opportunities for tourism but also opening the doors for the positive international community and cultural relations. Sports events are designed to effect both socially and environmentally.

Categories of the Sport Events

Sports events can be categorized in many ways. There are different useful stages of planning and these stages include the types of event, choice, size, bidding and situation. The purpose of the categorization of sports events is to classify the use of sport tourism. Sports tourism activity can be categorized as Mega events are very great and very significant events. They bring a huge number of tourists and economic benefits. This mega-events has a further two categories first one is an International major multisport game and the other is an International single sports event (SHUKLA, 2016). Games are also multisport event and these are also very large events as it includes a number of games hosted by one country. Championships are seasonal sports events. These are single sports events and mostly arranged or organized by the sports association of schools, colleges and universities. Tournaments are the single sports event as championships they are also owned by one institute. They are arranged by local sport organizations, school and the religious authorities. They are arranged in the same season and same place each year.

Moreover, Training camps are organized to give training to athletes of the games, tournaments and championships. These training camps take advantage of the off-season of the good facilities and raise their potential work for the special events. Sports organizations and industries arrange different conferences and meetings to discuss the schedule and venue of the games. Trade shows and exhibitions also part of these conferences. These develop different types of courses for athletes, coaches and officials of all levels. These are helpful for them in participating and being selecting in games. Professional sports attract more participants and visitors from smaller markets to the larger community (Dr.T.S.Deveraja, 2016). These are helpful for sports industries in sponsorships and hosting strategies.

A BETTER APPROACH TO BUSINESS TOURISM FOR BUSINESS MANAGEMENT

Although all the approaches of business tourism have their importance depending upon the type of Business. But business tourism along with incentives and relaxation activities generates fruitful results. Incentive Travel reward encourages human beings and it enhances proficiency in the work. Therefore when a company arranges such types of business tours specifically for its employees, this will surely benefit itself. The employees will work more than regular efforts and try their level best to raise the company to a higher level (Saniaghi, 2008). The opportunity for holidays and enjoyment raises the hard work of workers for their company. Each employee will feel satisfied and will be contented. It will enhance the best performance of every worker. Not each company does so but some companies have such type of offering.

Moreover, Sport tourism and sports event result in the economic values, social benefits and civic pride for any country that ultimately helps in enhancing the business of sponsorships and sport event planners. This sports business provides the following economic benefits like Identify the value of the tournaments and sports that they give to the public. Recognize the strength and values of the sports and games to the community that increases the number of events to fulfill the demand of the public. Ascertain the events that bring maximum economic significance, for example, sports and games; and focus on the mega-events. Raising the opportunities for the tourists and their families so that they could stay extra days and enjoyed games with more leisure. Provide especial values to volunteers and also facilitate with paid job opportunities to volunteer. Stay in touch with the linking establishments and increase the interest in the demographics of a sports event. Sports tourism provides opportunities for re-stimulate the miserable parts of the urban areas.

ROLE OF BUSINESS AGENCIES IN BUSINESS TOURISM

Business agencies play an important role in providing business travel as business travel agents work with business agencies. Business travel is all about providing the services and facilities for business people who travel for a business meeting, conferences, convention, and different fairs and exhibitions.

Incentive travel that is is leisure tourism and short breaks also part of it. It also gives a bid to the visitors and participants as an incentive to attain the targeted goals (Oklobdžija, 2015). Big travel industries perform the task regularly on behalf of small companies. They provide the facilities according to the need of the customers and the customers they design the travel schedules. They work as an agent for the corporation and travel industries that supply the product and other services to the clients. Mostly these agent companies are known as the principal's airlines, hotels and pick and drop services from the airport. Business travel agencies are categorized into the five kinds that are independent agencies, national agencies, global agencies, implant, and E-agent.

These kinds are categorized concerning their specific characteristics and product. Many independent agencies organize different offerings for business travelers. They are private and do not relate to the national agencies they have their setups and a team of workers. Sometimes they work for business travels but they like to deal more with both business and leisure tourism (Stetic, 2012). They provide services to business clients freely with a wide number of services. Independent business travel agents provide services to business clients on their own without having any personal motives. National agencies are not independent to provide as many services. They are bound to facilitate but somehow they provide basic services that fulfill the basic needs of the customers. Implants are the agent type that operates with the business shreds of evidence so that they look after the requirements of the tourists. The agent may work alone or with the company to whom premises are concerned.

The demand for online business travel services by leisure tourists is expanding day by day. That's why e- agents perform a significant role to provide the online services to the clients. Business travel agencies mostly deal with the small corporations and send their individuals to attend small trips and tours in a manner to collaborate with clients who want all their business travel succeeded. Some basic services are accommodation booking, trip booking, transportation, car hire or rail tickets, indemnifications and car parking. The agencies find the accommodations, stay, flight bookings on specific dates, transferals to the hotels, etc. All business travel markets want to tap in with profitable hotels and airlines and continually want to bring out new goods and services to attract business travelers (Cope, 2003). Hotels are arranged with the facility of wireless internet for the ease of the tourists. The management of the travel expenses of the customer is also a big assignment that accomplishes by the business travel agents. They cooperate with the industries and provide twenty-four-hour facilities. The business travel agencies negotiate with the suppliers of travelings and business clients and sort out their problems. They negotiate on the airlines, accommodation, transport, hoteling and meeting schedules (Kaltsova, 2009).

CONCLUSION

Business travel is the progressively famous industry in the field of tourism and results in high economic value. It involves the substantial income of the manager, transport providers, caterers and host companies that provide these services. Some reasons make business tourism as a high-value industry as sometimes clients and business managers have to travel at one short notice that they can not take benefit of tourism with leisure point of view. Business travelers need a high level of services, for example, high-quality lodgings include the factor of entertainment. They often travel in high or upgraded services for example business class or first class.

REFERENCES

Cieślikowski, K. (2015). Business tourism as innovative product for destinations. Research Gate.

Commonwealth. (2015, November). *Commonwealth health of government meetings*. Retrieved from The commonwealth: https://thecommonwealth.org/sites/default/files/news-items/documents/CHOGM%20 2015%20Communique.pdf

Cope, D. a. (2003). *Business Travel: Conferences, Incentive Travel, Exhibitions, Corporate Hospitality and Corporate Travel*. Research Gate.

Corporation, B. (2013). Sport Tourism. *Tourism Business Essentials, 51.*

Deveraja. (2016). Tour operator's contribution towards development of tourism destinations. *International Journal in Management and Social Science.*

Etiosa, O. (2012). *The Impacts Of Event Tourism On Host Communities*. Pietarsaari.

Ficarelli, S. (2013). Congress and Meetings Tourism and Destination Brand. *The Fifth International Conference on Creative Content Technologies*, 15.

Forum, I. T. (2012). *Niches For Developing countries*. Retrieved from International Trade Forum: http://www.tradeforum.org/Business-Tourism/

Getz, D. (2007). Event Tourism: Definition, evalution and research. *Science Direct*, 428.

Gračan, D. (2011). *Business tourism – Modern form for improvement of the competitiveness*. Academic Press.

Iacuone & Zarrilli. (2018). Business tourism in China. The case of Guangzhou. *GeoJournal of Tourism and Geosites.*

Kaltsova, S. (2009). The Business of Tourism. *Academia (Caracas).*

Lau, D. C. (2013). *Meetings, Incentives, Conventions and Exhibitions*. MICE.

McKay, T. (2012). *Adventure tourism: opportunities and management challenges for SADC destination.* Research Gate.

McKercher, B., Ho, P. S., Cros, H. D., & So-Ming, B. C. (2002). Activities-based segmentation of the cultural tourism market. *Journal of Travel & Tourism Marketing, 12*(1), 23–46. doi:10.1300/J073v12n01_02

Morrison, A. M. (2013). *Marketing and managing tourism destinations*. Routledge. doi:10.4324/9780203081976

Neirotti, L. D. (2007). Sport and Adventure Tourism. Academic Press.

Oklobdžija, S. (2015). The role of events in tourism development. *Professional Paper*, 97.

Portal, E. (2017). *Tourism Notes*. Retrieved from Tourism Notes Eduction portal: https://tourismnotes.com/

Rydalmere. (2019). *Exhibition and Trade Fairs*. Retrieved from Exhibitions and Event association of Austrailia: https://eeaa.com.au/member/exhibitions-and-trade-fairs/

Saniaghi, R. (2008). Strategic positioning and performance of winter destination. *Tourism Review*, *63*(4), 40–57. doi:10.1108/16605370810912209

Santibáñez, D. B. (2017). *Business Plan for a travel agency and tour*. Academic Press.

Seebaluck, V. (2015). Prospects and challenges of business tourism. Academic Press.

Shukla, A. (2016). Growth of Travel and Tourism in India: A driver of EmploymentGeneration and Economic Development. *International Journal in Management and Social Science*.

Skoultsos, S. (2009). *Event tourism: Statements and questions about its impacts on rural areas*. Retrieved from Research Gate: https://www.researchgate.net/publication/47697270_Event_tourism_Statements_and_questions_about_its_impacts_on_rural_areas

Stetic, S. (2012). *The importance of business tourism to national economies in transition: The case of Serbia*. Research Gate.

Sung, H. H. (2004). Classification of adventures travelers. *Journel of Travel Research*, 343-356.

Swarbrooke, J., & Horner, S. (2012). *Business travel and tourism*. Routledge. doi:10.4324/9780080490601

Thomas, R. (2011). Understanding small firms in tourism. *Tourism Management*, 63–76.

Virgil Niculaa, P. R. (2014). *Business Tourism Market Developments*. Science Direct.

Wanhill, S. (2012). Small and medium tourism enterprises. *Annals of Tourism Research*, 132–146.

Woods, W. (2010). Performing rurality and practices in rural geography. *Progress in Human Geography*, *34*(6), 835–842. doi:10.1177/0309132509357356

Zima, K. (2012). *Event Tourism Economical and tourstic impacts on regional economy*. Academic Press.

KEY TERMS AND DEFINITIONS

Business Agencies: Play an important role in providing business travel as business travel agents work with business agencies.

Congress Tourism: This type of tourism is the provision of facilities and services to many ambassadors who attend meetings and congresses, exhibitions, business events, incentive travels, and corporate hospitality.

Convention Travels: Convention travels are also a congregation of managers with mutual goals and objectives.

Destination Management Companies (DMC): Comprises of the tourism services organization that includes ground transportation, logistics, organizing pre and post tours, and incentive trips.

Exhibition Organizer (EO): Have the opportunity to work with the motive of facilitating the business tourists more and more which will ultimately raise their position and value.

Exhibitions Trade: Types of traveling that are organized to introduced new products, services, and information to the business tourists.

Meetings Incentives Convention and Exhibition (MICE): A detailed and comprehensive type of business tourism with the aim to share knowledge and information related to a specific product.

Professional Conference Organizers (PCO): Do contracts with the meeting planner, handle on-site logistics, accommodation, hotels, registration, sponsorships, local conveyance, social program arrangements, entertainment and meeting preparation.

Compilation of References

"The Federal Law "On Freedom of Conscience and Religious Associations". (1997, September 26 N 125-F3).

Ababach, R. J. (2008) The Impact of knowledge management and organizational learning on organizational innovation. Proceedings of the knowledge management international conference on Transferring managing, maintaining knowledge for National Capacity Development.

Abbas, S., Shouping, L., Sidra, F., & Sharif, A. (2018). Impact of Crime on Socio-Economic Development: A Study of Karachi. *Malaysian Journal of Social Sciences and Humanities*, *3*(3), 148–159.

Abu, S. (2015). Fostering Knowledge Sharing Behaviour among Public Sector Managers: A Proposed Model for the Malaysian Public Service. *Journal of Knowledge Management*, *19*(1), 121–140. doi:10.1108/JKM-11-2014-0449

Acemoglu, D., Akcigit, U., & Kerr, W. R. (2016). Innovation network. *Proceedings of the National Academy of Sciences of the United States of America*, *113*(41), 11483–11488. doi:10.1073/pnas.1613559113 PMID:27681628

Adama, T. Y. M (2017). Facteurs clés de succès de la gestion des connaissances et capacités d'innovation des entreprises nigériennes. *La revue Gestion et Organisation*, *9*(1), 11-24.

Adler, P. S., & Kwon, S. W. (2002). Social capital: Prospects for a new concept. *Academy of Management Review*, *27*(1), 17–40. doi:10.5465/amr.2002.5922314

Afsar, B. (2016). The impact of person-organization fit on innovative work behavior: The mediating effect of knowledge sharing behavior. *International Journal of Health Care Quality Assurance*, *29*(2), 104–122. doi:10.1108/IJHCQA-01-2015-0017 PMID:26959893

Aggestam, L., Durst, S., & Persson, A. (2014). Critical success factors in capturing knowledge for retention in IT-supported repositories. *Information*, *5*(4), 558–569. doi:10.3390/info5040558

Ahmad, H., Öztürk, M., Ahmad, W., & Khan, S. M. (2015). Status of natural resources in the uplands of the Swat Valley Pakistan. In *Climate change impacts on high-altitude ecosystems* (pp. 49–98). Springer. doi:10.1007/978-3-319-12859-7_2

Ahuja, G. (2000). Collaboration Networks, Structural Holes, and Innovation: A Longitudinal Study. *Administrative Science Quarterly*, *45*(3), 425–455. doi:10.2307/2667105

Airey, D., & Johnson, S. (1999). The content of tourism degree courses in the UK. *Tourism Management*, *20*(2), 229–235. doi:10.1016/S0261-5177(98)00079-X

Akhmetova, E. (2013). *Islam in Russia: Historical Facts and Modern Developments*. SIS Publications.

Akram, T., Lei, S. Haider, M. J. & Hussain, S. T. (2019). The impact of organizational justice on employee innovative work behavior: Mediating role of knowledge sharing. *Journal of Innovation & Knowledge*, 1-13.

Aksüzek, S. (2008). İşletmelerde Rekabet Avantajı Olarak Yaratıcı Düşünceden Yararlanılması Üzerine Bir Araştırma (Yayımlanmamış yüksek lisans tezi). Gazi Üniversitesi Sosyal Bilimler Enstitüsü, Ankara.

Al Hawamdeh, N., & Al-edenat, M. (2019). Determinants of Barriers to Knowledge Sharing in the Jordanian Hospitality Industry. *International Business Research, 12*(7), 121–132. doi:10.5539/ibr.v12n7p121

Alam, A., Almotairi, M., & Gaadar, K. (2013). Nation branding: An effective tool to enhance fore going direct investment (FDI) in Pakistan. *Research Journal of International Studies, 25*, 134–141.

Alasttal, M. N. M., & Burdey, M. B. (2017). An Exploratory Analysis Of Pakistan Tourism Market: Its Present Scenario, Issues, Challenges And Future Prospects. *Grassroots, 50*(3).

Alaverdov, E. (2013). *Islamic Factor in History and Modern Russia*. Georgian Technical University.

Alavi, M., & Leidner, D. E. (2001). Knowledge Management and Knowledge Management Systems: Conceptual Foundations and Research Issues. *Management Information Systems Quarterly, 25*(1), 107–136. doi:10.2307/3250961

Albdour, A. A., & Altarawneh, I. I. (2014). Employee engagement and organizational commitment: Evidence from Jordan. *Int. J. Bus., 19*(2), 192–212.

Aldrich, H. E., & Zimmer, C. (1985). Entrepreneurship through social network. In D. L. Sexton & R. W. Smilor (Eds.), *The Art and Science of Entrepreneurship* (pp. 3–24). Balling Publishing Company.

Alexandrova, Y. (2004). *International Tourism*. Aspect Press.

Algieri, B., Aquino, A., & Succurro, M. (2018). International competitive advantages in tourism : An eclectic view. *Tourism Management Perspectives, 25*, 41–52. doi:10.1016/j.tmp.2017.11.003

Ali, A. (2010). *Investigating project management practices in public sector organizations of a less developed country*. Academic Press.

Ali, A. (2010). Socio-Economic cost of terrorism: A case study of Pakistan. *Strategic Studies, 57*, 1–14.

Ali, F. (2015). Service quality as a determinant of customer satisfaction and resulting behavioural intentions: A SEM approach towards Malaysian resort hotels. *Tourism – International Journal (Toronto, Ont.), 63*(1), 37–51.

Ali, K. A. M., & Buang, M. (2016). Study on Factors that Influence Innovation in Malaysian Public Sector. *Journal of Advanced Research in Business and Management Studies, 4*(1), 60–73.

Allee, V. (2009). *The future of knowledge*. Routledge. doi:10.4324/9780080507156

Almeida, M. V., & Soares, A. L. (2014). Knowledge sharing in project-based organizations: Overcoming the informational limbo. *International Journal of Information Management, 34*(6), 770–779. doi:10.1016/j.ijinfomgt.2014.07.003

Alov, A., & Vladimirov, N. (2001). *Islam in Russia*. Moscow: Russian Research Institute of Cupid Heritage.

Al-Refaie, A. (2015). Effects of human resource management on hotel performance using structural equation modeling. *Computers in Human Behavior, 43*, 293–303. doi:10.1016/j.chb.2014.11.016

Altinay, L., Sigala, M., & Waligo, V. (2016). Social value creation through tourism enterprise. *Tourism Management, 54*, 404–417. doi:10.1016/j.tourman.2015.12.011

Amabile, T. (1988). A model of creativity and innovation in organizations. *Research in Organizational Behavior, 10*(1), 123–167.

Amabile, T. M., Schatzel, E. A., Moneta, G. B., & Kramer, S. J. (2004). Leader behaviors and the work environment for creativity: Perceived leader support. *The Leadership Quarterly, 15*(1), 5–32. doi:10.1016/j.leaqua.2003.12.003

Ambler, T., Witzel, M., & Xi, C. (2016). *Doing business in China*. Routledge.

Anand, A. (2011). Understanding Knowledge Management: A literature review. *International Journal of Engineering Science and Technology, 3*(2), 926–939.

Anatoliivna, V. (2013). *Absorptive Capacity in Organizational Theories: Learning, Innovation*. Managerial Cognition. Marketing and Management Innovation.

Anderson, J. R., Reder, L. M., & Simon, H. A. (1996). Situated learning and education. *Educational Researcher, 25*(4), 5–11. doi:10.3102/0013189X025004005

Anderson, N., De Dreu, C., & Nijstad, B. (2004). The routinization of innovation research: A constructively critical review of the state-of-the-science. *Journal of Organizational Behavior, 25*(2), 147–173. doi:10.1002/job.236

Anderson, N., Potočnik, K., & Zhou, J. (2014). Innovation and creativity in organizations: A state-of-the-science review, prospective commentary, and guiding framework. *Journal of Management, 40*(5), 1297–1333. doi:10.1177/0149206314527128

Andreeva, T., & Kianto, A. (2012). Does knowledge management really matter? Linking knowledge management practices, competitiveness and economic performance. *Journal of Knowledge Management, 16*(4), 617–636. doi:10.1108/13673271211246185

Angle, H. L., & Perry, J. L. (1981). An empirical assessment of organizational commitment and organizational effectiveness. *Administrative Science Quarterly, 26*(1), 1–14. doi:10.2307/2392596

Aquino, R. S., Lück, M., & Schänzel, H. A. (2018). A conceptual framework of tourism social entrepreneurship for sustainable community development. *Journal of Hospitality and Tourism Management, 37*(July), 23–32. doi:10.1016/j.jhtm.2018.09.001

Arab News. (n.d.). Retrieved from https://www.arabnews.com/node/1538466/saudi-arabia

Aral, N. (1999). Sanat Eğitimi - Yaratıcılık Etkileşimi. *Hacettepe Üniversitesi Eğitim Fakültesi Dergisi Ankara, 15*, 11–17.

Aral, S., Dellarocas, C., & Godes, D. (2013). Introduction to the special issue—social media and business transformation: A framework for research. *Information Systems Research, 24*(1), 3–13. doi:10.1287/isre.1120.0470

Aramendia Muneta, M. E., & Ollo López, A. (2013). ICT Impact on tourism industry. *International Journal of Management Cases, 15*(2), 87–98.

Arshad, M. I., Iqbal, M. A., & Shahbaz, M. (2017). Pakistan tourism industry and challenges: A review. *Asia Pacific Journal of Tourism Research*. doi:10.1080/10941665.2017.1410192

Aslan, E. (2001a). *Kavram Boyutunda Yaratıcılık*. Türk Psikolojik Danışma ve Rehberlik Dergisi 16.

Aslan, E. (2001b). Torrance Yaratıcı Düşünce Testi'nin Türkçe Versiyonu. *M.Ü. Atatürk Eğitim Fakültesi Eğitim Bilimleri Dergisi, 14*, 19–40.

Aslan, E., & Puccio, G. J. (2006). Developing and Testing A Turkish Version of Torrance's Tests of Creative Thinking: A Study Of Adults. *The Journal of Creative Behavior, 40*(3), 163–178. doi:10.1002/j.2162-6057.2006.tb01271.x

Atay, Z. (2009). *Okul öncesi eğitim kurumlarına devam eden 5-6 yaş öğrencilerinin yaratıcılık Düzeylerinin yaş, cinsiyet ve ebeveyn eğitim Durumlarına göre incelenmesi: Ereğli örneği. (Yayımlanmamış yüksek lisans tezi)*. Konya: Selçuk Üniversitesi Sosyal Bilimler Enstitüsü.

Ateljevic, J., & Li, L. (2017). Tourism entrepreneurship–concepts and issues. In *Tourism and entrepreneurship* (pp. 30–53). Routledge. doi:10.4324/9780080942728

Atilla, O. (2017). Müziğin Yaratıcı Mutfak Uygulamaları Üzerindeki Etkilerinin Belirlenmesi (Yayımlanmamış yüksek lisans tezi). Anadolu Üniversitesi Sosyal Bilimler Enstitüsü, Eskişehir.

Atitumpong, A., & Badir, Y. F. (2018). Leader-member exchange, learning orientation and innovative work behavior. *Journal of Workplace Learning*, *30*(1), 32–47. doi:10.1108/JWL-01-2017-0005

Atmojo, M. (2015). The influence of transformational leadership on job satisfaction, organizational commitment, and employee performance. *Int. Res. J. Bus. Stud.*, *5*(2), 113–128. doi:10.21632/irjbs.5.2.113-128

Audretsch, D., & Feldman, M. (1995). R&D Spillovers and the Geography of Innovation and Production. *The American Economic Review*, *86*(3), 630–640.

Avdimiotis, S. (2016). Tacit knowledge management within hospitality establishments: Revealing the body of the iceberg. *International Journal of Knowledge Management*, *12*(3), 15–29. doi:10.4018/IJKM.2016070102

Avolio, B., & Gardner, W. (2005). Authentic leadership development: Getting to the root of positive forms of leadership. *The Leadership Quarterly*, *16*(3), 315–338. doi:10.1016/j.leaqua.2005.03.001

Avraham, E., & Ketter, E. (2016). *Tourism marketing for developing countries: Battling stereotypes and crises in Asia, Africa and the Middle East*. Springer. doi:10.1057/9781137342157

Ayan, B. E. (2017). *Okul Öncesi ve İlkokul Öğretmenlerinin Yaratıcılığa İlişkin Görüşleri ve Öğrencilerinin Yaratıcılık Performansları Arasındaki İlişki* (Yayımlanmamış yüksek lisans tezi). Kocaeli Üniversitesi Sosyal Bilimler Enstitüsü, Kocaeli.

Babkin, A. (2008). *Special Types of Tourism*. Rostov-don.

Badilescu-Buga, E. (2013). Knowledge behaviour and social adoption of innovation. *Information Processing & Management*, *49*(4), 902–911. doi:10.1016/j.ipm.2013.02.001

Baerenholdt, J. O., & Haldrup, M. (2006). Mobile networks and place making in cultural tourism: Staging Viking ships and rock music in Roskilde. *European Urban and Regional Studies*, *13*(3), 209–224. doi:10.1177/0969776406065431

Baggio, R., & Cooper, C. (2010). Knowledge transfer in a tourism destination: The effects of a network structure. *Service Industries Journal*, *30*(10), 1757–1771. doi:10.1080/02642060903580649

Bagnoli, C., & Giachetti, C. (2014). Aligning Knowledge Strategy and Competitive Strategy in Small Firms. *Journal of Business Economics and Management*, *16*(3), 571–598. doi:10.3846/16111699.2012.707623

Baker, T. L., Hunt, T. G., & Andrews, M. C. (2006). Promoting ethical behavior and organizational citizenship behaviors: The influence of corporate ethical values. *Journal of Business Research*, *59*(7), 849–857. doi:10.1016/j.jbusres.2006.02.004

Baki, S. (2004). *Küresel Ekonomik Sistemde Rekabet Gücünün Geliştirilmesine Yönelik Yaratıcılık Stratejilerinin İncelenmesi* (Yayımlanmamış yüksek lisans tezi). Kocaeli Üniversitesi Sosyal Bilimler Enstitüsü, Kocaeli.

Baldwin-Evans, K. (2006). Hilton highlights link between staff loyalty and e-learning. *Human Resource Management International Digest*, *14*(1), 36–38. doi:10.1108/09670730610643990

Baloch, Q. B. (2007). Managing tourism in Pakistan: A case study of Chitral valley. *Journal of Managerial Sciences*, *2*(2), 169–190.

Baltacı. (2013). Çocuğun Yaratıcılık Performansında Görsel Değerlerin Yeri (Yayımlanmamış doktora tezi). Ondokuz Mayıs Üniversitesi Eğitim Bilimleri Enstitüsü, Samsun.

Bandura, A. (1997). *Self-efficacy: The exercise of control.* Macmillan.

Bani-Melhem, S., Zeffane, R., & Albaity, M. (2018). Determinants of employees' innovative behavior. *International Journal of Contemporary Hospitality Management, 30*(3), 1601–1620. doi:10.1108/IJCHM-02-2017-0079

Baptista, R., & Swann, P. (1998). Do firms in clusters innovate more? *Research Policy, 27*(4), 525–540. doi:10.1016/S0048-7333(98)00065-1

Barão, A., Vasconcelos, J. B. D., Rocha, Á., & Pereira, R. (2017). A knowledge management approach to capture organizational learning networks. *International Journal of Information Management Processing, 37*(5), 735–740.

Baret, C., & Maciel, S. O. (2004). *Apports et limites de mesure du capital social en recherche en gestion des ressources humaines.* Communication au XV ème Congrès AGRH, Montréal.

Bari, M. W., & Fanchen, M. (2017). Personal Interaction Drives Innovation: Instrumental Guanxi-Based Knowledge Café Approach. In Handbook of research on tacit knowledge management for organizational success (Vol. 1, pp. 177–200). Hershey, PA: IGI Publishers.

Bari, M. W., Abrar, M., Bashir, M., Baig, S. A., & Fanchen, M. (2019). Soft Issues During Cross-Border Mergers and Acquisitions and Industry Performance, China–Pakistan Economic Corridor Based View. *SAGE Open, 9*(2). doi:10.1177/2158244019845180

Barley, W. C., Treem, J. W., & Kuhn, T. (2018). Valuing multiple trajectories of knowledge: A critical review and agenda for knowledge management research. *The Academy of Management Annals, 12*(1), 278–317. doi:10.5465/annals.2016.0041

Batet, M., Moreno, A., Sánchez, D., Isern, D., & Valls, A. (2012). Turist@: Agent-based personalized recommendation of touristic activities. *Expert Systems with Applications, 39*(8), 7319–7329. doi:10.1016/j.eswa.2012.01.086

Batjargal, B. (2003). Social capital and entrepreneurial performance in Russia: A longitudinal study. *Organization Studies, 24*(4), 535–556. doi:10.1177/0170840603024004002

Battour, M., & Ismail, M. N. (2016). Halal tourism: Concepts, practises, challenges and future. *Tourism Management Perspectives, 19*(1), 150–154. doi:10.1016/j.tmp.2015.12.008

Baumol, W. J. (2002). *The free-market innovation machine: Analyzing the growth miracle of capitalism.* Princeton University Press. doi:10.1515/9781400851638

Beaverstock, J. V, & Budd, L. (2013). *International business travel in a digital world economy.* Academic Press.

Beaverstock, J., Derudder, B., Faulconbridge, J., & Witlox, F. (2009). International business travel: Some explorations. *Geografiska Annaler. Series B, Human Geography, 91*(3), 193–202. doi:10.1111/j.1468-0467.2009.00314.x

Bedi, P., Agarwal, S. K., & Jindal, V. (2014). MARST: Multi-Agent recommender system for e-tourism using reputation-based collaborative filtering. In Proceedings of databases in networked information system (Vol. 8381, pp. 189-201), Berlin: Springer.

Benedict, M., Valoggia, P., Rousseau, A., & Schmitt, C. (2009). Gestion des connaissances etcapital social: Quelles interrelations? *Management & Avenir, 14*(27), 152–170. doi:10.3917/mav.027.0152

Benn, S. H., Bubna-Litic, D. C., & Eckstein, D. (2001). Is the MBA sustainable? Degrees of change. In *Australian and New Zealand Academy of Management Conference.* Massey University.

Beritelli, P. (2011). Cooperation among prominent actors in a tourist destination. *Annals of Tourism Research, 38*(2), 607–629. doi:10.1016/j.annals.2010.11.015

Berry, M. J. A. (1997). *Linoff, G. Data mining techniques: for marketing, sales, and customer support.* New York: Wiley Computer Publishing.

Bessick, J., & Naicker, V. (2013). Barriers to tacit knowledge retention: An understanding of the perceptions of the knowledge management of people inside and outside the organisation. *South African Journal of Information Management, 15*(2), 1–8. doi:10.4102ajim.v15i2.556

Bevort, A., & Lallement, M. (2006). Le capital social. Performance, équité et réciprocité. La Découverte.

Bigné, J. E., Alcañiz, J. E. B., Font, X., & Andreu, L. (2000). *Marketing de destinos turísticos: análisis y estrategias de desarrollo* [Marketing of Tourist Destinations; analysis and development strategies]. Madrid: Esic editorial.

Biletska, I. (2011). *Business tourism in the context of international tourism development.* Academic Press.

Binimelis, A., & Ordinas, A. (2003). El turismo de reuniones y negocios en Mallorca [Marketing and business tourism in Mallorca]. *Cuadernos de Turismo, 12*, 35–52.

Bird, J., Wang, C., Watson, J., & Murray, L. (2009). Relationships among principal authentic leadership and teacher trust and engagement levels. *Journal of School Leadership, 19*(2), 153–171. doi:10.1177/105268460901900202

Blackler, F. (2002). Knowledge, knowledge work and organizations. *The Strategic Management of Intellectual Capital and Organizational Knowledge*, 47–62.

Blackshaw, P. (2004). *Consumer-generated media (CGM) 101: Word-of-mouth in the age of the web-fortified consumer.* Http://Www. Nielsen-Online. Com/Downloads/Us/Buzz/Nbzm_wp_CGM101. Pdf

Blau, P. M. (1964). *1964 Exchange and power in social life New York.* Wiley.

Bloodgood, J. M. (2019). Knowledge acquisition and fi rmcompetitiveness: The role of complements and knowledge source. *Journal of Knowledge Management, 23*(1), 46–66. doi:10.1108/JKM-09-2017-0430

Bock, G., Zmud, R., Kim, Y., & Lee, J. (2005). Behavioral intention formation in knowledge sharing: Examining the roles of extrinsic motivators, social-psychological factors, and organizational climate. *Management Information Systems Quarterly, 29*(1), 87–111. doi:10.2307/25148669

Boddez, Y., Houwer, J. D., & Beckers, T. (2017). The inferential reasoning theory of causal learning: Towards a multi-process propositional account. In *The Oxford handbook of causal reasoning* (p. 53). Oxford University Press.

Bojnordi, E., & Moradi, P. (2012). A novel collaborative filtering model based on the combination of the correlation method with matrix completion technique. In *16th CSI international symposium on artificial intelligence and signal processing (AISP).* IEEE. 10.1109/AISP.2012.6313742

Boni, F. (2019). *Sino-Pakistani Relations: Politics.* Military and Regional Dynamics. doi:10.4324/9780429431982

Boniface, B., Cooper, R., & Cooper, C. (2016). *Worldwide destinations: The geography of travel and tourism.* Routledge. doi:10.4324/9781315697666

Borgatti, S. P., & Cross, R. (2003). A relational view of information seeking and learning in social networks. *Management Science, 49*(4), 432–445. doi:10.1287/mnsc.49.4.432.14428

Boschma, R. A. (2005). Proximity and Innovation: A critical Assessment. *Regional Studies, 39*(1), 61–74. doi:10.1080/0034340052000320887

Bos-Nehles, A. C., & Veenendaal, A. A. R. (2017). Perceptions of HR practices and innovative work behavior: The moderating effect of an innovative climate. *International Journal of Human Resource Management, 5192*, 1–23. doi:10.1080/09585192.2017.1380680

Bouncken, R. (2002). Knowledge Management for Quality Improvement in Hotels. In R. Bouncken & P. Sungsoo (Eds.), *Knowledge Management in Hospitality and Tourism* (pp. 25–59). New York: The Haworth Hospitality Press. doi:10.1300/J162v03n03_03

Bourdieu, P. (1995). La misère des médias. *Télérama*, 8–12.

Bourdieu, P. (1980). Le capital social, notes provisoires. *Actes de la Recherche en Sciences Sociales, 31*(31), 2–3.

Bowen, J., & Ford, R. C. (2004). What experts say about managing hospitality service delivery systems. *International Journal of Contemporary Hospitality Management, 16*(7), 394–401. doi:10.1108/09596110410559078

Bradley, A., Hall, T., & Harrison, M. (2002). Selling cities: Promoting new images for meetings tourism. *Cities (London, England), 19*(1), 61–70. doi:10.1016/S0264-2751(01)00046-4

Brass, D., Butterfield, K., & Skaggs, B. (1998). Relationships and unethical behavior: A social network perspective. *Academy of Management Review, 23*(1), 14–31. doi:10.5465/amr.1998.192955

Bredin, K., & Söderlund, J. (2006). Perspectives on human resource management: An explorative study of the consequences of projectification in four firms. *International Journal of Human Resources Development and Management, 6*(1), 92–113. doi:10.1504/IJHRDM.2006.009751

Brief, A. P., & Motowidlo, S. J. (1986). Prosocial organizational behaviors. *Academy of Management Review, 11*(4), 710–725. doi:10.5465/amr.1986.4283909

Brookfield, S. D., & Holst, J. D. (2010). *Radicalizing learning: Adult education for a just world.* John Wiley & Sons.

Brooking, A. (1996). *Intellectual Capital: Core Asset for the Third Millennium Enterprise.* Thomson Business Press.

Brown, L., Edwards, J., & Hartwell, H. (2010). A taste of the unfamiliar. Understanding the meanings attached to food by international postgraduate students in England. *Appetite, 54*(1), 202–207. doi:10.1016/j.appet.2009.11.001 PMID:19896514

Buhalis, D. (1998). Strategic use of information technologies in the tourism industry. *Tourism Management, 19*(5), 409–421. doi:10.1016/S0261-5177(98)00038-7

Buhalis, D., & Costa, C. (2006). *Tourism management dynamics: trends, management and tools.* Routledge. doi:10.4324/9780080455907

Buhalis, D., & Law, R. (2008). Progress in information technology and tourism management: 20 years on and 10 years after the Internet—The state of eTourism research. *Tourism Management, 29*(4), 609–623. doi:10.1016/j.tourman.2008.01.005

Bulut, Y. (2014). Meslek Yüksekokulu Öğrencilerinin Yaratıcılık Açısından Kendilerini Değerlendirmeleri İle Yaratıcılık Düzeyleri Arasındaki İlişkinin İncelenmesi (Yayımlanmamış yüksek lisans tezi). Gazi Üniversitesi Eğitim Bilimleri Enstitüsü, Ankara.

Bunderson, J. S., & Boumgarden, P. (2010). Structure and learning in self-managed teams: Why "bureaucratic" teams can be better learners. *Organization Science, 21*(3), 609–624. doi:10.1287/orsc.1090.0483

Burleson, W. (2005). Developing creativity, motivation, and self-actualization with learning systems. *International Journal of Human-Computer Studies, 63*(4-5), 436–451. doi:10.1016/j.ijhcs.2005.04.007

Burt, R. S. (1992). *Structural holes, the social structure of competition.* Cambridge: Harvard University Press.

Burt, R. S. (2000). The network structure of social capital. *Research in Organizational Behavior, 22*, 345–423. doi:10.1016/S0191-3085(00)22009-1

Burt, R. S. (2001). Structural Holes Versus Network Closure as Social Capital. 31-56. In N. Lin, K. S. Cook, & R. S. Burt (Eds.), *Social Capital: Theory and Research*. New Brunswick, NJ: Transaction.

Burt, R. S. (2004). Structural holes and good ideas. *American Journal of Sociology, 110*(2), 349–399. doi:10.1086/421787

Buxbaum, O. (2016). *Key insights into basic mechanisms of mental activity*. Springer International Publishing. doi:10.1007/978-3-319-29467-4

Büyükbaykal, C. I. (2015). Communication technologies and education in the information age. *Procedia: Social and Behavioral Sciences, 174*, 636–640. doi:10.1016/j.sbspro.2015.01.594

Byukusenge, E., Munene, J., & Orobia, L. (2016). Does knowledge management lead to innovation? An empirical study on SMEs in Rwanda. *International Journal of Management Science and Business Administration, 2*(9), 7–19. doi:10.18775/ijmsba.1849-5664-5419.2014.29.1001

Cabrera, A., Collins, W. C., & Salgado, J. F. (2006). Determinants of individual engagement in knowledge sharing. *International Journal of Human Resource Management, 17*(2), 245–264. doi:10.1080/09585190500404614

Cabrera, E. F., & Cabrera, A. (2005). Fostering knowledge sharing through people management practices. *International Journal of Human Resource Management, 16*(5), 720–735. doi:10.1080/09585190500083020

Cabrera, E. F., & Cabrera, E. F. (2002). Knowledge-sharing dilemmas. *Organization Studies, 23*(5), 687–710. doi:10.1177/0170840602235001

Cainelli, G., Evangelista, R., & Savona, M. (2005). Innovation and economic performance in services: A firm-level analysis. *Cambridge Journal of Economics, 30*(3), 435–458. doi:10.1093/cje/bei067

Çakır, M. (2010). Otel İşletmelerinin Mutfak Bölümünde İstihdam Edilen Personelin Eğitim Sürecinin Değerlendirilmesi: İstanbul'daki 5 Yıldızlı Zincir Otellere Yönelik Bir Alan Çalışması (Yayımlanmamış yüksek lisans tezi). İstanbul Üniversitesi Sosyal Bilimler Enstitüsü, İstanbul.

Camelo-Ordaz, C., García-Cruz, J., Sousa-Ginel, E., & Valle-Cabrera, R. (2011). The Influence of Human Resource Management on Knowledge Sharing and Innovation in Spain: The Mediating Role of Affective Commitment. *International Journal of Human Resource Management, 22*(07), 1442–1463. doi:10.1080/09585192.2011.561960

Cameron, K., & Dutton, J. (2003). *Positive organizational scholarship: Foundations of a new discipline*. San Francisco, CA: Berrett-Koehler Publishers.

Carluer, F. (2006). Enterprises' networks and territorial dynamics: a strategic analysis » (Trad. « Réseaux d'entreprises et dynamiques territoriales: une analyse stratégique. *Geography Economy Society, 8*(2), 193–214.

Carmeli, A., Meitar, R., & Weisberg, J. (2006). Self-leadership skills and innovative behavior at work. *International Journal of Manpower, 27*(1), 75–90. doi:10.1108/01437720610652853

Carolina, N. (2006). Building Collegial Information Networks : A Theory of Advice Network Generation. *Academy of Management Review, 31*(3), 615–637. doi:10.5465/amr.2006.21318921

Carson, S. H., Peterson, J. B., & Higgins, D. M. (2005). Reliability, Validity, and Factor Structure of the Creative Achievement Questionnaire. *Creativity Research Journal, 17*(1), 37–50. doi:10.120715326934crj1701_4

Carter, C., & Williams, B. (1957). *Industry and Technical Progress*. Oxford: Oxford Univ. Press.

Caruana, R., & Niculescu-Mizil, A. (2003). An empirical comparison of supervised learning algorithms. In *Proceedings of the twenty-third international conference*. ACM.

Casidy, R., Wymer, W., & O'Cass, A. (2018). Enhancing hotel brand performance through fostering brand relationship orientation in the minds of consumers. *Tourism Management, 66*(June), 72–84. doi:10.1016/j.tourman.2017.11.008

Castillo, L., Armengol, E., Onaindia, E., Sebastia, L., González-Boticario, J., Rodriguez, A., ... Borrajo, D. (2008). Samap: A user-oriented adaptive system for planning tourist visits. *Expert Systems with Applications, 34*(2), 1318–1332. doi:10.1016/j.eswa.2006.12.029

Centre for the Promotion of Imports from Developing Countries. (2016). *CBI product factsheet: MICE from the EU*. Author.

Cepeda-Carrion, G., & Cillo, V. (2019). Tips to use partial least squares structural equation modelling (PLS-SEM) in knowledge management. *Journal of Knowledge Management, 23*(1), 67–89. doi:10.1108/JKM-05-2018-0322

Černe, M., Jaklič, M., & Škerlavaj, M. (2013). Authentic leadership, creativity, and innovation: A multilevel perspective. *Leadership, 9*(1), 63–85. doi:10.1177/1742715012455130

Chawla, M. I. (2017). One belt one road summit 2017 and its implications for CPEC: An overview. *South Asian Studies, 32*(2), 277–284.

Chen, C. F. (2006). Applying the analytical hierarchy process (AHP) approach to convention site selection. *Journal of Travel Research, 45*(2), 167–174. doi:10.1177/0047287506291593

Chen, W. J., & Cheng, H. Y. (2012). Factors affecting the knowledge sharing attitude of hotel service personnel. *International Journal of Hospitality Management, 31*(2), 468–476. doi:10.1016/j.ijhm.2011.07.005

Chen, W.-J. (2013). Factors influencing internal service quality at international tourist hotels. *International Journal of Hospitality Management, 35*, 152–160. doi:10.1016/j.ijhm.2013.06.004

Cheraghalizadeh, R., & Tümer, M. (2017). Journal of Hospitality and Tourism Management The effect of applied resources on competitive advantage in hotels : Mediation and moderation analysis. *Journal of Hospitality and Tourism Management, 31*, 265–272. doi:10.1016/j.jhtm.2017.04.001

Chesbrough, H. (2003). *Open Innovation: The new imperative for creating and profiting from technology*. Boston: Harvard Business School Press.

Chiang, C. C., King, B., & Nguyen, T. H. (2012). Taiwan's MICE visitors: Business, leisure and education dimensions. *International Journal of Culture, Tourism and Hospitality Research, 6*(1), 21–33. doi:10.1108/17506181211206225

Choi, B., Poon, S. K., & Davis, J. G. (2008). Effects of knowledge management strategy on organizational performance: A complementarity theory-based approach. *Omega, 36*(2), 235–251. doi:10.1016/j.omega.2006.06.007

Chossat, V., & Gergaud, O. (2003). Expert opinion and gastronomy: The recipe for success. *Journal of Cultural Economics, 27*(2), 127–141. doi:10.1023/A:1023432502059

Chouhan, V. S., & Srivastava, S. (2014). Understanding competencies and competency modelling—A literature survey. *IOSR Journal of Business and Management, 16*(1), 14–22. doi:10.9790/487X-16111422

Çidem, G. (2016). Mutfak Çalışanlarının Örgüt Kültürü Algıları ve Yaratıcılık Süreçleri Arasındaki İlişkiler: Kemer Bölgesi Örneği (Yayımlanmamış yüksek lisans tezi). Süleyman Demirel Üniversitesi Sosyal Bilimler Enstitüsü, Isparta.

Cieslikowski, K. (2015). Business Tourism as Innovative Product For Destinations. *Conference proceedings, Innovative (Eco-) Technology, Entrepreneurship and Regional Development. Conference Proceedings.*

Cieślikowski, K. (2015). Business tourism as innovative product for destinations. Research Gate.

Çifci, M. A. (2014). İş Ortamının Yaratıcılık Performansına Etkisi Mutfak Çalışanları Üzerine Bir Uygulama (Yayımlanmamış yüksek lisans tezi). Akdeniz Üniversitesi Sosyal Bilimler Enstitüsü, Antalya.

Clark, G. (2008). *Local development benefits from staging global events*. Publications de l'OCDE.

Claus, R., & Martin, R. (2004) Personalized multi-criteria decision strategies in location-based decision support. In *Proceedings of a journal of the association of Chinese professionals in geographic information systems* (Vol. 10, pp. 149-156). 10.1080/10824000409480666

Coakes, E., Willis, D., & Clarke, S. (2001). *Knowledge management in the sociotechnical world: the graffiti continues*. Springer Science & Business Media.

Cohen, E. (1988). Authenticity and commoditization in tourism. *Annals of Tourism Research, 15*(3), 371–386. doi:10.1016/0160-7383(88)90028-X

Cohen, W. M., & Levinthal, D. A. (1990). Absorptive capacity: A new perspective on learning and innovation. *Administrative Science Quarterly, 35*(1), 128–152. doi:10.2307/2393553

Coleman, J. S. (1988). Social capital in the creation of human capital. In E. Lesser (Ed.), *Knowledge and social capital*. Chicago: University of Chicago Press.

Commonwealth. (2015, November). *Commonwealth health of government meetings*. Retrieved from The commonwealth: https://thecommonwealth.org/sites/default/files/news-items/documents/CHOGM%202015%20Communique.pdf

Connolly, D. J., Sigala, M., & Buhalis, D. (2001). Major trends and IT issues facing the hospitality industry in the new economy: A review of the 5th annual pan-European hospitality technology exhibition and conference (EURHOTEC 2000). *International Journal of Tourism Research, 3*(4), 325–327. doi:10.1002/jtr.308

Connolly, T., Thorn, B. K., & Heminger, A. (1992). Discretionary databases as social dilemmas. *Social Dilemmas: Theoretical Issues and Research Findings, 7*, 199–207.

Constitution of the Russian Federation. (1993).

Cooke, P., Uranga, M. G., & Etxebarriab, G. (1997). Regional innovation systems: Institutional and organizational dimensions. *Research Policy, 26*(4-5), 475–491. doi:10.1016/S0048-7333(97)00025-5

Coombs, C. R. (2009). Improving retention strategies for IT professionals working in the public sector. *Information & Management, 46*(4), 233–240. doi:10.1016/j.im.2009.02.004

Cooper, C. (2006). Knowledge management and tourism. *Annals of Tourism Research, 33*(1), 47–64. doi:10.1016/j.annals.2005.04.005

Cooper, C. (2015). Managing tourism knowledge. *Tourism Recreation Research, 40*(1), 107–119. doi:10.1080/02508281.2015.1006418

Cope, D. a. (2003). *Business Travel: Conferences, Incentive Travel, Exhibitions, Corporate Hospitality and Corporate Travel*. Research Gate.

Corporation, B. (2013). Sport Tourism. *Tourism Business Essentials, 51*.

Correia, A., Mesquita, A. & Paulos, A. (2010). Innovation through virtual communities of practice: motivation and constraints in the knowledge-creation process. *Projectics / Proyéctica / Projectique, 4*(1), 39–49.

Crawford, J., Dawkins, S., Martin, A., & Lewis, G. (2017). Understanding the organizational climate of unethical leadership in the Australian Football League. *The Journal of Leadership Studies*, *11*(2), 52–54. doi:10.1002/jls.21525

Crawford, J., Dawkins, S., Martin, A., & Lewis, G. (2019). Putting the leader back into authentic leadership: Reconceptualising and rethinking leaders. *Australian Journal of Management*; Advanced Online Publication. doi:10.1177/0312896219836460

Crawford, J., Kelder, J.-A., & Knox, M. (in-press). What does it take to be a social entrepreneur? Authentic leaders and their effect on innovation. In Ç. Dogru (Ed.), *Leadership styles, innovation, and social entrepreneurship in the era of digitalization*. Hershey, PA: IGI Global.

Crippen, C. (2012). Enhancing authentic leadership– followership: Strengthening school relationships. *Management in Education*, *26*(4), 192–198. doi:10.1177/0892020612439084

Crnjar, K. (2013). *Drivers and Barriers to Introducing Knowledge Management Initiatives in Hotel Enterprises in Croatia*. Academic Press.

Crossan, M. M., Lane, H. W., & White, R. E. (1999). An organizational learning framework: From intuition to institution. *Academy of Management Review*, *24*(3), 522–537. doi:10.5465/amr.1999.2202135

Csikszentmihalyi, M. (1996). *Creativity: Flow and the Psychology of Discovery and Invention*. HarperCollinsPublishers. https://books.google.com.tr/books?id=K0buAAAAMAAJadresindenerişildi

CWT Meetings & Events (2019). *Informe 2019 de CWT Meetings & Events*. Available from https://www.cwt-meetings-events.com/es/es/insights/meetings-and-events-trends/

Czamanski, S., & Ablas, L. A. (1979). Identication of Idustrial clusters and Complexes: A Commarison of Methods and Findings. *Urban Studies (Edinburgh, Scotland)*, *16*(1), 61–80. doi:10.1080/713702464

Czernek, K. (2017). Tourism features as determinants of knowledge transfer in the process of tourist cooperation. *Current Issues in Tourism*, *20*(2), 204–220. doi:10.1080/13683500.2014.944107

Dahou, K., Hacini, I., & Burgoyne, J. (2019). Knowledge management as a critical success factor in developing international companies' organizational learning capability. *Journal of Workplace Learning*, *31*(1), 2–16. doi:10.1108/JWL-12-2017-0118

Dalkir, K. (2013). *Knowledge management in theory and practice*. Routledge. doi:10.4324/9780080547367

Dallen, T. J., & Olsen, D. H. (2006). *Religion and Spiritual Journeys*. Routledge.

Darr, E. D., Argote, L., & Epple, D. (1995). The acquisition, transfer, and depreciation of knowledge in service organizations: Productivity in franchises. *Management Science*, *41*(11), 1750–1762. doi:10.1287/mnsc.41.11.1750

Darroch, J., & Mcnaughton, R. (2002). Examining the link between knowledge management practices and types of innovation. *Journal of Intellectual Capital*, *3*(3), 210–222. doi:10.1108/14691930210435570

Dasarathy, B. (1991). *Nearest neighbor (NN) norms: NN pattern classification techniques*. IEEE Computer Society Press.

Das, T., & Teng, B. (1998). Between trust and control: Developing confidence in partner cooperation in alliances. *Academy of Management Review*, *23*(3), 491–512. doi:10.5465/amr.1998.926623

Davenport, T. H., & Prusak, L. (1998). *Working knowledge: How organizations manage what they know*. Harvard Business Press.

Davenport, T. H., & Prussak, L. (1998). *Working Knowledge: how organizations manage what they know*. Boston: Harvard Business School Press.

Davenport, T., & Prusak, L. (1998). *Working Knowledge – How Organizations Manage What They Know*. Boston, MA: Harvard Business School Press.

Davidson, R. (1994). *Business Travel*. Addison Wesley Longman.

Davidson, R., & Cope, B. (2003). *Business travel: Conferences, incentive travel, exhibitions, corporate hospitality and corporate travel*. Pearson Education.

Davila, T., Epstein, M., & Shelton, R. (2012). Making innovation work: How to manage it, measure it, and profit from it. Upper Saddle River, NJ: FT Press.

De Jong, J., & Den Hartog, D. (2010). Measuring innovative work behaviour. *Creativity and Innovation Management*, *19*(1), 23–36. doi:10.1111/j.1467-8691.2010.00547.x

Del Chiappa, G., & Baggio, R. (2015). Knowledge transfer in smart tourism destinations: Analyzing the effects of a network structure. *Journal of Destination Marketing & Management*, *4*(3), 145–150. doi:10.1016/j.jdmm.2015.02.001

Demarest, M. (1997). Understanding knowledge management. *Long Range Planning*, *30*(3), 374–384. doi:10.1016/S0024-6301(97)90250-8

Demirci, C. (2007). Fen Bilgisi Öğretiminde Yaratıcılığın Erişi Ve Tutuma Etkisi. *Hacettepe Üniversitesi Eğitim Fakültesi Dergisi*, (32), 65-75.

Deng, C., Xiaofe, H., Ji-Rong, W., & Wei-Ying, M. (2004). Block-level link analysis. *Proceedings of the 27th annual international ACM SIGIR conference on research and development in information retrieval*, 440–7.

Denicolai, S., Cioccarelli, G., & Zucchella, A. (2010). Resource-based local development and networked core-competencies for tourism excellence. *Tourism Management*, *31*(2), 260–266. doi:10.1016/j.tourman.2009.03.002

Department of Statistics. Malaysia. (2018, January 29). *Economic census 2016: Tourism statistics*. Retrieved 4 November 2018 from https://www.dosm.gov.my/v1/ index.php?r=column/ctheme&menu_id=b0pIV1E3RW40VWRTUkZochyZ1pLUT09&bul_id=VmZsbTU4NDlFcFZRdVF6ZDF3OW4zZz09#

Deveraja. (2016). Tour operator's contribution towards development of tourism destinations. *International Journal in Management and Social Science*.

Dhar, R. L. (2015). Service quality and the training of employees: The mediating role of organizational commitment. *Tourism Management*, *46*, 419–430. doi:10.1016/j.tourman.2014.08.001

Díaz, C. G. Q. (1995). El mercado de reuniones, congresos e incentivo en España [The market for meetings, conferences, and incentives in Spain]. *Estudios Turísticos*, *126*, 179–190.

Dibiaggio, L., & Ferrary, M. (2003). Communautés de pratiques et réseaux sociaux dans la dynamique de fonctionnement des clusters de hautes technologies. *Revue d'Economie Industrielle*, *103*(1), 111–130. doi:10.3406/rei.2003.3112

Dilsiz, B. (2010). Türkiye'de Gastronomi ve Turizm (İstanbul Örneği) (Yayımlanmamış yüksek lisans tezi). İstanbul Üniversitesi Sosyal Bilimler Enstitüsü, İstanbul.

Dimitrios, B., Ioannis, R., Efstathios, V., Christos, A., Dimitrios, T., & Labros, S. (2018). Successful and Efficient Knowledge Management in the Greek Hospitality Industry: Change the Perspective! *Academic Journal of Interdisciplinary Studies*, *7*(1), 185–191. doi:10.2478/ajis-2018-0019

Donaldson, B., & O'Toole, T. (2007). *Strategic market relationships: From strategy to implementation*. John Wiley & Sons.

Donate, M. J., & Guadamillas, F. (2011). Organizational factors to support knowledge management and innovation. *Journal of Knowledge Management, 15*(6), 890–914. doi:10.1108/13673271111179271

Dossi, G. (1988). Sources, procedures, and microeconomic effects of innovation. *Journal of Economic Literature, 26*(3), 1120–1171.

Dracheva, E. (2010). Economics and Tourism Organization. Moscow: Knorus.

Dragicevic, V., Jovicic, D., Belsic, I., Stankov, U., & Boskovic, D. (2012). Business tourism destination competitiveness: A case of vojvodina province (Serbia). *Economic Research-. Ekonomska Istrazivanja, 25*(2), 311–332. doi:10.1080/1331677X.2012.11517510

Drazin, R., Glynn, M., & Kazanjian, R. (1999). Multilevel theorizing about creativity in organizations: A sensemaking perspective. *Academy of Management Review, 24*(2), 286–307. doi:10.5465/amr.1999.1893937

Dredge, D., Benckendorff, P., Day, M., Gross, M. J., Walo, M., Weeks, P., & Whitelaw, P. (2011). *Conceptualising the perfect blend in the tourism and hospitality education curriculum space*. Academic Press.

Dretske, F. (2004). Psychological vs. biological explanations of behaviour. *Behavior and Philosophy, §§§*, 167–177.

Drucker. (1993). *Post-capitalist Society*. Butterworth-Heinemann.

Du Plessis, M. (2007). The role of knowledge management in innovation. *Journal of Knowledge Management, 11*(4), 20–29. doi:10.1108/13673270710762684

Dunham, M. (2002). *Data mining: Introductory and advanced topics*. Prentice Hall.

Dunning, J. (2014). *The globalization of business (routledge revivals): The challenge of the 1990s*. Abingdon, UK: Routledge. doi:10.4324/9781315743691

Durukan, H. (2006). Okul Yöneticisinin Vizyoner Liderlik Rolü. *Ahi Evran Üniversitesi Kırşehir Eğitim Fakültesi Dergisi, 7*(2), 277–286.

Dutta, S., & Mia, I. (2011). The global information technology report 2010–2011. In *World Economic Forum* (Vol. 24). Academic Press.

Dvir, R., & Pasher, E. (2004). Innovation engines for knowledge cities: An innovation ecology perspective. *Journal of Knowledge Management, 8*(5), 16–27. doi:10.1108/13673270410558756

Dwyer, L., & Kim, C. (2003). Destination competitiveness: Determinants and indicators. *Current Issues in Tourism, 6*(5), 369–414. doi:10.1080/13683500308667962

Edú-Valsania, S., Moriano, J., & Molero, F. (2016). Authentic leadership and employee knowledge sharing behavior: Mediation of the innovation climate and workgroup identification. *Leadership and Organization Development Journal, 37*(4), 487–506. doi:10.1108/LODJ-08-2014-0149

Egginton, W. E. (2010). *Project management education, training, working & learning: A longitudinal study into the experiences of British army officers in UK defence projects* (Doctoral dissertation). University of Bristol.

Eklinder-Frick, J., Eriksson, L. T., & Hallen, L. (2012). Effects of social capital on processes in a regional strategic network. *Industrial Marketing Management, 41*(5), 800–806. doi:10.1016/j.indmarman.2012.06.007

Ekvall, G. (1996). Organizational climate for creativity and innovation. *European Journal of Work and Organizational Psychology, 5*(1), 105–123. doi:10.1080/13594329608414845

Elliott, J. (2002). *Tourism: Politics and public sector management*. Routledge. doi:10.4324/9780203416136

Epignosis, L. L. C. (2014). *E-learning concepts, trends, applications*. Author.

Epstein, R. M., & Hundert, E. M. (2002). Defining and assessing professional competence. *Journal of the American Medical Association, 287*(2), 226–235. doi:10.1001/jama.287.2.226 PMID:11779266

Erden, Z., von Krogh, G., & Nonaka, I. (2008). The quality of group tacit knowledge. *The Journal of Strategic Information Systems, 17*(1), 4–18. doi:10.1016/j.jsis.2008.02.002

Eriksen, M. (2009). Authentic leadership: Practical reflexivity, self-awareness, and self-authorship. *Journal of Management Education, 33*(6), 747–771. doi:10.1177/1052562909339307

Espejo, C., & Ponce, M. D. (1999). Turismo de congresos y reuniones en la ciudad de Murcia [Congress and meeting tourism in the city of Murcia]. *Cuadernos de Turismo, 3*, 45–60.

Etaati, L., & Sundaram, D. (2015). *Adaptive tourist recommendation system: conceptual frameworks and implementations. In Proceedings of journal of computer science* (Vol. 2, pp. 95–107). Springer.

Etiosa, O. (2012). *The Impacts Of Event Tourism On Host Communities*. Pietarsaari.

Eurostat. (2000). *Methodological Manual for Satistics on Congresses and Conferences*. Brussels: Author.

Eva, E., & Maria, E. (2011). Organizational culture As the driver of dense intra-orgnaizational networks. *Journal of Competitiveness, 2*, 28–42.

Evangelista, R., Lucchese, M., & Meliciani, V. (2013). Business services, innovation, and sectoral growth. *Structural Change and Economic Dynamics, 25*(June), 119–132. doi:10.1016/j.strueco.2012.02.005

Evans, M., Dalkir, K., & Bidian, C. (2014). A holistic view of the knowledge life cycle: The knowledge management cycle (KMC) model. *Electronic Journal of Knowledge Management, 12*(2), 85–97.

Fakhar, A. H. (2010). Factors affecting tourism, tourism potential and strategies For development as an Industry in Pakistan.

Fekete, A. C., Fekete, M. C., & Greculescu, A. (2014). Factors influencing psychosocial communication in the reason/emotion society. *Procedia: Social and Behavioral Sciences, 128*, 123–127. doi:10.1016/j.sbspro.2014.03.129

Feldman, M., & Francis, J. (2002). Entrepreneurs and the formation of industrial clusters. 3rd day on proximity, Paris.

Fenich, G. (1992). Convention center development: Pros, cons and unanswered questions. *International Journal of Hospitality Management, 11*(3), 183–196. doi:10.1016/0278-4319(92)90017-P

Ferraris, A., Santoro, G., & Dezi, L. (2017). How MNC's subsidiaries may improve their innovative performance? The role of external sources and knowledge management capabilities. *Journal of Knowledge Management, 21*(3), 540–552. doi:10.1108/JKM-09-2016-0411

Ferris, G., Treadway, D., Perrewe, P., Brouer, R., Douglas, C., & Luxm, S. (2007). Political skills in organizations. *Journal of Management, 33*(3), 290–320. doi:10.1177/0149206307300813

Fess, C. (2019). Beginners Guide to Starting a Tourism Business. Tourism Tattler. Retrieved from https://www.tourismtattler.com/articles/business-and-finance/beginners-guide-to-starting-a-tourism-business/6681

Ficarelli, S. (2013). Congress and Meetings Tourism and Destination Brand. *The Fifth International Conference on Creative Content Technologies*, 15.

Fidgeon, P. R. (2010). Tourism education and curriculum design: A time for consolidation and review? *Tourism Management, 31*(6), 699–723. doi:10.1016/j.tourman.2010.05.019

Filipatali, T. (2013). *Learning Theories. Their Influence on Teaching Methods*. Munich: GRIN Verlag. https://www.grin.com/document/293498

Filiz, F. (2013). Kimya Dersleri İçin Bilimsel Yaratıcılık Ölçeğinin Geliştirilmesi Ve Genel Yaratıcılık İle Bilimsel Yaratıcılık Arasındaki İlişkinin Belirlenmesi (Yayımlanmamış yüksek lisans tezi). Balıkesir Üniversitesi Fen Bilimleri Enstitüsü, Balıkesir.

Finnegan, D., & Willcocks, L. (2006). Knowledge sharing issues in the introduction of a new technology. *Journal of Enterprise Information Management*, *19*(6), 200–221. doi:10.1108/17410390610708472

Flamarich, M., & Duro, J. A. (2011). Business (events) tourism in Spain: Analysis and prospects. *Papers de Turisme*, *49/50*, 59–75.

Florida, R. (2006). The flight of the creative class: The new global competition for talent. *Liberal Education*, *92*(3), 22–29.

Ford, C. (1996). A theory of individual creative action in multiple social domains. *Academy of Management Review*, *21*(4), 1112–1142. doi:10.5465/amr.1996.9704071865

Forum, I. T. (2012). *Niches For Developing countries*. Retrieved from International Trade Forum: http://www.tradeforum.org/Business-Tourism/

Foss, N. J., & Pedersen, T. (2002). Transferring knowledge in MNCs: The role of sources of subsidiary knowledge and organizational context. *Journal of International Management*, *8*(1), 49–67. doi:10.1016/S1075-4253(01)00054-0

FoundationsI. (n.d.). Retrieved from https://indigenousfoundations.arts.ubc.ca/un_declaration_on_the_rights_of_indigenous_peoples/

Fraj, E., Matute, J., & Melero, I. (2015). Environmental strategies and organizational competitiveness in the hotel industry: The role of learning and innovation as determinants of environmental success. *Tourism Management*, *46*, 30–42. doi:10.1016/j.tourman.2014.05.009

Framke, W. (2002). The destination as a concept: A discussion of the business-related perspective versus the socio-cultural approach in tourism theory. *Scandinavian Journal of Hospitality and Tourism*, *2*(2), 92–108. doi:10.1080/15022250216287

Franzidis, A. (2018). An examination of a social tourism business in Granada, Nicaragua. *Tourism Review*.

Freydouni, S. G., & Woods, P. (2010). Barriers of knowledge sharing in Malaysian hotels. In 5th *Knowledge Management International Conference (KMICe)* (pp. 461-466). Academic Press.

Friedman, N., Geiger, D., & Goldszmidt, M. (1997). Bayesian network classifiers. *Mach Learn, 29*(2-3), 131–63.

Fries-Britt, S., & Snider, J. (2015). Mentoring outside the line: The importance of authenticity, transparency, and vulnerability in effective mentoring relationships. *New Directions for Higher Education*, *2015*(171), 3–11. doi:10.1002/he.20137

Fritsch, M., & Kauffeld-Monz, M. (2009). The impact of network structure on knowledge transfer: An application of social network analysis in the context of regional innovation networks. *The Annals of Regional Science*, *44*(1), 21–38. doi:10.100700168-008-0245-8

Fukey, L. N., & Issac, S. S. (2014). Connect among green, sustainability and hotel industry: A prospective simulation study. *Energy Conservation, 6*, 8.

Gainutdin, R. (1997). *Islam Belief in Charity*. Moscow: Russian Muftis Council.

Gall, J., & Strezo, M. (2019). Quantitative Analysis of Environment Potential for Cluster Development in Tourist Regions of Slovak Republic. *Geographica Pannonica*, *23*(3), 195–203. doi:10.5937/gp23-21375

Garcia, I., Sebastia, L., Onaindia, E., & Guzman, C. (2009). A group recommender system for tourist activities. In Proceedings of an international conference on electronic commerce and web technologies (Vol. 5692, pp. 26-37). Springer.

Garcia, I., Sebastia, L., & Onaindia, E. (2011). *On the design of Individual and Group Recommender Systems for Tourism. In Proceedings of expert systems with applications* (Vol. 38, pp. 7683–7692). Elsevier.

Gavalas, D., & Kenteris, M. (2011). A web-based pervasive recommendation system for mobile tourist guides. *Personal and Ubiquitous Computing*, *15*(7), 759–770. doi:10.100700779-011-0389-x

Gavalas, D., Konstantopoulos, C., Mastakas, K., & Pantziou, G. (2014). *Mobile recommender system in tourism. In Proceedings of the journal of network and computer applications* (Vol. 39, pp. 319–333). Elsevier.

Getz, D. (2007). Event Tourism: Definition, evalution and research. *Science Direct*, 428.

Getz, D., & Page, S. J. (2016). Progress and prospects for event tourism research. *Tourism Management*, *52*, 593–631. doi:10.1016/j.tourman.2015.03.007

Ghapar, A., Othman, N., & Jamal, S. (2014). Community resilience and knowledge on vulnerability in tourism development. Tourism. *Leisure and Global Change.*, *1*(1), 140–148.

Giap, T. K., Gopalan, S., & Ye, Y. (2016). Drivers of growth in the travel and tourism industry in Malaysia a geweke causality analysis. *Economies*, *4*(1), 1–15. doi:10.3390/economies4010003

Giddens, A. (1990). Structuration theory and sociological analysis. *Anthony Giddens: Consensus and controversy.*

Glasser, H., Calder, W., & Fadeeva, Z. (2005). *Definition: research in higher education for sustainability*. Halifax, Nova Scotia: Document Prepared for the Halifax Consultation.

Global Business Travel Association. (2013). *Global Business Travel Association (GBTA)*. Retrieved from: http://www.gbta.org.

Global Business Travel Association. (2019). Available from https://www.gbta.org/

Global Muslim Travel Index (GMTI). (2018). Retrieved from https://www.crescentrating.com/reports/global-muslim-travel-index-2019.html

Gloet, M., & Terziovski, M. (2004). Exploring the relationship between knowledge management practices and innovation performance. *Journal of Manufacturing Technology Management*, *15*(5), 402–409. doi:10.1108/17410380410540390

Goffman, E. (1978). The presentation of self in everyday life. London, UK: Harmondsworth.

Goh, S. C. (2002). Managing effective knowledge transfer: An integrative framework and some practice implications. *Journal of Knowledge Management*, *6*(1), 23–30. doi:10.1108/13673270210417664

Gold, A. H., Malhotra, A., & Segars, A. H. (2001). Knowledge management: An organizational capabilities perspective. *Journal of Management Information Systems*, *18*(1), 185–214. doi:10.1080/07421222.2001.11045669

Gonzalez, R. V. D., & de Melo, T. M. (2018). The effects of organization context on knowledge exploration and exploitation. *Journal of Business Research*, *90*, 215–225. doi:10.1016/j.jbusres.2018.05.025

Gonzalez, R. V. D., & Meli, T. M. (2017). Linkage between dynamics capability and knowledge management factors: A structural equation model. *Management Decision*, *55*(10), 2256–2276. doi:10.1108/MD-03-2017-0180

Gooroochurn, N., & Sugiyarto, G. (2005). Competitiveness indicators in the travel and tourism industry. *Tourism Economics*, *11*(1), 25–43. doi:10.5367/0000000053297130

Gouldner, A. W. (1960). The norm of reciprocity: A preliminary statement. *American Sociological Review, 25*(2), 161–178. doi:10.2307/2092623

Gouvêa, M., & dos Santos, R. M. (2005). Marketing de servicios en el contexto del turismo de negocios [Marketing services in the context of business tourism]. *Estudios y Perspectivas en Turismo, 14*(2).

Gračan, D. (2011). *Business tourism – Modern form for improvement of the competitiveness*. Academic Press.

Graen, G. B., Hui, C., & Taylor, E. A. (2006). Experience-based learning about LMX leadership and fairness in project teams: A dyadic directional approach. *Academy of Management Learning & Education, 5*(4), 448–460. doi:10.5465/amle.2006.23473205

Graham, B., & Vowles, T. M. (2006). Carriers within carriers: A strategic response to low-cost airline competition. *Transport Reviews, 26*(1), 105–126. doi:10.1080/01441640500179377

Granovetter, M. (1973). The strength of weak ties. *American Journal of Sociology, 78*(6), 1360–1380. doi:10.1086/225469

Granovetter, M. (2005). The impact of social structure on economic outcomes. *The Journal of Economic Perspectives, 19*(1), 33–50. doi:10.1257/0895330053147958

Grant, R. M. (2000). Shifts in the world economy: the drivers of knowledge management. *Knowledge Horizons: The Present and the Promise of Knowledge Management*, 27–53.

Grant, R. M. (1996). Prospering in dynamically-competitive environments: Organizational capability as knowledge integration. *Organization Science, 7*(4), 375–387. doi:10.1287/orsc.7.4.375

Grant, R. M. (1996). Toward a knowledge-based theory of the firm. *Strategic Management Journal, 17*(S2), 109–122. doi:10.1002mj.4250171110

Grant, R. M. (2005). *Contemporary Strategy Analysis*. Oxford: Blackwell.

Grant, R. M., & Baden-Fuller, C. (2004). A knowledge accessing theory of strategic alliances. *Journal of Management Studies, 41*(1), 61–84. doi:10.1111/j.1467-6486.2004.00421.x

Granvotter, M. (1985). Economics action and social structure: The problem of embeddedness. *American Journal of Sociology, 91*(3), 481–510. doi:10.1086/228311

Gray, P. H., & Meister, D. B. (2004). Knowledge sourcing effectiveness. *Management Science, 50*(6), 821–834. doi:10.1287/mnsc.1030.0192

Greenspoon, J. S. (2008). *Sustainability in Graduate Business Curriculum: Implications for Business Schools*. ProQuest.

Gretzel, U., & Yoo, K. H. (2008). Use and impact of online travel reviews. *Information and Communication Technologies in Tourism, 2008*, 35–46. doi:10.1007/978-3-211-77280-5_4

Grissemann, U., Plank, A., & Brunner-Sperdin, A. (2013). Enhancing business performance of hotels: The role of innovation and customer orientation. *International Journal of Hospitality Management, 33*(1), 347–356. doi:10.1016/j.ijhm.2012.10.005

Gubbins, C., & Dooley, L. (2014). Exploring Social Network Dynamics Driving Knowledge Management for Innovation. *Journal of Management Inquiry, 23*(2), 162–185. doi:10.1177/1056492613499203

Gülözge, G. (2006). *Sınıf Öğretmeni Adaylarının Yaratıcılık Düzeylerinin Çeşitli Değişkenler Açısından İncelenmesi* (Yayımlanmamış yüksek lisans tezi). Pamukkale Üniversitesi Sosyal Bilimler Enstitüsü.

Gumuchian, H., & Pecqueur, B. (2007). *La ressource territoriale*. Paris, France: Économica Editions Anthropos.

Gumusluoglu, L., & Ilsev, A. (2009). Transformational leadership, creativity, and organizational innovation. *Journal of Business Research, 62*(4), 461–473. doi:10.1016/j.jbusres.2007.07.032

Gunduz, M., & Yahya, A. M. A. (2018). Analysis of project success factors in construction industry. *Technological and Economic Development of Economy, 24*(1), 67–80. doi:10.3846/20294913.2015.1074129

Güneştekin, F. (2011). İlköğretim 1-5. Sınıf Öğrencilerinin Yaratıcılığının Bazı Demografik Değişkenler Açısından İncelenmesi (Yayımlanmamış yüksek lisans tezi). Akdeniz Üniversitesi Sosyal Bilimler Enstitüsü, Antalya.

Gustafson, P. (2012). Managing business travel: Developments and dilemmas in corporate travel management. *Tourism Management, 33*(2), 276–284. doi:10.1016/j.tourman.2011.03.006

Halawi, L., Aronson, J., & McCarthy, R. (2005). Resource-based view of knowledge management for competitive advantage. *Electronic Journal of Knowledge Management, 3*(2), 75–86. Retrieved from https://commons.erau.edu/publication/319/

Hall, C. M. (2012, February). Boosting food and tourism-related regional economic development. In Food and the Tourism Experience: The OECD-Korea Workshop; OECD Studies on Tourism (pp. 49-62). OECD.

Hallin, C. A., & Marnburg, E. (2007). Knowledge management in the hospitality industry: A review of empirical research. *Tourism Management.*

Hall, M. C., & Williams, A. (2008). *Tourism and innovation.* Routledge. doi:10.4324/9780203938430

Hamill, J. (1997). The Internet and international marketing. *International Marketing Review, 14*(5), 300–323. doi:10.1108/02651339710184280

Han & Kamber. (2007). *Data Mining Concepts & Techniques.* Morgan Kaufmann Publishers.

Hankinson, G. (2005). Destination brand images: A business tourism perspective. *Journal of Services Marketing, 19*(1), 24–32. doi:10.1108/08876040510579361

Hawkins, C., & Bonney, M. (2019). Lean thinking in leisure: Continuously improving event volunteering and management. *Annals of Leisure Research, 22*(3), 362–372. doi:10.1080/11745398.2019.1568267

Hegarty, J. A., & Barry O'Mahony, G. (2001). Gastronomy: A phenomenon of cultural expressionism and an aesthetic for living. *International Journal of Hospitality Management, 20*(1), 3–13. doi:10.1016/S0278-4319(00)00028-1

Hemsley, J., & Mason, R. M. (2012). The nature of knowledge in the social media age: Implications for knowledge management models. In *2012 45th Hawaii International Conference on System Sciences* (pp. 3928–3937). IEEE.

Henderson, J. C. (2016). Halal food, certification, and halal tourism: Insights from Malaysia and Singapore. *Tourism Management Perspectives, 19*(1), 160–164. doi:10.1016/j.tmp.2015.12.006

Hennessey, B. A. & Amabile, T. M. (1987). *Creativity and Learning: What Research Says to the Teacher.* ERIC Document Reproduction Service No. ED 312-835. Advance online publication. https://www.eric.ed.gov/PDFS/ED312835.pdf

Hinds, P., & Pfeffer, J. (2003). Why organizations don't "know what they know": Cognitive and motivational factors affecting the transfer of expertise. In M. Volker (Ed.), *Sharing expertise: Beyond knowledge management* (pp. 3–26). Boston, MA: MIT Press.

Hirst, P., Thompson, G., & Bromley, S. (2015). *Globalization in question.* Hoboken, NJ: John Wiley & Sons.

Hjalager, A.-M. (1996). Tourism and the environment: The innovation connection. *Journal of Sustainable Tourism, 4*(4), 201–218. doi:10.1080/09669589608667268

Hjalager, A.-M. (2002). Repairing innovation defectiveness in tourism. *Tourism Management, 23*(5), 465–474. doi:10.1016/S0261-5177(02)00013-4

Hjalager, A.-M. (2010). A review of innovation research in tourism. *Tourism Management, 31*(1), 1–12. doi:10.1016/j.tourman.2009.08.012

Hollan, P. C. (2008). Cognitive versus stimulus-response theories of learning. *Learning & Behavior, 36*(3), 227–241. doi:10.3758/LB.36.3.227 PMID:18683467

Holman, D., Epitropaki, O., & Fernie, S. (2001). Understanding learning strategies in the workplace: A factor analytic investigation. *Journal of Occupational and Organizational Psychology, 74*(5), 675–681. doi:10.1348/096317901167587

Holman, D., Totterdell, P., Axtell, C., Stride, C., Port, R., Svensson, R., & Zibarras, L. (2012). Job Design and the Employee Innovation Process: The Mediating Role of Learning Strategies. *Journal of Business and Psychology, 27*(2), 177–191. doi:10.100710869-011-9242-5

Holste, J. S., & Fields, D. (2010). Trust and tacit knowledge sharing and use. *Journal of Knowledge Management, 14*(1), 128–140. doi:10.1108/13673271011015615

Hong Trinh, T. (2016). The industry cluster approach for the development of central Vietnam. *International Journal of Business and Management, 11*(5), 167–178.

Hong, S. W. C. (2008). *Competitiveness in the tourism sector: a comprehensive approach from Economic and Management points.* Springer Science & Business Media.

Horng, J.-S. & Hu, M.-L. (2008). The Mystery in the Kitchen: Culinary Creativity (C. 1-2, C. 20). *Creativity Research Journal.* doi:10.1080/10400410802060166

Hosteltur. (2019). *Las tres tendencias que mueven el turismo MICE en 2018* [The three trends that move MICE tourism in 2018]. Retrieved from https://www.hosteltur.com/126296_tres-tendencias-mueven-turismo-mice-2018.html

Hu, H.-H., Kandampully, J., & Juwaheer, T. D. (2009). Relationships and impacts of service quality, perceived value, customer satisfaction, and image: An empirical study. *Service Industries Journal, 29*(2), 111–125. doi:10.1080/02642060802292932

Hu, L. T., & Bentler, P. M. (1999). Cutoff criteria for fit indexes in covariance structure analysis: Conventional criteria versus new alternatives. *Structural Equation Modeling, 6*(1), 1–55. doi:10.1080/10705519909540118

Hyder, S., & Lussier, R. N. (2016). Why businesses succeed or fail: A study on small businesses in Pakistan. *Journal of Entrepreneurship in Emerging Economies, 8*(1), 82–100. doi:10.1108/JEEE-03-2015-0020

Iacuone & Zarrilli. (2018). Business tourism in China. The case of Guangzhou. *GeoJournal of Tourism and Geosites.*

IBTM Americas. (2019). *MICE Travel Report.* Available from https://puntomice.com/tag/ibtm-americas/

IBTM. (2019). *Trends Watch Report 2019.* Retrieved from https://www.ibtmworld.com/PageFiles/397712/IBTM%20World%202018%20Trends%20Watch%20Report.pdf?v=636789096955805051

Ilies, R., Morgeson, F., & Nahrgang, J. (2005). Authentic leadership and eudaemonic well-being: Understanding leader–follower outcomes. *The Leadership Quarterly, 16*(3), 373–394. doi:10.1016/j.leaqua.2005.03.002

Ilya, S. (2011). *Eight-window sampling model for the research interview.* Available at: http://fieldsociology.blogspot.ru/2011/12/blog-post.html

In pictures: Prince William, Kate bid farewell to Pakistan after action-packed 5-day tour. (2019). Retrieved from https://www.dawn.com/news/1511580

INE. (2019). *Estatísticas do Turismo 2018*. Lisbon, Portugal: Instituto Nacional de Estatística.

Instituto Nacional de Estadística de España. (2019). www.ine.es

Instituto Nacional de Estatística. (2019). https://www.ine.pt/xportal/xmain?xpid=INE&xpgid=ine_publicacoes&PUBLICACOEStipo=ea&PUBLICACOEScoleccao=107668&selTab=tab0&xlang=pt

International Congress and Convention Association. (2019). https://www.iccaworld.org/

Inui, Y., Wheeler, D., & Lankford, S. (2006). Rethinking tourism education: What should schools teach. *Journal of Hospitality, Leisure, Sport and Tourism Education*, *5*(2), 25–35. doi:10.3794/johlste.52.122

Isa, S. M., Chin, P. N., & Mohammad, N. U. (2018). Muslim tourist perceived value: A study on Malaysia Halal tourism. *Journal of Islamic Marketing*, *9*(2), 402–420. doi:10.1108/JIMA-11-2016-0083

Isinkaye, F. O., Folajimi, Y. O., & Ojokoh, B. A. (2015). *Recommendation systems: Principles, methods, and evaluation. Egyptian Informatics Journal* , 16, 261–273.

Iszatt-White, M., Whittle, A., Gadelshina, G., & Mueller, F. (2018). The 'Corbyn phenomenon': Media representations of authentic leadership and the discourse of ethics versus effectiveness. *Journal of Business Ethics*; Advanced Online Publication.

Jackson, S. W. (2017). *A case study of hospitality and tourism challenges and opportunities in Malawi*. Academic Press.

Jackson, J., & Murphy, P. (2006). Clusters in regional tourism An Australian case. *Annals of Tourism Research*, *33*(4), 1018–1035. doi:10.1016/j.annals.2006.04.005

James, R. (2002). *Academic standards and the assessment of student learning: Some current issues in Australian Higher Education*. Centre for the Study of Higher Education.

Janssen, O. (2000). Job demands, perceptions of effort-reward fairness and innovative work behaviour. *Journal of Occupational and Organizational Psychology*, *73*(3), 287–302. doi:10.1348/096317900167038

Janssen, O. (2004). How fairness perceptions make innovative behavior more or less stressful. *Journal of Organizational Behavior*, *25*(2), 201–215. doi:10.1002/job.238

Janssen, O. (2005). The joint impact of perceived influence and supervisor supportiveness on employee innovative behaviour. *Journal of Occupational and Organizational Psychology*, *78*(4), 573–579. doi:10.1348/096317905X25823

Javalgi, R. G., & Steven White, D. (2002). Strategic challenges for the marketing of services internationally. *International Marketing Review*, *19*(6), 563–581. doi:10.1108/02651330210451926

Jensen, C. F., Hansen, J., & Metz, L. (1993). *Turistvirksomhed*. Copenhagen: Akademisk Forlag.

Jeppesen, L. B., & Frederiksen, L. (2006). Why do users contribute to firm-hosted user communities? The case of computer-controlled music instruments. *Organization Science*, *17*(1), 45–63. doi:10.1287/orsc.1050.0156

Jiménez-Zarco, A. I., González-González, I., Martínez-Ruíz, M. P., & Izquierdo-Yusta, A. (2014). New service innovation success: Analyzing the influence of performance indicator nature. *Computers in Human Behavior*, *51*, 1024–1031. doi:10.1016/j.chb.2014.09.046

Jitenev, S. (2016). Religious Tourism and Pilgrimage in Russian federation: Current State and Prospects. *Heritage Institute Journal, 1*(4).

Jitenev, S. (2019). *Educational Tourism in Russia*. Moscow: Yurayt.

Jitenev, S., & Abakumov, S. (2010). *Current State of Religious Tourism and its Social Role in Russian and Abroad.* Moscow: International Tourism Academy.

Joo, B.-K., & Yoon, H. J. (2012). The Effects of Core Self Evaluations and Transformational Leadership on Organizational Commitment. *Leadership and Organization Development Journal, 33*(6), 564–582. doi:10.1108/01437731211253028

Joong Kim, Y., & Hancer, M. (2010). The effect of knowledge management resource inputs on organizational effectiveness in the restaurant industry. *Journal of Hospitality and Tourism Technology, 1*(2), 174–189. doi:10.1108/17579881011065065

Jost, P. M., & Sandhu, H. S. (2000). *The hawala alternative remittance system and its role in money laundering*. Interpol Lyon.

Kahn, W. (1990). Psychological conditions of personal engagement and disengagement at work. *Academy of Management Journal, 33*(4), 692–724.

Kaltsova, S. (2009). The Business of Tourism. *Academia (Caracas).*

Kamarulzaman, N., Jomhari, N., Raus, N., & Yusoff, M. Z. M. (2015). Applying the Fuzzy Delphi Method to analyze the user Requirement for user Centered Design Process in Order to Create Learning Applications. *Indian Journal of Science and Technology, 8*(32), 1–7. doi:10.17485/ijst/2015/v8i32/92146

Kang, Y. J., Jin, Y. L., & Kim, H. W. (2017). A psychological empowerment approach to online knowledge sharing. *Computers in Human Behavior, 74*, 175–18. doi:10.1016/j.chb.2017.04.039

Kanwal, S., Pitafi, A. H., Pitafi, A., Nadeem, M. A., Younis, A., & Chong, R. (2019). China–Pakistan Economic Corridor (CPEC) development projects and entrepreneurial potential of locals. *Journal of Public Affairs.*

Kareiva, P. (2005). Is the key to conservation changing ethical values or policing unethical behavior? *Current Biology, 15*(2), 40–42. doi:10.1016/j.cub.2004.12.048 PMID:15668150

Karlsson, C., & Rodriguez, A. (2015). *Knowledge sharing in an open innovation collaboration. A case study of the SEVS Project* (Master's Thesis). Chalmers University of Technology, Gothenburg, Sweden.

Karlsson, C., Johansson, B., & Stough, R. R. (2005). Industrial Clusters and Inter-firm Networks: An Introduction. In Industrial Clusters and Inter-Firm Networks. Edward Elgar.

Keating, E., Oliva, R., Repenning, N., Rockart, S., & Sterman, J. (1999). Overcoming the improvement paradox. *European Management Journal, 17*(2), 120–134. doi:10.1016/S0263-2373(98)00072-3

Keleşoğlu, S. (2017). Öğretmen Eğitiminde Yaratıcı Düşünme ve İnovasyon Eğitim Programının Tasarımı, Denenmesi ve Değerlendirilmesi (Yayımlanmamış doktora tezi). Gazi Üniversitesi Eğitim Bilimleri Enstitüsü, Ankara.

Kesíci, M. & Önçel, S. (2015). Aşçıların Mesleki Özerklikleri İle Yaratıcılık Süreci İlişkisi. *İşletme Fakültesi Dergisi, 16*(1), 23-45.

Khan, K. A., & Sheikh, R. A. (2005). Project management in Pakistan government. *Proceedings of the PMI Asia-Pacific Global Congress.*

Khan, K., Turner, J. R., & Maqsood, T. (2013, June). Factors that influence the success of public sector projects in Pakistan. In *Proceedings of IRNOP 2013 Conference* (pp. 17-19). Academic Press.

Khan, I. A., & Khan, M. S. (2018). *Developing Sustainable Agriculture in Pakistan.* CRC Press. doi:10.1201/9781351208239

Khan, S. A. (2013). Geo-economic imperatives of Gwadar Sea Port and Kashgar economic zone for Pakistan and China. *IPRI Journal, 13*(2), 87–100.

Khristov, T. (2005). *Religious Tourism.* Moscow: Publishing Center Avademy.

Kim, H. J., Chen, M. H., & Jang, S. C. S. (2006). Tourism expansion and economic development: The case of Taiwan. *Tourism Management, 27*(5), 925–933. doi:10.1016/j.tourman.2005.05.011

Kim, S., Im, H. H., & King, B. (2015). Muslim travellers in Asia: The destination preferences and brand perceptions of Malaysian tourists. *Journal of Vacation Marketing, 21*(1), 3–21. doi:10.1177/1356766714549648

Kim, T. H., Lee, J. N., Chun, J. U., & Benbasat, I. (2014). Understanding the effect of knowledge management strategies on knowledge management performance: A contingency perspective. *Information & Management, 51*(4), 398–416. doi:10.1016/j.im.2014.03.001

Kim, T. T., Kim, W. G., Park, S. S.-S., Lee, G., & Jee, B. (2012). Intellectual Capital and Business Performance: What Structural Relationships Do They Have in Upper-Upscale Hotels? *International Journal of Tourism Research, 14*(4), 391–408. doi:10.1002/jtr.1868

Kim, T. T., & Lee, G. (2013). Hospitality employee knowledge-sharing behaviors in the relationship between goal orientations and service innovative behavior. *International Journal of Hospitality Management, 34*(1), 324–337. doi:10.1016/j.ijhm.2013.04.009

Klima, R. (1972). Theoretical pluralism, methodological dissension and the role of the sociologist: The West German case. *Social Sciences Information. Information Sur les Sciences Sociales, 11*(3-4), 69–108. doi:10.1177/053901847201100303

Kline, S., & Rosenberg, N. (1986). An overview of innovation. In R. Landau & N. Rosenberg (Eds.), *The Positive Sum Strategy. Harnessing Technology for Economic Growth* (pp. 275–306). Washington, DC: National Academy of Sciences.

Kmieciak, R., Michna, A., Kmieciak, R., & Michna, A. (2018). Knowledge management orientation, innovativeness, and competitive intensity : evidence from Polish SMEs intensity : evidence from Polish SMEs. *Knowledge Management Research and Practice, 16*(4), 559–572. doi:10.1080/14778238.2018.1514997

Knoema. (2018). *Contribution of travel and tourism to GDP as a share of GDP.* Retrieved from https://knoema.com/atlas/topics/Tourism/Travel-and-Tourism-Total-Contribution-to-GDP/Contribution-of-travel-and-tourism-to-GDP-percent-of-GDP

Knox, M., Crawford, J., & Kelder, J.-A. (2019). Developing authentic leadership and fostering social entrepreneurial innovation. In A. Ayandibu (Ed.), *Reshaping entrepreneurship education with strategy and innovation.* Hershey, PA: IGI Global.

Knox, M., Crawford, J., & Young, S. (in press). Ecotourism Emergence in Tasmania through Social Entrepreneurs and Authentic Leaders. In D. Batabyal (Ed.), *Global Entrepreneurial Trends in the Tourism and Hospitality Industry.* Hershey, PA: IGI Global.

Ko, D. G., Kirsch, L. J., & King, W. R. (2005). Antecedents of knowledge transfer from consultants to clients in enterprise system implementations. *Management Information Systems Quarterly, 29*(1), 59–85. doi:10.2307/25148668

Koens, K. (2012). Competition, cooperation and collaboration: business relations and power in township tourism. In *Slum Tourism* (pp. 101–118). Routledge.

Konak, A. (2008). İlköğretim 6. Sınıf Öğrencilerinin Sanatsal Yaratıcılık Düzeyleri (Yayımlanmamış yüksek lisans tezi). Süleyman Demirel Üniversitesi Sosyal Bilimler Enstitüsü, Isparta.

Koren, Y., Bell, R., & Kolinsky, C. (2009). *Matrix factorization techniques for recommender systems* (Vol. 8). IEEE Comput.

Kortsch, T., Schulte, E. M., & Kauffeld, S. (2019). Learning @ work: Informal learning strategies of German craft workers. *European Journal of Training and Development*, *43*(5–6), 418–434. doi:10.1108/EJTD-06-2018-0052

Kostova, T., & Roth, K. (2003). Social Capital in Multinational Corporations and Micro-Macro Model of its Formation. *Academy of Management Review*, *28*(2), 297–317. doi:10.5465/amr.2003.9416356

Kronstadt, K. A. (2009). *Pakistan-US relations*. Library of Congress Washington Dc Congressional Research Service.

Kucharska, W., & Wildowicz-Giegiel, A. (2017). Company Culture, Knowledge Sharing and Organizational Performance: The Employee's Perspective. *Proceedings of the 18th European Conference on Knowledge Management*, *1*, 524–531.

Kumar, A. A. (2017). Knowledge Retention: A Key attribute in Organizational growth. Advances in Applied Science Research, 1-9.

Kurata, Y., & Hara, T. (2013). CT-planner4: Toward a more user-friendly interactive day-tour planner. In *Information and communication technologies in tourism 2014* (pp. 73–86). Springer International Publishing. doi:10.1007/978-3-319-03973-2_6

Kurmanaliyeva, Rysbekova, Dussenbayeva, & Izmailov. (2014). Religious Tourism as a Sociocultural Phenomenon of the Present The Unique Sense Today is a Universal Value Tomorrow. This is the Way Religions are Created and Values are Made. *Pricedia -Social and Behavioral Sciences*.

Kuura, A. (2011). *Policies for projectification: support, avoid or let it be?* Academic Press.

Kwakman, K. (2003). Factors affecting teachers' participation in professional learning activities. *Teaching and Teacher Education*, *19*(2), 149–170. doi:10.1016/S0742-051X(02)00101-4

Ladkin, A. (2006). Conference tourism–MICE market and business tourism. In *Tourism business frontiers* (pp. 78–88). Routledge. doi:10.1016/B978-0-7506-6377-9.50014-6

Lai, G., Lin, N., & Leung, S. Y. (1998). Network resources, contact resources, and status attainment. *Social Networks*, *20*(2), 159–178. doi:10.1016/S0378-8733(97)00012-9

Landry, R., Amara, N., & Lamari, M. (2002). Does social capital determine innovation? To what extent? *Technological Forecasting and Social Change*, *69*(7), 681–701. doi:10.1016/S0040-1625(01)00170-6

Lane, L. P., Koka, B., & Pathak, S. (2006). The reification of Absorptive Capacity: A critical review and rejuvenation of the Construct. *Academy of Management Review*, *31*(4), 833–863. doi:10.5465/amr.2006.22527456

Larose, T. D. (2005). *Discovering knowledge in data*. Hoboken, NJ: John Wiley.

Larsen, J., Axhausen, K., & Urry, J. (2006). Geographies of social networks: Meetings, travel and communications. *Mobilities*, *1*(2), 261–283. doi:10.1080/17450100600726654

Latilla, V. M., Frattini, F., Petruzzelli, A. M., Berner, M., Latilla, V. M., Frattini, F., ... Berner, M. (2018). Knowledge management, knowledge transfer and organizational performance in the arts and crafts industry : A literature review. *Journal of Knowledge Management*, *22*(6), 1310–1331. doi:10.1108/JKM-08-2017-0367

Lau, R. K.-S. (2020). Africa-China Relations in the Context of Belt and Road Initiative: Realizing African-Chinese Dreams for Common Development? In China's Globalization and the Belt and Road Initiative (pp. 157–178). Springer.

Lau, C. (2009). *Manual on elective. I, Meetings, incentives, conventions and exhibitions (MICE)*. Hong Kong: Education Bureau.

Lau, D. C. (2013). *Meetings, Incentives, Conventions and Exhibitions*. MICE.

Lazega, E. (2008). Théorie de la coopération entre concurrents: Interdépendance, discipline sociale et processus sociaux. *Le Libellio d'Aegis*, *4*(3), 1–5.

Lee, C. S., Chang, Y. C., & Wang, M. H. (2009). Ontological recommendation multi-agent for Tainan city travel. *Expert Systems with Applications*, *36*(3), 6740–6753. doi:10.1016/j.eswa.2008.08.016

Lee, C. Y. (2018). Geographical clustering and firm growth: Differential growth performance among clustered firms. *Research Policy*, *47*(6), 1173–1184. doi:10.1016/j.respol.2018.04.002

Lee, D. Y., & Tsang, E. W. (2001). The effects of entrepreneurial personality, background and network activities on venture growth. *Journal of Management Studies*, *38*(4), 583–602. doi:10.1111/1467-6486.00250

Lee, H., & Choi, B. (2003). Knowledge management enablers, processes, and organizational performance: An integrative view and empirical examination. *Journal of Management Information Systems*, *20*(1), 179–228. doi:10.1080/07421222.2003.11045756

Lee, M. J., & Back, K. J. (2005). A review of convention and meeting management research 1990–2003: Identification of statistical methods and subject areas. *Journal of Convention & Event Tourism*, *7*(2), 1–20. doi:10.1300/J452v07n02_01

Lee, P. T.-W., Hu, Z.-H., Lee, S.-J., Choi, K.-S., & Shin, S.-H. (2018). Research trends and agenda on the Belt and Road (B&R) initiative with a focus on maritime transport. *Maritime Policy & Management*, *45*(3), 282–300. doi:10.1080/03088839.2017.1400189

Lee, S., Gon, B., Hoyal, K., Lee, S., Kim, B. G., & Kim, H. (2012). An integrated view of knowledge management for performance. *Journal of Knowledge Management*, *16*(2), 183–203. doi:10.1108/13673271211218807

Lefika, P. T., & Mearns, M. A. (2015). Adding knowledge cafés to the repertoire of knowledge sharing techniques. *International Journal of Information Management*, *35*(1), 26–32. doi:10.1016/j.ijinfomgt.2014.09.005

Leiper, N. (1979). The framework of tourism. *Annals of Tourism Research*, *6*(4), 390–407. doi:10.1016/0160-7383(79)90003-3

Leiper, N. (1990). Tourist attraction systems. *Annals of Tourism Research*, *17*(3), 367–384. doi:10.1016/0160-7383(90)90004-B

Levin, D. Z., Walter, J., Appleyard, M. M., & Cross, R. (2016). Relational Enhancement: How the Relational Dimension of Social Capital Unlocks the Value of Network-Bridging Ties. *Group & Organization Management*, *41*(4), 415–457. doi:10.1177/1059601115574429

Levitt, T. (1993). The globalization of markets. In R. Aliber & R. Click (Eds.), *Readings in international business: a decision approach* (pp. 249–267). Boston, MA: MIT Press.

Lewin, A. Y., Massini, C., & Peeters, S. (2011). Microfoundations of Internal and External Absorptive Capacity Routines. *Organization Science*, *22*(1), 81–98. doi:10.1287/orsc.1100.0525

Lickorish, L. J., Jefferson, A., Bodlender, J., & Jenkins, C. L. (1991). *Developing tourism destinations: policies and perspectives.* London: Financial Times Prentice Hall.

Lieven, A. (2017). Counter-insurgency in Pakistan: The role of legitimacy. *Small Wars & Insurgencies*, *28*(1), 166–190. doi:10.1080/09592318.2016.1266128

Liu, C. H. S. (2018). Examining social capital, organizational learning and knowledge transfer in cultural and creative industries of practice. *Tourism Management*, *64*(February), 258–270. doi:10.1016/j.tourman.2017.09.001

Li, Y. Q., & Liu, C. H. (2019). The power of coworkers in service innovation: The moderating role of social interaction. *International Journal of Human Resource Management, 30*(12), 1956–1976. doi:10.1080/09585192.2017.1314310

Loh, Lorenzi, Saldaña, & Licthnow. (2004). A tourism recommender system based on collaboration and text analysis. In *Proceedings of cognizant comm. corp.* (Vol. 6, pp. 157-165). Academic Press.

Lohmann, M. (2004). New demand factors in tourism. In *European Tourism Forum* (Vol. 10). Academic Press.

Longhi, C., & Keeble, D. (2000). *High-Technology Clusters and Evolutionary Trends in the 1990s, in the High-Technology Clusters, Networking and Collective Learning in Europe*. Ashgate.

Lucas, J., Luz, N., Moreno, M., Anacleto, R., Figueiredo, A., & Martins, C. (2013). A hybrid recommendation approach for a tourism system. *Expert Systems with Applications, 40*(9), 3532–3550. doi:10.1016/j.eswa.2012.12.061

Luke, P., Maxine, R., Kamal, M., David, D., & Andy, N. (2004). Networking and innovation: A systematic review of the evidence. *International Journal of Management Reviews, 5–6*(3–4), 137–168. doi:10.1111/j.1460-8545.2004.00101.x

Lundvall, G. A. (1993). Explaining interfirm cooperation-the limits of transaction cost approach. In G. Grabher (Ed.), *The Embedded Firm; On the Socioeconomics of Industrial Networks* (pp. 52–64). London: Routledge.

Lundvall, G. A. (1995). *National Systems of Innovation. Pinter Publishers*. London: Editions.

Luthans, F. (2002). The need for and meaning of positive organizational behavior. *Journal of Organizational Behavior: The International Journal of Industrial, Occupational and Organizational Psychology and Behavior, 23*(6), 695–706. doi:10.1002/job.165

Luthans, F., & Avolio, B. (2009). Inquiry unplugged: Building on Hackman's potential perils of POB. *Journal of Organizational Behavior, 30*(2), 323–328. doi:10.1002/job.590

Lyons, G. (2013). Business travel—The social practices surrounding meetings. *Research in Transportation Business & Management, 9*, 50–57. doi:10.1016/j.rtbm.2013.03.001

MAH. (2018). *Hotel Statistics, Malaysian Association of Hotels*. Available at www.hotels.org.my/ home.asp?hdnMRef1/457

Mahapa, M. (2013). Impact of knowledge management strategies on organizational performance in the hospitality industry of Zimbabwe. *Public Administration Review, 2*(1), 76.

Mainga, W. (2017). Examining project learning, project management competencies, and project efficiency in project-based firms (PBFs). *International Journal of Managing Projects in Business, 10*(3), 454–504. doi:10.1108/IJMPB-04-2016-0035

Makarenko, V. (2000). Who are the allies of Russia? Moscow: The Mental and Geopolitics: Paradoxes of Russia's Security Policy.

Malashenko, A. (1998). *Islamic Rebirth in Present-Day Russia*. Moscow: Carnegie Moscow Center.

Malik, A. R. (2018). The China–Pakistan Economic Corridor (CPEC): A Game Changer for Pakistan's Economy. In China's Global Rebalancing and the New Silk Road (pp. 69–83). Springer.

Malmberg, J., Järvelä, S., Holappa, J., Haataja, E., Huang, X., & Siipo, A. (2019). Going beyond what is visible: What multichannel data can reveal about interaction in the context of collaborative learning? *Computers in Human Behavior, 96*, 235–245. doi:10.1016/j.chb.2018.06.030

Mangion, M. L., Durbarry, R., & Sinclair, M. T. (2005). Tourism competitiveness: Price and quality. *Tourism Economics, 11*(1), 45–68. doi:10.5367/0000000053297202

Mansour, N., & Gaha, G. (2009). L'échec d'un projet de gestion des connaissances: Cas d'une entreprise de conseil. *Management International*, *13*(4), 91–101. doi:10.7202/038587ar

Manuti, A., Pastore, S., Scardigno, A. F., Giancaspro, M. L., & Morciano, D. (2015). Formal and informal learning in the workplace: A research review. *International Journal of Training and Development*, *19*(1), 1–17. doi:10.1111/ijtd.12044

Marco-Lajara, B., Claver-Cortés, E., Úbeda-García, M., García-Lillo, F., & Zaragoza-Sáez, P. C. (2019). The role of internal knowledge generation and external knowledge acquisition in tourist districts. *Journal of Business Research*, *101*, 767–776. doi:10.1016/j.jbusres.2018.12.045

Margana, G. S., Widodo, S. E., & Mukhtar, M. (2019). *A Study of the Influence of Transformational Leadership, Organizational Learning on Employees Innovative Behavior and Work Engagement at Sekolah Bahasa Polri.* Academic Press.

Marshall, A. (1919). *Industry and Trade: A Study of Industrial Technique and Business Organization and of Their Influences on the Condition of Various Classes and Nations*. London: MacMillian and Co. Limited.

Martin, A., Sanderson, K., Warnecke, E., Dawkins, S., Bartlett, L., Memish, K., & Peebles, D. (2017). *An integrated approach to workplace mental health.* Hobart, Australia: University of Tasmania.

Martínez, L., Rodriguez, R. M., & Espinilla, M. (2009). REJA: A georeferenced hybrid recommender system for restaurants. In *Proceedings of IEEE/WIC/ACM international conference on web intelligence and intelligent agent technology* (pp. 187–190). 10.1109/WI-IAT.2009.259

Martinez-Conesa, I., Soto-Acosta, P., & Carayannis, E. G. (2017). On the path towards open innovation: Assessing the role of knowledge management capability and environmental dynamism in SMEs. *Journal of Knowledge Management*, *21*(3), 553–570. doi:10.1108/JKM-09-2016-0403

Martínez-Martínez, A., Cegarra-Navarro, J. G., & García-Pérez, A. (2015). Environmental knowledge management: A long-term enabler of tourism development. *Tourism Management*, *50*, 281–291. doi:10.1016/j.tourman.2015.03.006

Martínez-Santiago, F., Ariza-López, F., Montejo-Ráez, A., & Ureña-López, A. (2012). GeOasis: A knowledge-based geo-referenced tourist assistant. *Expert Systems with Applications*, *39*(14), 11737–11745. doi:10.1016/j.eswa.2012.04.080

Maskell, P. (2000). Social Capital, Innovation, and Competitiveness. In S. Baron, J. Field, & T. Schuller (Eds.), *Social Capital: Critical Perspectives* (pp. 111–123). Oxford: Oxford University Press.

Maskell, P., & Malmberg, A. (1999). Localised learning and industrial competitiveness. *Cambridge Journal of Economics*, *23*(2), 167–185. doi:10.1093/cje/23.2.167

Maskell, P., & Malmberg, A. (2007). Myopia, knowledge development and cluster evolution. *Journal of Economic Geography*, *7*(5), 603–618. doi:10.1093/jeg/lbm020

Mason, D., & Pauleen, D. J. (2003). Perceptions of knowledge management: A qualitative analysis. *Journal of Knowledge Management*, *7*(4), 38–48. doi:10.1108/13673270310492930

Mastercard and CrescentRating. (2018). *Global Muslim travel index 2018*. Available at www. crescentrating.com/halal-muslim-travel-market-reports.html

Mathieson, A., & Wall, G. (1982). *Tourism, economic, physical and social impacts*. Longman.

Matsuo, M. (2019). Effect of learning goal orientation on work engagement through job crafting: A moderated mediation approach. *Personnel Review*, *48*(1), 220–233. doi:10.1108/PR-11-2017-0346

Mayer, D. M., Kuenzi, M., Greenbaum, R., Bardes, M., & Salvador, R. B. (2009). How low does ethical leadership flow? Test of a trickle-down model. *Organizational Behavior and Human Decision Processes, 108*(1), 1–13. doi:10.1016/j.obhdp.2008.04.002

Mayer, M. (2009). Innovation as a success factor in tourism: Empirical evidence from western Austrian cable-car companies. *Erdkunde, 63*(2), 123–139. doi:10.3112/erdkunde.2009.02.02

McAdam, R., & McClelland, J. (2002). Sources of new product ideas and creativity practices in the UK textile industry. *Technovation, 22*(2), 113–121. doi:10.1016/S0166-4972(01)00002-5

McDermott, R., & O'Dell, C. (2001). Overcoming cultural barriers to sharing knowledge. *Journal of Knowledge Management, 5*(1), 76–85. doi:10.1108/13673270110384428

McIver, D., Lengnick-Hall, C. A., Lengnick-Hall, M. L., & Ramachandran, I. (2013). Understanding work and knowledge management from a knowledge-in-practice perspective. *Academy of Management Review, 38*(4), 597–620. doi:10.5465/amr.2011.0266

McKay, T. (2012). *Adventure tourism: opportunities and management challenges for SADC destination.* Research Gate.

McKercher, B. (2002). The future of tourism education: An Australian scenario? *Tourism and Hospitality Research, 3*(3), 199–210. doi:10.1177/146735840200300302

McKercher, B., Ho, P. S., Cros, H. D., & So-Ming, B. C. (2002). Activities-based segmentation of the cultural tourism market. *Journal of Travel & Tourism Marketing, 12*(1), 23–46. doi:10.1300/J073v12n01_02

McLeod, S. A. (2017). *Kolb - learning styles.* Retrieved from www.simplypsychology.org/learning-kolb.html

Medina-Muñoz, D., & García-Falcón, J. M. (2000). Successful relationships between hotels and agencies. *Annals of Tourism Research, 27*(3), 737–762. doi:10.1016/S0160-7383(99)00104-8

Meehan, K., Lunney, T., Curran, K., & McCaughey, A. (2013). Context-aware intelligent recommendation system for tourism. In *Proceedings of the 11th IEEE international conference on pervasive computing and communications* (pp. 328–331). IEEE.

Melton, H. L., & Hartline, M. D. (2013). Employee collaboration, learning orientation, and new service development performance. *Journal of Service Research, 16*(1), 67–81. doi:10.1177/1094670512462139

Meutia, M. (2013). Entrepreneurial Social Competence and Entrepreneurial Orientation to Build SME's Business Network and Business Performance. *International Journal of Social Science and Humanity, 3*(4), 381–385. doi:10.7763/IJSSH.2013.V3.266

Meyer, J. P., & Allen, N. J. (1991). A three-component conceptualization of organizational commitment. *Human Resource Management Review, 1*(1), 61–89. doi:10.1016/1053-4822(91)90011-Z

Mia, M.B., & Ramage, M. (2011). *IT project management in developing countries: Approaches and factors affecting success in the microfinance sector of Bangladesh.* Academic Press.

Milhem, W., Abushamsieh, K., & Pérez Aróstegui, M.N. (2014). Training Strategies, Theories and Types. *Journal of Accounting, Business & Management, 21*(1).

Miller, K. D., Fabian, F., & Lin, S. (2009). Strategies for online communities. *Strategic Management Journal, 30*(3), 305–322. doi:10.1002mj.735

Mill, R. C., & Morrison, A. M. (1985). *The Tourism System.* Prentice Hall.

Minbaeva, D. B., Makela, K., & Rabbiosi, L. (2007). *Absorptive capacity and knowledge transfer: Individual level.* unpublished paper, Copenhagen Business School.

Miralles, P. (2010). Technological innovation, a challenge for the hotel sector. *UPGRADE: The European Journal for the Informatics Professional, 11*(2), 33–38.

Mir, F. A., & Pinnington, A. H. (2014). Exploring the value of project management: Linking project management performance and project success. *International Journal of Project Management, 32*(2), 202–217. doi:10.1016/j.ijproman.2013.05.012

Miron, E., Erez, M., & Naveh, E. (2004). Do personal characteristics and cultural values that promote innovation, quality, and efficiency compete or complement each other? *Journal of Organizational Behavior, 25*(2), 175–199. doi:10.1002/job.237

Mitchell, R., Burgess, J., & Waterhouse, J. (2010). Proximity and knowledge sharing in clustered firms. *International Journal of Globalisation and Small Business, 4*(1), 5–24. doi:10.1504/IJGSB.2010.035328

Mohajan, H. (2016). *A Comprehensive Analysis of Knowledge Management Cycles.* Academic Press.

Mohammad, A. A., Rashid, B. B., & Tahir, S. B. (2013). Assessing the influence of customer relationship management (CRM) dimensions on organisation performance: An empirical study in the hotel industry. *Journal of Hospitality and Tourism, 4*(3), 228–247. doi:10.1108/JHTT-01-2013-0002

Mohr, J., & Spekman, R. (1994). Characteristics of partnership success: Partnership attributes, communication behavior, and conflict resolution techniques. *Strategic Management Journal, 15*(2), 135–152. doi:10.1002mj.4250150205

Molina-Morales, F. X., & Martinez-Fernandez, M. T. (2010). Social networks: Effects of social capital on firm innovation. *Journal of Small Business Management, 48*(2), 258–279. doi:10.1111/j.1540-627X.2010.00294.x

Molz, J. G. (2013). Social networking technologies and the moral economy of alternative tourism: The case of couchsurfing.org. *Annals of Tourism Research, 43*, 210–230. doi:10.1016/j.annals.2013.08.001

Montani, F., Battistelli, A., & Odoardi, C. (2017). Proactive Goal Generation and Innovative Work Behavior: The Moderating Role of Affective Commitment, Production Ownership and Leader Support for Innovation. *The Journal of Creative Behavior, 51*(2), 107–127. doi:10.1002/jocb.89

Montes, J. C., & Pronovost, S. (2007). *La conversation sur les notions de capital social et de réseau: À la recherche d'une canne blanche. 16th AIMS Conference*, Montreal, Canada.

Morrison, A. M. (2013). *Marketing and managing tourism destinations.* Routledge. doi:10.4324/9780203081976

Mowery, D. C., & Rosenberg, N. (1979). The influence of market demand upon innovation: A critical review of some recent empirical studies. *Research Policy, 8*(2), 102–153. doi:10.1016/0048-7333(79)90019-2

Muhammad, S. B. (2014). Dynamics of the nationalist insurgency in the Balochistan province of Pakistan. Institute of World Economy and International Relations, 69.

Mukhametshin, F., & Dubkov, A. (2001). The State, Religion, and Church in Russia and Abroad. Information-analytical bulletin, no. 2. Muslim Organizations in the Russian Federation.

Muniandy, K., Rahim, S. A., Ahmi, A., & Rahman, N. A. A. (2019). Factors that influence customers' intention to visit green hotels in Malaysia. *International Journal of Supply Chain Management., 8*(3), 994–1003.

Musulin, J., Gamulin, J., & Crnojevac, I. H. (2011). Knowledge management in tourism: The importance of tacit knowledge and the problem of its elicitation and sharing. In *2011 Proceedings of the 34th International Convention MIPRO*, (pp. 981- 987). IEEE.

Muthuveloo, R., Shanmugam, N., & Teoh, A. P. (2017). The impact of tacit knowledge management on organisational performance: Evidence from Malaysia. *Asia Pacific Management Review*, *22*(4), 192–201. doi:10.1016/j.apmrv.2017.07.010

Myers, S., & Marquis, D. G. (1996). *Successful Industrial Innovation*. Washington, DC: National Science Foundation.

Naeem, A., Mirza, N. H., Ayyub, R. M., & Lodhi, R. N. (2019). HRM practices and faculty's knowledge sharing behavior: Mediation of affective commitment and affect based trust. *Studies in Higher Education*, *44*(3), 499–512. doi:10.1080/03075079.2017.1378635

Nahapiet, J., & Goshal, S. (1998). Social capital, intellectual capital and the organizational advantage. *Academy of Management Review*, *23*(2), 242–266. doi:10.5465/amr.1998.533225

Najafi-Tavani, S., Najafi-Tavani, Z., Naudé, P., Oghazi, P., & Zeynaloo, E. (2018). How Collaborative innovation networks affect new product performance: Product innovation capability, process innovation capability, and absorptive capacity. *Industrial Marketing Management*, *73*(25), 193–205. doi:10.1016/j.indmarman.2018.02.009

Nambisan, S., & Sawhney, M. (2011). Orchestration Processes in Network-Centric Innovation: Evidence from the Field. *The Academy of Management Perspectives*, *25*(3), 40–57.

Nan, N. (2008). A principal-agent model for incentive design in knowledge sharing. *Journal of Knowledge Management*, *12*(3), 101–113. doi:10.1108/13673270810875895

National Research Council. (2000). *How people learn: Brain, mind, experience, and school: Expanded edition*. National Academies Press.

Neirotti, L. D. (2007). Sport and Adventure Tourism. Academic Press.

Nelson, R. R. (Ed.). (1993). National Innovation Systems: A Comparative Analysis. Oxford University Press.

Newstead, T., Dawkins, S., Macklin, R., & Martin, A. (2019). *We don't need more leaders–We need more good leaders. Advancing a virtues-based approach to leader (ship) development. Leadership Quarterly*; Advanced Online Publication.

Nguyen, T. P. L., Nguyen, K. N., Do, T. D., & Nguyen, T. T. M. (2019). Knowledge sharing and innovative work behavior: The case of Vietnam. *Uncertain Supply Chain Management*, *7*, 619–634. doi:10.5267/j.uscm.2019.5.001

Niessen, C., Müller, T., Hommelhoff, S., & Westman, M. (2018). The impact of preventive coping on business travelers' work and private life. *Journal of Organizational Behavior*, *39*(1), 113–127. doi:10.1002/job.2215

Nieves, J., & Diaz-Meneses, G. (2018). Knowledge sources and innovation in the hotel industry: Empirical analysis on Gran Canaria Island, a mature mass-tourism destination. *International Journal of Contemporary Hospitality Management*, *30*(6), 2537–2561. doi:10.1108/IJCHM-07-2016-0341

Nieves, J., & Segarra-Ciprés, M. (2015). Management innovation in the hotel industry. *Tourism Management*, *46*, 51–58. doi:10.1016/j.tourman.2014.06.002

Nikolova, I., Van Ruysseveldt, J., De Witte, H., & Syroit, J. (2014). Work-based learning: Development and validation of a scale measuring the learning potential of the workplace (LPW). *Journal of Vocational Behavior*, *84*(1), 1–10. doi:10.1016/j.jvb.2013.09.004

Noguera, J. M., Barranco, M. J., Segura, R. J., & Martínez, L. (2012). A mobile 3D-GIS hybrid recommender system for tourism. *Information Sciences*, *215*, 37–52. doi:10.1016/j.ins.2012.05.010

Nonaka, I., & Takeuchi, H. (1995). *The knowledge-creating company: How Japanese companies create the dynamics of innovation*. Oxford University Press.

Nonaka, I. (1991). The knowledge-creating company. *Harvard Business Review*, *69*, 96–104.

Nonaka, I., & Konno, N. (1998). The concept of "Ba": Building a foundation for knowledge creation. *California Management Review*, *40*(3), 40–54. doi:10.2307/41165942

Nonaka, I., & Takeuchi, H. (1995). *The Knowledge Creating Company*. New York: Oxford University Press.

Nonaka, I., & Takeuchi, H. (1995). *The knowledge-creating company: how Japaneses companies create the dynamics of innovation*. New York: Oxford University Press, Inc.

Nonaka, I., & Takeuchi, H. (1996). The knowledge-creating company: How Japanese companies create the dynamics of innovation. *Long Range Planning*, *4*(29), 592. doi:10.1016/0024-6301(96)81509-3

Nonaka, I., & Toyama, R. (2005). The theory of the knowledge-creating firm: Subjectivity, objectivity, and synthesis. *Industrial and Corporate Change*, *14*(3), 419–436. doi:10.1093/icc/dth058

Nordin, S. (2003). *Tourism Clustering & Innovation*. European Tourism Research institute. [Ostersund, Sweden.]. *Mid-Sweden University.*, *SE-831*, 25.

Nunes, S., Estevão, C., & Filipe, M. N. (2018). Determinant factors of competitiveness in the hotel sector : The case of Portugal. *Competitiveness Review*, *28*(2), 122–135. doi:10.1108/CR-07-2016-0043

O'Sullivan, K. (Ed.). (2007). *Strategic knowledge management in multinational organizations*. IGI Global.

Oklobdžija, S. (2015). The role of events in tourism development. *Professional Paper*, 97.

Oktadiana, H., Pearce, P. L., & Chon, K. (2016). Muslim travellers' needs: What don't we know? *Tourism Management Perspectives*, *20*(1), 124–130. doi:10.1016/j.tmp.2016.08.004

Ologbo, A. C., & Noor, K. M. (2015). Knowledge management processes and firm innovation capability: A theoretical model. *Asian Social Science*, *11*(8), 10–17. doi:10.5539/ass.v11n18p10

Öncü, T. (1992). Yaratıcılığın Betimlenmesi Ve Yaratıcılık Üzerine Çevresel Etkiler. *Ankara Üniversitesi Dil ve Tarih-Coğrafya Fakültesi Felsefe Bölümü Dergisi*, *14*(1), 255–264.

Öncü, T. (2000). Anasınıfı (6 Yaş) Düzeyindeki Çocukların Şekilsel Yaratıcılıklarının Cinsiyet Değişkeni Açısından Karşılaştırılması. *Ankara Üniversitesi Dil ve Tarih-Coğrafya Fakültesi Dergisi*, *40*(1-2), 25–34.

Ottenbacher, M. C. (2007). Innovation management in the hospitality industry: Different strategies for achieving success. *Journal of Hospitality & Tourism Research (Washington, D.C.)*, *31*(4), 431–454. doi:10.1177/1096348007302352

Özden, Y. (2003). *Öğrenme ve Öğretme*. Ankara: Pegem Yayıncılık.

Özer, Z. (1991). Yaratıcılığa Giden Yolda Beyin Fırtınası. *Bilim ve Teknik*, (348), 50.

Özerbaş, M. A. (2011). Yaratıcı düşünme öğrenme ortamının akademik başarı ve bilgilerin kalıcılığa etkisi. *Gazi Üniversitesi Gazi Eğitim Fakültesi Dergisi*, *31*(3), 675–705.

Ozigbo, N. C. (2015). Knowledge Management Implementation in the Nigerian Hospitality Industry. *Review Pub Administration Manage*, *3*(147), 2. doi:10.4172/2315-7844.1000148

Paine, L. (1994). Managing for organizational integrity. *Harvard Business Review*, *72*(2), 106–117.

Pak, Y. S., Ra, W., & Lee, J. M. (2015). An integrated multi-stage model of knowledge management in international joint ventures: Identifying a trigger for knowledge exploration and knowledge harvest. *Journal of World Business*, *50*(1), 180–191. doi:10.1016/j.jwb.2014.02.006

Palacios Marqués, D., & José Garrigós Simón, F. (2006). The effect of knowledge management practices on firm performance. *Journal of Knowledge Management*, *10*(3), 143–156. doi:10.1108/13673270610670911

Pan, W., & Hawryszkiewycz, I. T. (2004). A method of defining learning processes. In *Annual Conference of the Australasian Society for Computers in Learning in Tertiary Education*. ASCILITE.

Pandey, A., Gupta, V., & Gupta, R. K. (2019). Spirituality and innovative behaviour in teams: Examining the mediating role of team learning. *IIMB Management Review*, *31*(2), 116–126. doi:10.1016/j.iimb.2019.03.013

Park, D., Kim, J., Kim, W. G., & Park, H. (2019). Does distance matter? Examining the distance effect on tourists' multi-attraction travel behaviors. *Journal of Travel & Tourism Marketing*, *36*(6), 693–710. doi:10.1080/10548408.2019.1624243

Park, Y. K., Song, J. H., Yoon, S. W., & Kim, J. (2014). Learning organization and innovative behavior: The mediating effect of work engagement. *European Journal of Training and Development*, *38*(1–2), 75–94. doi:10.1108/EJTD-04-2013-0040

Paton, R. A., & McLaughlin, S. (2008). Services innovation: Knowledge transfer and the supply chain. *European Management Journal*, *26*(April), 77–83. doi:10.1016/j.emj.2008.01.004

Pearson, A. W., Carr, J. C., & Shaw, J. C. (2008). Toward a theory of familiness: A social capital perspective. *Entrepreneurship Theory and Practice*, *32*(6), 949–969. doi:10.1111/j.1540-6520.2008.00265.x

Pedersen, L. B. (2012). *Creativity in Gastronomy: exploring the connection between art and craft* Unpublished Thesis). Copenhagen.

Penfold, P. (2008). Learning through the world of second life-a hospitality and tourism experience. *Journal of Teaching in Travel & Tourism*, *8*(2–3), 139–160. doi:10.1080/15313220802634224

Petersen, J., & Belchambers, K. (1990). Business Travel – a Boom Market. In M. Quest (Ed.), *Horwath Book of Tourism* (pp. 56–65). London: Palgrave Macmillan. doi:10.1007/978-1-349-11687-4_6

Pettenati, M. C., & Cigognini, M. E. (2007). Social networking theories and tools to support connectivist learning activities. *International Journal of Web-Based Learning and Teaching Technologies*, *2*(3), 42–60. doi:10.4018/jwltt.2007070103

Peyman, A. M. J. (2006). Critical issues for knowledge management implementation at a national level. *Vine*, *36*(1), 52–66. doi:10.1108/03055720610667372

Philmore, A. (2016). The influence of organisational commitment and corporate ethical values on non-public accountants' whistle-blowing intentions in Barbados. *Journal of Applied Accounting Research, Emerald Group Publishing*, *17*(2), 190–210. doi:10.1108/JAAR-12-2013-0118

Pieterse, J. N. (2019). *Globalization and culture: Global mélange*. Rowman & Littlefield Publishers.

Pillai, R., Schriesheim, C. A., & Williams, E. S. (1999). Fairness perceptions and trust as mediators for transformational and transactional leadership: A two-sample study. Journal of Management, 25. doi:10.1177/014920639902500606

Piller, F. T., & Reichwald, R. (2009). Wertschöpfungsprinzipien von Open Innovation. In *Kommunikation als Erfolgsfaktor im Innovationsmanagement* (pp. 105–120). Springer. doi:10.1007/978-3-8349-8242-1_5

Pinch, S., Henry, N., Jenkins, M., & Tallman, S. (2003). From 'industrial districts' to 'knowledge clusters': A model of knowledge dissemination and competitive advantage in industrial agglomerations. *Journal of Economic Geography, 3*(4), 373–388. doi:10.1093/jeg/lbg019

Pine, B. J., & Gilmore, J. H. (2011). *The experience economy*. Harvard Business Press.

Pinho, J., da Silva, B., Moreno, M., de Almeida, A. M., & Lopes, C. (2011). *Applying recommender methodologies in the Tourism sector. In Highlights in practical applications of agents and multi-agent systems in advances in Intelligent and Soft Computing* (pp. 101–108). Springer.

Polanyi, M. t. (1966). *1966 The tacit dimension*. London: Routledge & Kegan Paul.

Ponce, M. D. (2006). Las administraciones ante el turismo de reuniones y congresos en Murcia [The administrations before the tourism of meetings and congresses in Murcia]. *Papeles de Geografía, 44*, 95–118.

Portal, E. (2017). *Tourism Notes*. Retrieved from Tourism Notes Eduction portal: https://tourismnotes.com/

Porter, B. A., Orams, M. B., & Lück, M. (2018). Sustainable entrepreneurship tourism: An alternative development approach for remote coastal communities where awareness of tourism is low. *Tourism Planning & Development, 15*(2), 149–165. doi:10.1080/21568316.2017.1312507

Porter, M. (1998). *On Competition*. Boston: Harvard Business Review.

Porter, M. E. (1998). Clusters and the new economics of competition. *Harvard Business Review, 76*(6), 77-90.

Prabakaran, R. (2003). Good practices in planning and management of integrated commercial poultry production in South Asia. Food and Agriculture Organization.

Puri, S. (2017). Brinkmanship, not COIN, in Pakistan's post-9/11 Internal War. *Small Wars & Insurgencies, 28*(1), 218–232. doi:10.1080/09592318.2016.1263471

Putnam, R. D. (1995). Bowling alone: America's declining social capital. *Journal of Democracy, 6*(1), 65–78. doi:10.1353/jod.1995.0002

Qammach, N. I. J. (2016). The Mediating Role of Knowledge Sharing on Relationship between IT Capability and IT Support as Predictors of Innovation Performance: An Empirical Study on Mobile Companies in Iraq. *Procedia Economics and Finance, 39*, 562–570. doi:10.10162212-5671(16)30300-8

Quandt, C., Ferraresi, A., Kudlawicz, C., Martins, J., & Machado, A. (2017). Social innovation practices in the regional tourism industry: Case study of a cooperative in Brazil. *Social Enterprise Journal, 13*(1), 78–94. doi:10.1108/SEJ-12-2015-0038

Quintane, E., Mitch Casselman, R., Sebastian Reiche, B., & Nylund, P. (2011). Innovation as a knowledge-based outcome. *Journal of Knowledge Management, 15*(6), 928–947. doi:10.1108/13673271111179299

Rabasa, A., Waxman, M., Larson, E. V., & Marcum, C. Y. (2004). *The Muslim world after 9/11*. Rand Corporation. doi:10.7249/RB151

Ragab, M. A. F., & Arisha, A. (2013). Knowledge management and measurement: A critical review. *Journal of Knowledge Management, 17*(6), 873–901. doi:10.1108/JKM-12-2012-0381

Rajapathirana, R. P. J., & Hui, Y. (2018). Relationship between innovation capability, innovation type, and firm performance. *Journal of Innovation & Knowledge, 14*(3), 44–55.

Rao, Y., Yang, M., & Yang, Y. (2018). Knowledge Sharing, Organizational Learning and Service Innovation in Tourism. *Journal of Service Science and Management, 11*(05), 510–526. doi:10.4236/jssm.2018.115035

Rauterberg, M. (2019). *Expert Opinion*. Retrieved April 24, 2019, from http://www.idemployee.id.tue.nl/g.w.m.rauterberg/lectures.htm

Refworld. (n.d.). Retrieved from https://www.refworld.org/docid/4ae8611155.html

Rego, A., Sousa, F., Marques, C., & Cunha, M. (2012). Authentic leadership promoting employees' psychological capital and creativity. *Journal of Business Research, 65*(3), 429–437. doi:10.1016/j.jbusres.2011.10.003

Reilly, A. J. 1998. *Three approaches to organizational learning*. The Pfeiffer Library.

Reinholt, M., & Foss, N. J. (2011). Why a central network position isn't enough: The role of motivation and ability for knowledge sharing in employee networks. *Academy of Management Journal, 54*(6), 1277–1297. doi:10.5465/amj.2009.0007

Rein, I., Kotler, P., & Haider, D. (1993). *Marketing places: attracting investment, industry, and tourism to cities, states, and nations*. Free Press.

Ricci, F., & Werthner, H. (2002). Case base querying for travel planning recommendation. *Information Technology & Tourism, 4*(3), 215–226. doi:10.3727/109830501108751001

Richards, B. (1992). Tourism blues. *Leisure Management, 12*(11), 30–32.

Richards, G. (2011). Creativity and tourism. The state of the art. *Annals of Tourism Research, 38*(4), 1225–1253. doi:10.1016/j.annals.2011.07.008

Richards, G., & Wilson, J. (2006). Developing creativity in tourist experiences: A solution to the serial reproduction of culture? *Tourism Management, 27*(6), 1209–1223. doi:10.1016/j.tourman.2005.06.002

Richards, G., & Wilson, J. (Eds.). (2007). *Tourism, creativity and development* (Vol. 10). Routledge. doi:10.4324/9780203933695

Ritchie, J. R. B., & Crouch, G. I. (2003). *The competitive destination: A sustainable tourism perspective*. Cabi. doi:10.1079/9780851996646.0000

Robbins, S., & Judge, T. A. (2007). *Organizational Behavior* (12th ed.). Upper Saddle River, NJ: Prentice-Hall.

Robinson, P. (2012). *Tourism: The key concepts*. Abingdon, UK: Routledge. doi:10.4324/9780203104910

Rogers, T. (2013). Conferences and conventions: A global industry (3rd ed.). Abingdon, UK: Routledge. doi:10.4324/9780203119402

Rogers, E. M. (1995). Diffusion of Innovations: modifications of a model for telecommunications. In *Die diffusion von innovationen in der telekommunikation* (pp. 25–38). Springer. doi:10.1007/978-3-642-79868-9_2

Rogerson, C. M. (2015). Unpacking business tourism mobilities in sub-Saharan Africa. *Current Issues in Tourism, 18*(1), 44–56. doi:10.1080/13683500.2014.898619

Rogers, T. (2008). Forewords. In T. Rogers (Ed.), *Conferences and Conventions: A global industry* (pp. 13–17). Oxford: Butterworth Heinemann.

Rosanas, J. M., & Velilla, M. (2005). The ethics of management control systems: Developing technical and moral values. *Journal of Business Ethics, 57*(1), 83–96. doi:10.100710551-004-3826-1

Rosenberg, N. (1982). *Inside the Black Box: Technology and Economics*. New York: Cambridge Univ. Press.

Roth, C. (2008). Réseaux épistémiques: Formaliser la cognition distribuée. *Sociologie du Travail, 50*(3), 353–371. doi:10.4000dt.19477

Rudez, H. (2010). Knowledge management in the hotel industry before and after the entry in the EU: The case of Slovenia. *Organizacija, 43*(4), 197–201. doi:10.2478/v10051-010-0019-9

Ruotsalo, T., Haav, K., Stoyanov, A., Rochee, S., Fanid, E., & Deliaic, R. (2013). SMART MUSEUM: A mobile recommender system for the web of data. *Journal of Web Semantics, 20*, 50–67. doi:10.1016/j.websem.2013.03.001

Russion Tourism Market Report. (n.d.). Retrieved from https://mtu.gov.hu/documents/prod/RMAA-Group--ndash--Russian-Tourism-Market-Report--Trends---Analysis-and-Statistics.pdf

Rydalmere. (2019). *Exhibition and Trade Fairs*. Retrieved from Exhibitions and Event association of Austrailia: https://eeaa.com.au/member/exhibitions-and-trade-fairs/

Sadikov, M., & Khanbabayev, K. (2009). *Religious and Political Extremism (a textbook for high schools). Makhachkala* Institute of Theology.

Saifi, S. A. A., Dillon, S., & Mcqueen, R. (2016). The relationship between face to face social networks and knowledge sharing: An exploratory study of manufacturing firms. *Journal of Knowledge Management, 20*(2), 308–326. doi:10.1108/JKM-07-2015-0251

Sa, L. M. L., Choon-Yon, S., Chai, Y. K., & Joo, J., H, A.(2019). Knowledge creation process, customer orientation and firm performance: Evidence from small hotels in Malaysia, Asia Pacific. *Management Review*. doi:10.1016/j.apmrv.2019.07.002

Salem, I. E. B. (2014). Toward better understanding of knowledge management: Correlation to hotel performance and innovation in five-star chain hotels in Egypt. *Tourism and Hospitality Research, 14*(4), 176–196. doi:10.1177/1467358414542265

Salsali, M., Cheraghi, M. A., & Ahmadi, F. (2009). Organizational factors influencing knowledge transfer into practice in Iranian nursing context: A grounded theory approach. *International Journal of Nursing Practice, 15*(5), 426–436. doi:10.1111/j.1440-172X.2009.01775.x

Samori, Salleh, & Khalid. (2015). Current trends on Halal tourism; Cases on selected Asian countries. Tourism Management perspective. doi:10.1016/j.tmp.2015.12.011

Sanchez, R., & Mahoney, J. T. (1996). Modularity, flexibility, and knowledge management in product and organization design. *Strategic Management Journal, 17*(SI), 63–76. doi:10.1002mj.4250171107

Sánchez, D., & Moreno, A. (2008a). Pattern-based automatic taxonomy learning from the web. *AI Communications, 21*, 27–48.

Sánchez, D., & Moreno, A. (2008b). Learning non-taxonomic relationships from web documents for domain ontology construction. *Data and Knowledge Engineering Journal, 64*(3), 600–623. doi:10.1016/j.datak.2007.10.001

Sanchez-Lezama, A. P., Cavazos-Arroyo, J., & Albavera-Hernandez, C. (2014). Applying the Fuzzy Delphi Method for determining socio-ecological factors that influence adherence to mammography screening in rural areas of Mexico. *Cadernos de Saude Publica, 30*(2), 245–258. doi:10.1590/0102-311X00025113 PMID:24627054

Sanderson, K., Martin, A., Warnecke, E., Dawkins, S., Peebles, D., Bartlett, L., . . . LaMontagne, A. (2017). Workplace mental health: Co-production of an action plan for Australia. In Work, Stress, and Health Conference: Contemporary Challenges and Opportunities (p. 2). Academic Press.

San, İ. (1979). Yaratıcılık, Iki Düşünce Biçimi Ve Çocuğun Yaratıcılık Eğitimi. *Ankara Ünversitesi Eğitim Bilimleri Fakültesi Dergisi, 12*(1-4), 177–190.

Saniaghi, R. (2008). Strategic positioning and performance of winter destination. *Tourism Review, 63*(4), 40–57. doi:10.1108/16605370810912209

Santibáñez, D. B. (2017). *Business Plan for a travel agency and tour*. Academic Press.

Santich, B. (2004). The Study of Gastronomy and Its Relevance to Hospitality Education and Training. *International Journal of Hospitality Management, 23*(1), 15–24. doi:10.1016/S0278-4319(03)00069-0

Schilling, J., & Kluge, A. (2009). Barriers to organizational learning: An integration of theory and research. *International Journal of Management Reviews, 11*(3), 337–360. doi:10.1111/j.1468-2370.2008.00242.x

Schmitz, H., & Nadvi, K. (1999). Clustering and Industrialization: Introduction. *World Development, 27*(9), 1503–1514. doi:10.1016/S0305-750X(99)00072-8

Schmookler, J. (1966). *Invention and economic growth*. Academic Press.

Schmookler, J. (1966). *Invention and Economic Growth*. Cambridge, MA: Harvard University Press. doi:10.4159/harvard.9780674432833

Schumpeter, J. (1934). The theory of economic development, translated by Redvers Opie. *Harvard: Economic Studies, 46*.

Schweisfurth, T. G., & Raasch, C. (2018). Absorptive capacity for need knowledge: Antecedents and effects for employee innovativeness. *Research Policy, 47*(4), 687–699. doi:10.1016/j.respol.2018.01.017

Scott, B. R., & Lodge, G. C. (1985). US competitiveness in the world economy. *The International Executive, 27*(1), 26–26. doi:10.1002/tie.5060270112

Scott, N., & Ding, P. (2008). Management of tourism research knowledge in Australia and China. *Current Issues in Tourism, 11*(6), 514–528. doi:10.1080/13683500802475802

Scott, N., & Laws, E. (2006). Knowledge sharing in tourism and hospitality. *Journal of Quality Assurance in Hospitality & Tourism, 7*(1-2), 1–12. doi:10.1300/J162v07n01_01

Scott, S. G., Bruce, R. A., Scott, S. G., & Bruce, R. A. (2013). Determinants of innovative behavior: a path model of individual innovation in the workplace. *Academy of Management Journal, 37*(3), 580–607. doi:10.2307/256701

Sebastia, L., Garcia, I., Onaindia, E., & Guzman, C. (2009). E-Tourism: A Tourist recommendation and planning application. *International Journal on Artificial Intelligence Tools, 18*, 717-738.

Sebastia, L., Garcia, I., Onaindia, E., & Guzman, C. (2008). e-Tourism: a tourist recommendation and planning application. *20th IEEE International Conference on Tools with Artificial Intelligence (ICTAI)*. 10.1109/ICTAI.2008.18

Seebaluck, V. (2015). Prospects and challenges of business tourism. Academic Press.

Seggelen - Damen, I., & Dam, K. (2016). Self-reflection as a mediator between self-efficacy and well-being. *Journal of Managerial Psychology, 31*(1), 18–33. doi:10.1108/JMP-01-2013-0022

Senge, P. M. (1990). *The Fifth Discipline: The Art and Practice of the Learning Organization*. New York: Doubleday Currency.

Set, K. (2013). Tourism small and medium enterprises (TSMEs) in Malaysia. *International Journal of Business and Social Science, 4*(16), 58–66.

Sevimli, A. D. (2015). Yaratıcılık ve Sanat Bağlamında Sinema Bir Örnek: Akira Kurosawa (Yayımlanmamış doktora tezi). İstanbul Üniversitesi Sosyal Bilimler Enstitüsü, İstanbul.

Shafaei, F. (2017). The relationship between involvement with travelling to Islamic destinations and Islamic brand equity: A case of Muslim tourists in Malaysia. *Asia Pacific Journal of Tourism Research, 22*(3), 255–271. doi:10.1080/10941665.2016.1232741

Shafaei, F., & Mohamed, B. (2017). Malaysia's branding as an Islamic tourism hub: An assessment. *Geografia: Malaysian Journal of Society and Space, 11*(1), 97–106.

Shah, s. N. U., & Baloch, D. Q. B. (2017). *Comparative Analysis on Service Quality & Customer Satisfaction: Empirical Evidence from Pakistan Hotel Industry*. Academic Press.

Shalley, C., Zhou, J., & Oldham, G. (2004). The effects of personal and contextual characteristics on creativity: Where should we go from here? *Journal of Management, 30*(6), 933–958. doi:10.1016/j.jm.2004.06.007

Shamim, S., Cang, S., & Yu, H. (2017). Supervisory orientation, employee goal orientation, and knowledge management among front line hotel employees. *International Journal of Hospitality Management, 62*, 21–32. doi:10.1016/j.ijhm.2016.11.013

Shankar, R., Acharia, S., & Baveja, A. (2009). Soft-system knowledge management framework for new product development. *Journal of Knowledge Management, 13*(1), 135–153. doi:10.1108/13673270910931224

Shanker, D. (2008). ICT and Tourism : Challenges and Opportunities. *Proceedings Ohe International Conference on Tourism in India-Challenges Ahead*, 51–58.

Sharpe, M. (2004). *Islam in Russia: The Politics of Identity and Security. Moscow*. Moscow: Cambridge University Press. Cambridge.

Shaw, G., & Williams, A. M. (2009). Knowledge Transfer and Management in Tourism Organisations: An emerging research agenda. *Tourism Management, 30*(3), 325–335. doi:10.1016/j.tourman.2008.02.023

Shaw, G., Williams, A. M., & Nelson, R. R. (1995). Critical issues in tourism: A geographical perspective. *The Canadian Geographer. Geographe Canadien, 39*(3), 283.

Sheldon, P. J. (2007). 18 Tourism information technology. *International Handbook on the Economics of Tourism*, 399.

Sheldon, P., Fesenmaier, D. R., & Tribe, J. (2009). *White paper: The Tourism Education Futures Initiative (TEFI): Activating change in tourism education*. Academic Press.

Sheldon, P. J., Pollock, A., & Daniele, R. (2017). Social entrepreneurship and tourism: Setting the stage. In *Social Entrepreneurship and Tourism* (pp. 1–18). Springer. doi:10.1007/978-3-319-46518-0_1

Shukla, A. (2016). Growth of Travel and Tourism in India: A driver of EmploymentGeneration and Economic Development. *International Journal in Management and Social Science*.

Sidani, Y., & Rowe, W. (2018). A reconceptualization of authentic leadership: Leader legitimation via follower-centered assessment of the moral dimension. *The Leadership Quarterly, 29*(6), 623–636. doi:10.1016/j.leaqua.2018.04.005

Siddiqa, A. (2019). Pakistan—From Hybrid-Democracy to Hybrid-Martial Law. *Journal of South Asian and Middle Eastern Studies, 42*(2), 52–72.

Siegel, A., Brickman, S., Goldberg, Z., & Pat-Horenczyk, R. (2019). Preventing Future Terrorism: Intervening on Youth Radicalization. In An International Perspective on Disasters and Children's Mental Health (pp. 391–418). Springer.

Sigala, M., & Chalkiti, K. (2015). Knowledge management, social media and employee creativity. *International Journal of Hospitality Management*, *45*, 44–58. doi:10.1016/j.ijhm.2014.11.003

Silantyev, R. (2007). *The Latest History of Islam in Russia*. Moscow: Algorithm.

Simon, H. A. (2000). Bounded rationality in social science: Today and tomorrow. *Mind & Society*, *1*(1), 25–39. doi:10.1007/BF02512227

Singh, P. K. (2018). Knowledge strategy, sharing behavior and performance. *Management Research Review*, *41*(3), 395–411. doi:10.1108/MRR-01-2017-0001

Skoultsos, S. (2009). *Event tourism: Statements and questions about its impacts on rural areas.* Retrieved from Research Gate: https://www.researchgate.net/publication/47697270_Event_tourism_Statements_and_questions_about_its_impacts_on_rural_areas

Skyrme, D., & Amidon, D. (1997). The knowledge agenda. *Journal of Knowledge Management*, *1*(1), 27–37. doi:10.1108/13673279710800709

Slaughter, S. A., & Kirsch, L. J. (2006). The effectiveness of knowledge transfer portfolios in software process improvement: A field study. *Information Systems Research*, *17*(3), 301–320. doi:10.1287/isre.1060.0098

Smallbone, D., Kitching, J., & Athayde, R. (2010). Ethnic diversity, entrepreneurship and competitiveness in a global city. *International Small Business Journal*, *28*(2), 174–190. doi:10.1177/0266242609355856

SME Corporation Malaysia. (2018). *Small and medium enterprises*. http://www. smecorp.gov.my/index.php/en/small-and-medium-enterprises

Smith, A. (1999). International briefing 4 Training and development in Australia. *International Journal of Training and Development*, *3*(4), 301–313. doi:10.1111/1468-2419.00088

Smith, H., & Schurink, W. (2005). The interface between knowledge management and human resources: A qualitative study. *SA Journal of Human Resource Management*, *3*(1), 6–13. doi:10.4102ajhrm.v3i1.58

Smith, S. L. J. (1989). Tourism analysis. *Tourism Analysis*.

Snowden, D. (2002). Complex acts of knowing: Paradox and descriptive self-awareness. *Journal of Knowledge Management*, *6*(2), 100–111. doi:10.1108/13673270210424639

Snyman, R., & Kruger, C. J. (2004). The interdependency between strategic management and strategic knowledge management. *Journal of Knowledge Management*, *8*(1), 5–19. doi:10.1108/13673270410523871

Söderlund, J. (2011). Theoretical foundations of project management. In The Oxford handbook of project management. doi:10.1093/oxfordhb/9780199563142.003.0003

Söderlund, J. (2008). Competence dynamics and learning processes in project-based firms: Shifting, adapting and leveraging. *International Journal of Innovation Management*, *12*(01), 41–67. doi:10.1142/S1363919608001911

Soliman, M., & Shou, M. A. A. (2016). Knowledge Management and Tourism: An Exploratory Study Applied to Travel Agents in Egypt. *Environments*, *22*, 31.

Sönmez, S. F. (1998). Tourism, terrorism, and political instability. *Annals of Tourism Research*, *25*(2), 416–456. doi:10.1016/S0160-7383(97)00093-5

Sozbilir, F. (2018). The interaction between social capital, creativity and efficiency in organizations. *Thinking Skills and Creativity*, *27*(March issue), 92–100. doi:10.1016/j.tsc.2017.12.006

Spencer, J. P., & Bavuma, Z. (2018). How important are mice to the tourism economy? *The Business & Management Review, 9*(4), 125–134.

Spender, J.-C. (2008). Organizational learning and knowledge management: Whence and whither? *Management Learning, 39*(2), 159–176. doi:10.1177/1350507607087582

Spreitzer, G. M. (1995). Psychological empowerment in the workplace: Dimensions, measurement, and validation. *Academy of Management Journal, 38*(5), 1442–1465.

Standford. (n.d.). Retrieved from https://web.stanford.edu/group/sjir/3.2.07_meyer.html

Statista. (2019). *Global Business Travel Industry – Statistics & facts.* Retrieved from https://www.statista.com/topics/2439/global-business-travel-industry/

Statista. (n.d.). Retrieved from https://www.statista.com/statistics/617696/saudi-arabia-total-hajj-pilgrims/

Stetic, S. (2012). *The importance of business tourism to national economies in transition: The case of Serbia.* Research Gate.

Storper, M. (1997). *The regional world: Territorial development in a global economy.* New York: Guilford Press.

Suanmali, S. (2014). Factors Affecting Tourist Satisfaction: An Empirical Study in the Northern Part of Thailand. In *Proceedings of 4th international conference on tourism research SHS web of the conference* (Vol. 12). EDP Sciences.

Sugarman, L. (1987). Experiential learning: Experience as the source of learning and development, David A. Kolb, Prentice-Hall International, Hemel Hempstead, Herts., 1984. No. of pages: xiii + 256. *Journal of Organizational Behavior, 8*(4), 359–360. doi:10.1002/job.4030080408

Suire, R., & Vicente, J. (2014). Récents enseignements de la théorie des réseaux en faveur de la politique et du management des clusters. *Revue d'Economie Industrielle, 152*(4), 91–119.

Sundbo, J. (1998). *The theory of innovation: enterpreneurs, technology and strategy.* Edward Elgar Publishing.

Sung, H. H. (2004). Classification of adventures travelers. *Journel of Travel Research*, 343-356.

Sungsoo, P. (2005). Knowledge map for tourist destinations needs and implications. *Tourism Management, 26*(4), 583–594. doi:10.1016/j.tourman.2004.03.001

Sünnetçioğlu, S. & Özkök, F. (2017). Güzel, Özgün ve İyi Yemeğin Peşindeki Yolculuk: Gastronomi. *Journal of Awareness, 2*, 585–596.

Suvorova, P., & Ovsyannikov, V. (2014). Cultural Code in Poetic Text of the XX century. *World Applied Sciences Journal, 29*(4), 489-192.

Sveiby, K. E., & Simons, R. (2002). Collaborative climate and effectiveness of knowledge work–an empirical study. *Journal of Knowledge Management, 6*(5), 420–433. doi:10.1108/13673270210450388

Swanson, J. R., & Edgell, D. L. Sr. (2013). *Tourism policy and planning: Yesterday, today, and tomorrow.* Routledge.

Swarbrooke, J., & Horner, S. (2007). Main concepts in consumer behaviour, including models of consumer behaviour adapted for tourism. *Consumer Behaviour in Tourism*, 40-49.

Swarbrooke, J., & Horner, S. (2012). *Business travel and tourism.* Routledge. doi:10.4324/9780080490601

Syukiyanen, L. (2015). *Islam and Human Rights (coursebook).* Moscow: St. Petersburg State University.

Szulanski, G. (1996). Exploring internal stickiness: Impediments to the transfer of best practice within the firm. *Strategic Management Journal, 17*(S2), 27–43. doi:10.1002mj.4250171105

Taigibova, T. (2011). Actual Issues of Economic Science. In *Materials of the International Scientific Conference* (pp. 125-128). Ufa.

Tang, C. F., & Tan, E. C. (2015). Does tourism effectively stimulate Malaysia's economic growth? *Tourism Management, 46*, 148–163. doi:10.1016/j.tourman.2014.06.020

Tapscott, D. (1996). *The digital economy: Promise and peril in the age of networked intelligence*. New York, NY: McGraw-Hill.

Taylor, C. (1992). *The ethics of authenticity*. Boston, MA: Harvard University Press.

Taylor, E. Z. (2006). The effect of incentives on knowledge sharing in computer mediated communication: An experimental investigation. *Journal of Information Systems, 20*(1), 103–116. doi:10.2308/jis.2006.20.1.103

Te Yang, J. (2015). Effect of internal marketing on knowledge sharing and organisational effectiveness in the hotel industry. *Total Quality Management & Business Excellence, 26*(1–2), 76–92. doi:10.1080/14783363.2012.661131

Te Yang, J., & Wan, C. S. (2004). Advancing organizational effectiveness and knowledge management implementation. *Tourism Management, 25*(5), 593–601. doi:10.1016/j.tourman.2003.08.002

Teece, D. J. (1998). Research Directions for Knowledge Management. *California Management Review, 40*(3), 289–293. doi:10.2307/41165957

Teece, D. J. (2007). Explicating dynamic capabilities: The nature and microfoundations of (sustainable) enterprise performance. *Strategic Management Journal, 28*(13), 1319–1350. doi:10.1002mj.640

Tekin, Ö. A., & Çidem, G. (2017). The Relationship between Creativity Processes and Organizational Culture in Five-Star Hotel Cuisines. *Journal of Tourism and Gastronomy Studies, 324*, 342.

Tekusheva, I., & Shevchenko, K. (2011). Islam in the North Caucasus: History and Present Medium Orient... Prague: Academic Press.

Tellis, G.J. (2017). *Interesting and impactful research: on phenomena, theory, and writing*. Academic Press.

Temizkalp, G. (2010). *Öğretmen Adaylarının Yaratıcılık Düzeyleri* (Yayımlanmamış yüksek lisans tezi). Mehmet Akif Ersoy Üniversitesi Sosyal Bilimler Enstitüsü, Burdur.

Thakur, R., Hsu, S., & Fontenot, G. (2012). Innovation in healthcare: Issues and future trends. *Journal of Business Research, 65*(4), 562–569. doi:10.1016/j.jbusres.2011.02.022

The National Center for Biotechnology Information. (n.d.). Retrieved from https://www.ncbi.nlm.nih.gov/pmc/articles/PMC3747831/

The News. (2016, March 17). 5,634 foreign tourists visited Pakistan in 2015, NA told.

Thomas, R. (2011). Understanding small firms in tourism. *Tourism Management*, 63–76.

Thomas, R. (2012). Business elites, universities and knowledge transfer in tourism. *Tourism Management, 33*(3), 553–561. doi:10.1016/j.tourman.2011.06.009

Tichy, N. M., Tushman, M. L., & Fombrun, C. (1979). Social network analysis for organizations. *Academy of Management Review, 4*(4), 507–519. doi:10.5465/amr.1979.4498309

Torrance, E. P. (2004). Great Expectations: Creative Achievements of the Sociometric Stars in a 30-Year Study. *The Journal of Secondary Gifted Education, 16*(1), 5-13. https://files.eric.ed.gov/fulltext/EJ682714.pdf

Torre, A. (2006). Clusters et systèmes locaux d'innovation: Un retour critique sur les hypothèses naturalistes de la transmission des connaissances à l'aide des catégories de l'économie de la proximité. *Région et Développement, 24*, 15–44.

Tounkara, T., & Arduin, P.-E. (2014). Aligning knowledge sharing strategy with organizational culture. *European Conference on Knowledge Management*, 983–993.

Tourism Research Australia. (2019a). *State tourism satellite accounts: 2017-2018*. Barton, Australia: Austrade.

Tourism Research Australia. (2019b). *International visitors in Australia: Year ending March 2019*. Barton, Australia: Austrade.

Tourism, M. (2017). *Tourism Malaysia-Statstics*. available at: http://corporate.tourism. gov.my/research.asp?page1/4facts_ figures

Tourism, M. (2018). *Tourism Malaysia-Statstics*. Available at: http://corporate.tourism. gov.my/research.asp?page1/4facts_ figures

TourismNotes. (2018). Tourism. Retrieved from https://tourismnotes.com/travel-tourism/

TourismR. (n.d.). Retrieved from https://www.russiatourism.ru/news/14078/

Tribe, J. (1999). The concept of tourism: Framing a wide tourism world and broad tourism society. *Tourism Recreation Research, 24*(2), 75–81. doi:10.1080/02508281.1999.11014879

Trilling, L. (2009). *Sincerity and authenticity*. Boston, MA: Harvard University Press. doi:10.2307/j.ctvjhzrdp

Trivellas, P., Akrivouli, Z., Tsifora, E., & Tsoutsa, P. (2015). The impact of knowledge sharing culture on job satisfaction in accounting firms. The mediating effect of general competencies. *Procedia Economics and Finance, 19*, 238–247. doi:10.1016/S2212-5671(15)00025-8

Tsai, W. P., & Ghoshal, S. (1998). Social capital and value creation: The role of intrafirm network. *Academy of Management Journal, 41*(4), 464–476.

Tsoukas, H. (1996). The firm as a distributed knowledge system: A constructionist approach. *Strategic Management, 17*(2), 11–25. doi:10.1002mj.4250171104

Tsoukas, H., & Vladimirou, E. (2001). What is Organizational Knowledge? *Journal of Management Studies, 38*(7), 973–993. doi:10.1111/1467-6486.00268

Tzortzaki, A. M., & Mihiotis, A. (2012). A three dimensional knowledge management framework for hospitality and tourism. *Foresight, 14*(3), 242–259. doi:10.1108/14636681211239773

Ucar, T., & Karahoca, A. (2015). Personalizing trip recommendations: A framework proposal. *Global Journal of Computer Sciences, 05*, 30–35.

Uit Beijerse, R. P. (1999). Questions in knowledge management: defining and conceptualising a phenomenon. *Journal of Knowledge Management, 3*(2), 94–109.

Ulyana, A., Vinokurova, & Dambaeva, L. (2014). Sacred Sites of the Sakha People: Custom and Law. *Anthropology & Archeology of Eurasia*.

Underwood, B. J. (1982). *Studies in Learning and Memory: Selected Paper*. Praeger Publishers.

United Nations. (2018). *About the Sustainable Development Goals*. Retrieved from https://bit.ly/2jHjQmD

UNTWO. (2019). *UNTWO Tourism Highlights, 2018 edition*. Retrieved from https://www.e-unwto.org/doi/pdf/10.18111/9789284419876

UNWTO. (2006). *Informe global sobre la industria de reuniones* [Global report on the meetings industry]. Retrieved from http://cf.cdn.unwto.org/sites/all/files/pdf/am_8_meetings_esp-red.pdf

UNWTO. (2008). *International recommendations for tourism statistics 2008*. United Nations World Tourism Organization.

UNWTO. (2014). *Global Report on the Meetings Industry*. Madrid: World Tourism Organization (UNWTO).

UNWTO. (2019). *International tourism highlights* (2019 edition). United Nations World Tourism Organization.

Usoro, A., & Abiagam, B. (2018). Culture effect on knowledge management adoption in Nigerian hospitality industry. *Journal of Information and Knowledge Management Systems*, *48*(3), 314–332. doi:10.1108/VJIKMS-11-2017-0080

Üstündağ, T. (2009). Yaratıcılığa yolculuk (6. Baskı.). Ankara: Pegem Akademi.

Uzzi, B. (1996). The Sources and Consequences of Embeddedness for the Economic Performance of Organizations: The Network Effect. *American Sociological Review*, *61*(4), 674–698. doi:10.2307/2096399

Valente, F., Dredge, D., & Lohmann, G. (2015). Leadership and governance in regional tourism. *Journal of Destination Marketing & Management*, *4*(2), 127–136. doi:10.1016/j.jdmm.2015.03.005

Van Dijk, M. P., & Arni, S. (2003). Enterprise clusters in developing countries: Mechanisms of transition and stagnation. *Entrepreneurship and Regional Development*, *15*(3), 183–206. doi:10.1080/08985620210159239

Veiga, P. M., Ambrósio, F., & Ferreira, R. R. (2020). Competitiveness of Hotel Industry: a Knowledge Management Approach. In S. J. Teixeira & J. J. Ferreira (Eds.), *Multilevel Approach to Competitiveness in the Global Tourism Industry* (pp. 9–25). IGI Global. doi:10.4018/978-1-7998-0365-2.ch002

Vera-Muñoz, S. C., Ho, J. L., & Chow, C. W. (2006). Enhancing knowledge sharing in public accounting firms. *Accounting Horizons*, *20*(2), 133–155. doi:10.2308/acch.2006.20.2.133

Villena Navarro, M. (2014). *Turismo de Negocios: Análisis del gasto diario de los turistas de negocios* [Business Tourism: Analysis of the daily expenditure of business tourists]. Cartagena: Universidad Politécnica de Cartagena.

Virgil Niculaa, P. R. (2014). *Business Tourism Market Developments*. Science Direct.

Voegeli, C. (2019). Understanding the Current State and Future Directions of Hotel Sector Knowledge Management: A Literature Review. *Business and Economic Review*, *9*(2), 118–145. doi:10.5296/ber.v9i2.14778

Voinescu, R., & Moisoiu, C. (2019). Belt and Road Initiative and Possible Implications for Central and Eastern Europe Countries. *Global Economic Observer*, *7*(1), 195–204.

Von Hippel, E. (1986). Lead users: A source of novel product concepts. *Management Science*, *32*(7), 791–805. doi:10.1287/mnsc.32.7.791

Von Hippel, E. (1988). *The Sources of Innovation*. Oxford: Oxford Univ. Press.

Walker, G., Kogut, B., & Shan, W. (1997). Social Capital, Structural Holes and the Formation of an Industrial Network. *Organization Science*, *8*(2), 109–125. doi:10.1287/orsc.8.2.109

Wang, C. (2004). The iñuence of ethical and self-interest concerns on knowledge sharing intentions among managers: An empirical study. *International Journal of Management*, *21*(3), 370–381.

Wang, C. L., Senaratne, C., & Rafiq, M. (2015). Success traps, dynamic capabilities and firm performance. *British Journal of Management, 26*(1), 26–44. doi:10.1111/1467-8551.12066

Wang, S., Noe, R. A., & Wang, Z. M. (2011). Motivating knowledge sharing in knowledge management systems: A quasi-field experiment. *Journal of Management, 40*(4), 978–1009. doi:10.1177/0149206311412192

Wang, W., Zeng, G., & Tang, D. (2011). Bayesian intelligent semantic mash-up for tourism. *Concurrency and Computation, 23*(8), 850–862. doi:10.1002/cpe.1676

Wang, Y., & Bramwell, B. (2012). Heritage protection and tourism development priorities in Hangzhou, China: A political economy and governance perspective. *Tourism Management, 33*(4), 988–998. doi:10.1016/j.tourman.2011.10.010

Wang, Y., & Fesenmaier, D. R. (2003). Assessing motivation of contribution in online communities: An empirical investigation of an online travel community. *Electronic Markets, 13*(1), 33–45. doi:10.1080/1019678032000052934

Wang, Z., Wang, N., Cao, J., & Ye, X. (2016). The impact of intellectual capital - knowledge management strategy fit on firm performance. *Management Decision, 54*(8), 1861–1885. doi:10.1108/MD-06-2015-0231

Wanhill, S. (2012). Small and medium tourism enterprises. *Annals of Tourism Research*, 132–146.

Wan, Y. K. P. (2011, April). Assessing the strengths and weaknesses of Macao as an attractive meeting and convention destination: Perspectives of key informants. *Journal of Convention & Event Tourism, 12*(2), 129–151. doi:10.1080/15470148.2011.565704

Wardi, Y., Abror, A., & Trinanda, O. (2018). Halal tourism: Antecedent of tourist's satisfaction and word of mouth (WOM). *Asia Pacific Journal of Tourism Research, 23*(5), 463–472. doi:10.1080/10941665.2018.1466816

Warr, P., & Downing, J. (2000). Learning strategies, learning anxiety and knowledge acquisition. *British Journal of Psychology, 91*(3), 311–333. doi:10.1348/000712600161853 PMID:10958577

Wasko, M., & Faraj, S. (2005). Why Should I Share? Examining Social Capital and Knowledge Contribution in Electronic Networks of Practice. *Management Information Systems Quarterly, 29*(1), 35–57. doi:10.2307/25148667

Wasserman, S., & Faust, K. (1994). *Social network analysis: methods and applications*. Cambridge University Press. doi:10.1017/CBO9780511815478

Wayne, J. H., Casper, W. J., Matthews, R. A., & Allen, T. D. (2013). Family-supportive organization perceptions and organizational commitment: The mediating role of work family conflict and enrichment and partner attitudes. *The Journal of Applied Psychology, 98*(4), 606–622. doi:10.1037/a0032491 PMID:23565896

WCED. (1987). World commission on environment and development. *Our Common Future, 17*, 1–91.

Weber, K., & Chon, K. (2002). *Convention tourism: International research and industry perspectives*. Hove, UK: Psychology Press.

Webster, J., & Watson, R. T. (2002). Analyzing the past to prepare for the future: Writing a literature review. MIS Quarterly, xiii-xxiii.

Weggeman, M. (1997). *Kennismanagement – inrichting en besturing van kennisintensieve organizations*. Scriptum Management.

Weidenfeld, A., & Hall, C. M. (2014). Tourism in the development of regional and sectoral innovation systems. *The Wiley Blackwell Companion to Tourism*, 578.

Weidenfeld, A., Williams, A. M., & Butler, R. W. (2010). Knowledge transfer and innovation among attractions. *Annals of Tourism Research, 37*(3), 604–626. doi:10.1016/j.annals.2009.12.001

Weiermair, K. (2006). *Product Improvement or Innovation*. Academic Press.

Wells, D., & Schminke, M. (2001). Ethical development and human resources training an integrative framework. *Human Resource Management Review, 11*(1-2), 135–158. doi:10.1016/S1053-4822(00)00044-9

Westlund, H., & Nilsson, E. (2005). *Measuring enterprises' investments in social capital-a pilot study*. The 43rd Congress of the European Regional Science Association, Jyvaskyla, Finland. 10.1080/00343400500328164

Widmann, A., Messmann, G., & Mulder, R. H. (2016). The Impact of Team Learning Behaviors on Team Innovative Work Behavior: A Systematic Review. *Human Resource Development Review, 15*(4), 429–458. doi:10.1177/1534484316673713

Wielenga-Meijer, E. G. A., Taris, T. W., Kompier, M. A. J., & Wigboldus, D. H. J. (2010). From task characteristics to learning: A systematic review. *Scandinavian Journal of Psychology, 51*(5), 363–375. doi:10.1111/j.1467-9450.2009.00768.x PMID:20180922

Wilkea, E. P., Costa, B. K., Freire, O. B. D. L., & Ferreira, M. P. (2019). Interorganizational cooperation in tourist destination : Building performance in the hotel industry. *Tourism Management, 72*, 340–351. doi:10.1016/j.tourman.2018.12.015

Williams, E., Pillai, R., Deptula, B., & Lowe, K. B. (2012). The effects of crisis, cynicism about change, and value congruence on perceptions of authentic leadership and attributed charisma in the 2008 presidential election. *The Leadership Quarterly, 23*(3), 324–341. doi:10.1016/j.leaqua.2011.07.003

Williams, M. K. (2017). John Dewey in the 21st Century. *Journal of Inquiry and Action in Education, 9*(1). Retrieved from https://digitalcommons.buffalostate.edu/jiae/vol9/iss1/7

Williamson, A. J., Battisti, M., Leatherbee, M., & Gish, J. J. (2019). Rest, Zest, and My Innovative Best: Sleep and Mood as Drivers of Entrepreneurs' Innovative Behavior. *Entrepreneurship Theory and Practice, 43*(3), 582–610. doi:10.1177/1042258718798630

Wilson, S. M., & Peterson, P. L. (2006). *Theories of Learning and Teaching: What Do They Mean for Educators?* Working Paper. National Education Association Research Department.

Wilson, S., Fesenmaier, D., Fesenmaier, J., & Van Es, J. C. (2001). Factors for success in rural tourism development. *Journal of Travel Research, 40*(2), 132–138. doi:10.1177/004728750104000203

Winter, S. (1987). Knowledge and Competence as Strategic Assets. In D. Teece (Ed.), *The Competitive Challenge: Strategies for Industrial Innovation and Renewal*. New York: Harper & Row, Ballinger Division.

Wong, T., French, L., & Wickham, M. (2017). An examination of the knowledge management process in the emerging Chinese hotel industry. *Knowledge Management Strategies and Applications*, 239.

Woodman, R., Sawyer, J., & Griffin, R. (1993). Toward a theory of organizational creativity. *Academy of Management Review, 18*(2), 293–321. doi:10.5465/amr.1993.3997517

Woods, W. (2010). Performing rurality and practices in rural geography. *Progress in Human Geography, 34*(6), 835–842. doi:10.1177/0309132509357356

Wooldridge, M. (2009). An introduction to multi-agent systems. John Wiley and Sons Yager, R.(1988).On ordered weighted averaging operators. Operators in multi-criteria decision making. *IEEE Transactions on Systems, Man, and Cybernetics, 18*, 183–190.

Wootton, G., & Stevens, T. (1995). Business tourism: A study of the market for hotel-based meetings and its contribution to Wales's tourism. *Tourism Management, 16*(4), 305–313. doi:10.1016/0261-5177(95)00020-O

World Tourism Organization Annual Report. (2016). World Tourism Organization.

WTO. (2007). *A practical guide to tourism destination management*. World Tourism Organization.

Wu, C., & Lin, C. (2009). Case study of knowledge creation facilitated by Six Sigma. *International Journal of Quality & Reliability Management, 26*(9), 911–932. doi:10.1108/02656710910995091

Wu, I.-L., & Lin, H.-C. (2009). A Strategy-Based Process for Implementing Knowledge Management: An Integrative View and Empirical Study. *Journal of the American Society for Information Science and Technology, 60*(4), 789–802. doi:10.1002/asi.20999

Wu, T.-P., & Wu, H.-C. (2019). The link between tourism activities and economic growth: Evidence from China's provinces. *Tourism and Hospitality Research, 19*(1), 3–14. doi:10.1177/1467358416683769

Wynen, J., Boon, J., Kleizen, B., & Verhoest, K. (2019). How Multiple Organizational Changes Shape Managerial Support for Innovative Work Behavior: Evidence From the Australian Public Service. *Review of Public Personnel Administration, §§§*, 1–25. doi:10.1177/0734371X18824388

Xerri, M., & Brunetto, Y. (2011). Fostering the innovative behaviour of SME employees: A social capital perspective. *Research and Practice in Human Resource Management, 19*(2), 43–59.

Xinhua. (2018). China's inbound, outbound tourism markets expand in 2017: reports. Retrieved from http://usa.chinadaily.com.cn/a/201806/27/WS5b3371e aa3 1033 49141df4cf.html

Xin, J. Y., Ramayah, T., Soto-Acosta, P., Popa, S., & Teoh, A. P. (2014). Analysing the use of Web 2.0 for brand awareness and competitive advantage: An empirical study in the Malaysian hospitality industry. *Information Systems Management, 31*(2), 96–103. doi:10.1080/10580530.2014.890425

Xu, C. (2010). *A Systematic Review of Knowledge Management Research in the Hospitality and Tourism Industry* (Dissertation). University of Nevada Las Vegas.

Yakubov, M. (2011). *Constructive Potentials of Islam on the Example of the Muslim Revival in Russia at the Kazan.* Academic Press.

Yakunin, V. (2015). Specific features and classification of tourism in the domestic historiography. *Science and Tourism: Strategies and Interactions, 4*(2).

Yaman, S. & Yalçın, N. (2005). Fen Bilgisi Öğretiminde Probleme Dayalı Öğrenme Yaklaşımının Yaratıcı Düşünme Becerisine Etkisi. *İlköğretim-Online, 4*(1), 42-52.

Yammarino, F., Dionne, S., Schriesheim, C., & Dansereau, F. (2008). Authentic leadership and positive organizational behavior: A meso, multi-level perspective. *The Leadership Quarterly, 19*(6), 693–707. doi:10.1016/j.leaqua.2008.09.004

Yang, Y., & Wong, K. K. F. (2012). The influence of cultural distance on China inbound tourism flows A panel data gravity model approach. *Asian Geographer, 29*(1), 21–37. doi:10.1080/10225706.2012.662314

Yanwen, W. (2012). The study on complex project management in developing countries. *Physics Procedia, 25*, 1547–1552. doi:10.1016/j.phpro.2012.03.274

Yasir, M. (2017). Is Crime Rate Increasing in Pakistan as Chinese Immigrants Increase? Propakistani. Retrieved from https://propakistani.pk/problog/welcome-china-crimecorruption-pakistan/

Yin-Fah, B. C., Foon, Y. S., Chee-Leong, L., & Osman, S. (2010). An exploratory study on turnover intention among private sector employees. *International Journal of Business and Management, 5*(8), 57–64.

Yiu, M., & Law, R. (2013). Review and application of knowledge management and knowledge sharing in tourism. *Asia Pacific Journal of Tourism Research, 19*(7), 737–759. doi:10.1080/10941665.2013.812128

Yong, S. H., Kim, B., Lee, H., & Kim, Y. G. (2013). The effects of individual motivations and social capital on employees' tacit and explicit knowledge sharing intentions. *International Journal of Information Management, 33*(2), 356–366. doi:10.1016/j.ijinfomgt.2012.10.009

Young, L. D. (2012). How to promote innovative behavior at work? The role of justice and support within organizations. *The Journal of Creative Behavior, 46*(3), 220–243. doi:10.1002/jocb.15

Yousaf, N., Rosnan, H., & Zamzuri, N. H. (2019). Internalisation process of the medical tourism industry in Malaysia, A sequential approach. *Revista Publicando, 6*(19), 448–462.

Yu, C. C., & Chang, H. P. (2009). Personalized location-based recommendation services for tour planning in mobile tourism applications. In *Proceedings of the 10th international conference on e-commerce and web technologies (EC-Web'09)* (pp. 38–49). 10.1007/978-3-642-03964-5_5

Yu, C., Yu-Fang, T., & Yu-Cheh, C. (2013). Knowledge sharing, organisational climate, and innovative behaviour: A cross-level analysis of effects. *Social Behavior and Personality, 41*(1), 143–156. doi:10.2224bp.2013.41.1.143

Zach, F. J., & Hill, T. L. (2017). Network, knowledge and relationship impacts on innovation in tourism destinations. *Tourism Management, 62*, 196–207. doi:10.1016/j.tourman.2017.04.001

Zaei, M. E., & Zaei, M. E. (2014). Knowledge management in hospitality and tourism industry: A KM research perspective. Information and Knowledge Management, 4(9), 114-122.

Zaheer, A., & George, V. P. (2004). Reach Out or Reach Within? Performance Implications of Alliances and Location in Biotechnology. *Managerial and Decision Economics, 25*(6-7), 437–452. doi:10.1002/mde.1200

Zaher, A., Gozubuyuk, R., & Milanov, H. (2010). It's the connections: The network perspective in interorganizational research. *The Academy of Management Perspectives, 24*(1), 62–77.

Zahra, S. A., & George, G. (2002). Absorptive capacity: A review, reconceptualization, and extension. *Academy of Management Review, 27*(2), 185–203. doi:10.5465/amr.2002.6587995

Zakariah, S. H., Zainal, A., & Shariff, F. M. (2018). Enhancing the Role of Innovation towards Employee Job Performance at Malaysian hotels. *International Journal of Academic Research in Business and Social Sciences, 8*(15), 146–159.

Zeitlin, J. H. (2008). Industrial districts and regional clusters. In G. Jones, & J. Zeitlin (Eds.), Oxford handbook of business history (pp. 219–243). Oxford University. doi:10.1093/oxfordhb/9780199263684.003.0010

Zerega, B. (2008). *Art of knowledge management.* http://www.accessmylibrary.com/article-1G1-20967956/art-knowledge-managementexperts

Zhang, Z., Song, F., & Song, Z. (2019). *Promoting knowledge sharing in the workplace: Punishment v. reward. Chaos, Solitons & Fractals.* Academic Press.

Zima, K. (2012). *Event Tourism Economical and tourstic impacts on regional economy.* Academic Press.

Zopiatis, A., Constanti, P., & Theocharous, A. L. (2014). Job involvement commitment, satisfaction and turnover: Evidence from hotel employees in Cyprus. *Tourism Management, 41*, 129–140. doi:10.1016/j.tourman.2013.09.013

Zvobgo, T. C. (2015). *A comparative study into the effectiveness of business rescue strategies in South Africa and the judicial management system in Zimbabwe* (Doctoral dissertation). University of Pretoria.

About the Contributors

Muhammad Waseem Bari is an assistant professor at Lyallpur Business school, Government College University, Faisalabad, Pakistan. Dr. Bari earned his Ph.D. degree from the Beijing Institute of Technology, China. He has ample insight into the topics of organizational sciences and published several papers in SSCI, ESCI, and EI journals. He has a vast experience in banking and the FMCG industries. His research areas are knowledge management, psychological contract, innovation, organizational changes. Dr. Bari is also serving as an editor at IGI Global and Sage Publications.

Sadia Shaheen earned her Ph.D. degree from Capital University of Technology, Pakistan. She has published several papers in the domain of human behavior and leadership. Her interest areas are organizational cronyism, organizational changes, employees' behaviors.

***

Filipe Abrósio has been an assistant professor since 2010 at the Polytechnic Institute of Guarda and since 2015 at the Polytechnic Institute of Viseu. Hospitality and food service Consultant. Certified Accountant.

Ali Ahsan received the masters and Ph.D. degrees in engineering management with a focus on software engineering/information systems, entrepreneurship, project management, technology, and management. He is currently the CEO/SBA/PM at the WISSEN Group, the Dean of Foundation University, and a Visiting Professor with Stratford University, USA. He specializes in the field of business and technology. He is a Gold Medalist and has represented many organizations at many national and international forums.

Emilia Alaverdov is an Associate Professor, Faculty of Law and International Relations, Georgian Technical University. She received her Bachelor's degree in languages: English and Spanish and was awarded the qualification of English and Spanish Philologist. She received her MA degree in Diplomatic Academy of Georgia, in the field of International Relation and Diplomacy. In 2013 she defended her Ph.D. thesis and became an academic doctor of Political Science. She has following editorial experience: Reviewer, Journal US-China Foreign Language, David Publishing Company, Member of Editorial Board, Dergi Parrrk Academic, International Journal of Social and Educational Sciences, Ankara, Turkey, Member of Editorial Board, IJOPEC Publication, London-Istanbul, Member of Editorial Board, European Journal of Economics, Law and Politics, ELP, Member of Editorial Board, "Journal of Education, Culture and Society" - Foundation Pro Scientia Publica, Wroclaw, Poland. Is an editor of the book „The Development of Religious Tourism in the World", publisher igi-global.com, (scheduled to be published in January 2021) Author of 2 book chapters and more than 20 scientific articles. Participant in 44 international scientific conferences, 2 scientific projects, and 3 E+Projects.

Abdullah Al Mamun is working as a Senior Lecturer in the Faculty of Entrepreneurship and Business, Universiti Malaysia Kelantan. His research interest includes development economics, entrepreneurship, and environmental economics. He is also the Founder and managing editor of Journal of Entrepreneurship and Business (E-ISSN: 2289-8298), publish by Faculty of Entrepreneurship and Business, Universiti Malaysia Kelantan.

Noelia Araújo Vila is PhD in Tourism Management and Planning (2013) and Bachelor in Business Administration and Management (2006). Since 2007 she collaborates in the coordination tasks of the University Master in Tourism Management and Planning and teaches as professor in the Departments of Business Organization and Marketing and Financial Economics and Accounting (University of Vigo). She participates in various research projects linked to the tourism and business sector. She is a member of the research group of the University of Vigo Emitur (Marketing and Tourism). She holds a Master in Management and Management of SMEs with a prize for the best business project.

Adil Azzahidi is a doctoral student and a researcher at the University of Ibn Zohr in Agadir, Morocco.

Sajjad Baig is an Associate Professor in the faculty of Management Sciences. Currently, He is working as Dean of faculty of Management science. Dr. Sajjad Baig has earned his Doctorate Degree in Total Quality Management from University of the Punjab, Lahore. His main research areas are Total Quality Management, Operation Management, Software Project Management, Quality Assurance and Engineering Management,, Organizational Behavior and HRM Dr. Sajjad baig has been a scholastic researcher whose valuable publications have contributed exceptionally in the discipline of Business and Management. His scholarly peer-reviewed articles, approximately, 34 have been published in the top Journals e.g. Quality & Quantity, Industria Textila, Journal of Managerial Sciences, NICE Research Journal, Materials Science, Journal of Industrial Engineering and Management, Cogent Economics & Finance, Pakistan Journal of Science, Theoretical Economics Letters, International Journal of Information, Business and Management, International Journal of Accounting and Financial Reporting, Business Management Dynamics, Amazonia Investiga, Actual Problems of Economics, etc.

Mustapha Bengrich is a doctor, a professor in Finance, and the Head of Department on Economics and Management at the University of Ibn Zohr in Agadir, Morocco.

Nurgül Boz was born in 1993 in Zonguldak, Turkey. She completed her high school education in Amasra Anatolian Hotel and Tourism Vocational High School. She completed her undergraduate studies in 2016 at Pamukkale University, Department of Gastronomy and Culinary Arts. She completed her master's degree at Pamukkale University Social Sciences Institute in 2019. Nurgül Boz is married.

Lucília Cardoso, PhD in Business Sciences and Tourism since 2010. Integrate Member of CITUR – Centre for Tourism Research, Development and Innovation, Portugal. Is postdoctoral researcher of Aveiro University, in Switzerland she's studding the Swiss Higher Education in Tourism & Hospitality. Is

International researcher of Favourite Destinations project of the Euro-Asian Tourism Studies Association. Has publications in destination brand-choice, destination image and imagery, food and mountains tourism and smart destinations. Is Jury member of Art&Tur International Tourism Film Festival since 2016.

Joseph Crawford has a PhD in leadership psychometrics from the University of Tasmania. His research involves understanding the parameters by which ethical and authentic leadership can be observed and developed. He is the Founder of men's ethical grooming company Fifth Estate Co., and takes pride in the opportunity to practice the evidence-based understanding of leadership he has. Dr Crawford was awarded Young Professional of the Year 2019 by the Launceston Chamber of Commerce and is currently serving as Associate Editor of the Journal of University Teaching and Learning Practice.

Arthur Filipe de Araújo is a PhD in Tourism Studies by the University of Aveiro. He has published works mainly in film tourism, slum tourists' motivations, destination imagery, food tourism and mountain tourism. In the industry, he has worked in several areas of Destination Management Organizations in Brazil and in an educational tourism initiative in Spain. Currently, Arthur is an Assistant Professor in the Lusophone University of Porto (Portugal) and a member of the Research Unit on Governance, Competitiveness and Public Policies of the University of Aveiro (Portugal). He also integrates the Euro-Asian Tourism Studies Association (EATSA)'s Favourite Destination project and the jury of the Art&Tur International Tourism Film Festival.

Francisco Dias holds a PhD degree in Tourism Sciences from the University of Perpignan and master degree in Social Psychology at University of Porto. Since 2009, he is associate professor in ESTM - Leiria Polytechnic. He was the editor-in-chief of the European Journal of Tourism, Hospitality and Recreation (EJTHR) until 2015. He also was the founder and now is the vice-president of EATSA – the Euro-Asia Tourism Studies Association. His empirical studies cover a large range of psychological and sociological aspects of tourism (imaginary, image, imagery, destination brand equity, tourist motivation, destination choice and consumer behaviour) and the development of indicators and data gathering methodologies for monitoring of tourism destinations. Currently, he is the coordinator of the NID - Research Group on Tourism Destination Management at CITUR, and he is the PI of two research projects: - The Favourite Destinations project, a worldwide multilingual survey that allows to assess the brand equity assigned to all tourism destinations from all over the world. - The Tourism Observatory of the Center of Portugal. Aside that, he is the general director of the ART&TUR – the International Tourism Film Festival, an annual event that is promoted by the Centro de Portugal Film Commission.

Afsheen Fatima received my PhD from SZABIST University, Islamabad. Currently I am working as a Lecturer at management sciences department of AAUR. My job responsibility include teaching, supervision and research. Before completing my graduation, I also held assistant director position in my department. Since 2009, I have been actively involved in management, teaching and research. My research interests have focused on leadership, employee engagement, emotional intelligence and innovation. My research on these and other topics has appeared in Impact factor, ISI, and HEC approved journals including AJBM, WASJ, PJPR, ... Moreover, I have also presented papers in various National and International conferences. For the same purpose, I have travelled to Thailand, Indonesia, Bangkok, Turkey, UK, Singapore and Malaysia. There are some research papers that are under consideration either for journal publications or for conference presentation.

Cristina Isabel Fernandes, PhD in Management from the University of Beira Interior, is currently assistant professor in the department of management and economics of the University of Beira Interior. She is currently the scientific coordinator of the Entrepreneurship, Competitiveness and Innovation research line of the research center NECE - Center for Studies in Business Sciences at the University of Beira Interior. In addition to management positions and degree coordination in other higher education institutions, she is part of the editorial board of several international journals (Management Decision; Journal of Knowledge and Innovation), has several dozen scientific articles published in international journals on entrepreneurship, innovation, Knowledge Intensive Business Services, strategic management and regional development. Actively participates in scientific meetings and international conferences on these topics, having been distinguished several times with awards for best article. Has participation in several international projects.

Sadia Hanif received her Master's degree in Development Studies and Master of Sciences Degree in Project Management from COMSATS Institute of Information Technology, Abbottabad, Pakistan, in 2007 and 2013 respectively. She started her Ph.D. Management Sciences in spring 2016 at Foundation University, Pakistan. She served the COMSATS Community Development Unit (CCDU) as an Assistant Trainer in the field of women empowerment and gender studies to reduce and overcome the trauma of Earthquake victims of 2005. She worked as a lecturer management sciences in COMWAVE, Sarhad University, Abbottabad Campus from 2012 to 2013. She worked as a visiting faculty for different universities in Rawalpindi/Islamabad. Currently, she is a full-time Ph.D. scholar only.

Naeem Hayat is a PhD scholar working in the business area at Univeristy Malayisa Kelatan. He completed his MBA and MSc HRM from UK. He has working experience of industry and academia. Now working in area of business management, economics of small businesses and methods. He is also worked in universities in Pakistan. Currently on study leave from GIFT University, Pakistan.

Matthew Knox is a recent graduate of the University of Tasmania with a passion for understanding how to create a better world. His current work is focused on understanding authentic followers and their role in enabling ethical leader behaviors.

Pedro Mota Veiga is an Adjunct Professor of Management and Marketing in the Polytechnic Institute of Viseu (Portugal). His current research interests include strategic management, innovation, entrepreneurship, knowledge management and applied quantitative methodologies to business, management and economics. He received his BS degree in Probabilities and Statistics from the Lisbon University (Portugal) and he gained his PhD in Management from Beira Interior (Portugal). His research has been appeared in several journals with Scopus/Web of Science indexation and in book chapters.

Hande Mutlu Öztürk was born in Ordu on 14.06.1982. She completed her primary education in France and secondary and high school education in Istanbul, Turkey. In 2005, she graduated from Pamukkale University Food Engineering Department. In 2007, she completed her master's degree at Pamukkale University, Institute of Science. In 2011, she completed her PhD at Ege University, Institute of Science and Technology. Between 2009-2013, she worked as a lecturer in Pamukkale University, School of Tourism and Hotel Management, Food and Beverage Management Department. In 2013, she started to work as Assistant Professor at Pamukkale University, Faculty of Tourism, Department of Gastronomy and Culinary Arts. She speaks fluent French and good English. She is married and has 2 children.

P. Victer Paul is working as Assistant Professor in the Department of Computer Science and Engineering at Indian Institute of Information Technology Kottayam, Kerala, India, an Institute of National Importance of India. He has completed his Ph.D. in the Department of Computer Science (2015) from Pondicherry Central University, Pondicherry, India. He has done his B.Tech. in Information Technology (2007) from SMVEC and M.Tech. in Network and Internet Engineering (2011) from Pondicherry Central University, Pondicherry, India. He is a recipient of eminent INSPIRE fellowship from Department of Science and Technology, New Delhi. He is a University Gold medalist and University Rank holder in M.Tech and B.Tech studies respectively. He has also received "TCS Best Student Award 2010" from TCS for academic excellence at Pondicherry Central University, Puducherry. He has 7+ years of academic & industrial experience and published over 90 research articles in various International Journals & Conferences. Currently, he is working in the fields of Optimization algorithms, Cloud Computing and Data Analytics.

Nabila Tanvir has done MA (English Literature), MBA (Executive), BEd. Currently, Nabila Tanvir is enrolled in MS in Business Administration in National Textile University, Faisalabad, Pakistan and working as Research Assistant at National Textile University, Faisalabad, Pakistan. Nabila Tanvir has been working as freelance writer and researcher, web content writer for 10 years.

Index

A

B

C

D

E

F

G

H

I

K

L

M

N

O

P

R

S

T

W

www.ingramcontent.com/pod-product-compliance
Lightning Source LLC
LaVergne TN
LVHW062135300926
849161LV00010B/2535